Chorology, Taxonomy and Ecology of the Floras of Africa and Madagascar

Chorology, Taxonomy and Ecology of the Floras of Africa and Madagascar

from the
Frank White
Memorial Symposium

held in the Plant Sciences Department,
Oxford University
on Sept 26-27th 1996

by the Linnean Society of London,
the Royal Botanic Gardens, Kew
and Wolfson College, Oxford

Edited by C.R. Huxley, J.M. Lock and D.F. Cutler

PREFACE

This book includes papers and posters invited and submitted to the Frank White Memorial Symposium which was held in the Plant Sciences Department of the University of Oxford on September 26–27th 1996. Many thanks go to all those who attended and contributed to the symposium.

**By the time the translation of the Bible
into Algonquin Indian was completed,
the race for whom it had been written had
been destroyed.
There was nobody left who could read it.**

George Steiner in 'After Babel'.

Frank White had this quotation written out by Rosemary Wise, the herbarium artist, and prominently displayed in his room in the Plant Sciences Department, University of Oxford.

Frontispiece. Frank White in Bagley Wood near Oxford in the early 1950s. Photo A. Angus.

CONTENTS

ADDRESSES OF CONTRIBUTORS

Angus, A., Rosebank, Boarhills, St Andrews, Fife, KY16 8PR, UK

Barthlott, W., Botanisches Institut und botanischer Garten der Universität Bonn, Meckenheimer Allee 170, 53115 Bonn, Germany

Berg, C. C., Botanical Institute, University of Bergen, the Norwegian Arboretum, N - 5067 Store Milde, Norway

Chapman, J. D., 11 Koromiko Crescent, R. D. I., Lyttleton, New Zealand

Clarke, G. P., The Hermitage, Crewkerne, Somerset, TA18 8ET, UK

Cronk, Q. C. B., Royal Botanic Garden, 20A Inverlieth Row, Edinburgh, EH3 5LR, UK

Cutler, D. F., Jodrell Laboratory, Royal Botanic Gardens, Kew, Richmond, Surrey, TW9 3AB, UK

Dowsett, R. J. and Dowsett-Lemaire, F., Rue des Lavandes 12, 34190 Ganges, France

Du Puy, D. J., Royal Botanic Gardens, Kew, Richmond, Surrey, TW9 3AB, UK

Friis, I., Centre for Tropical Biodiversity, Botanical Museum and Library, University of Copenhagen, 130, Gothersgade, DK-1123 Copenhagen K, Denmark

Grimshaw, J. M., 35 Wessex Way, Cox Green, Maidenhead, SL6 3BP, UK

Hedberg, I. and Hedberg O., Department of Systematic Botany, University of Uppasla, Villavägen 6, S-752 36 Uppsala, Sweden

Lawton, R. M., Mulberry House, Stanville Rd, Cumnor Hill, Oxford, OX2 9JF, UK

Lejoly, J., Laboratoire de Botanique systématique et de Phytosociologie, Université Libre de Bruxelles, CP 169, Av. F. Roosvelt 50, 1050 Bruxelles, Belgium

Linder, H. P., Dept. Botany, University of Cape Town, Private Bag Rondebosch 7700, South Africa

Lock, J. M., Royal Botanic Gardens, Kew, Richmond, Surrey, TW9 3AB, UK

Moat, J., Royal Botanic Gardens, Kew, Richmond, Surrey, TW9 3AB, UK

Nichol, A., 32 Elsdon Close, Peterlee, Co. Durham, SR8 1NE, UK

Okafor, J. C., P.O. Box 3856, Enugu, Nigeria

Pannell, C. M., Plant Sciences Department, University of Oxford, South Parks Rd, Oxford, OX1 3AB, UK

Porembski, S., Botanisches Institut und botanischer Garten der Universität Bonn, Meckenheimer Allee 170, 53115 Bonn, Germany

Prance, G. T., Royal Botanic Gardens, Kew, Richmond, Surrey, TW9 3AB, UK

Robbrecht, E., National Botanic Garden, Domein van Bouchout, B-1860 Meise, Belgium

Sonké, B., Département Biologiques, Ecole Normale Supérieure de Yaoundé, Université de Yaoundé I, B.P.047, Yaoundé, Cameroon

Tutin, C. E. G., S.E.G.C., B.P. 7847, Libreville, Gabon.

Prance, Sir G., Nichol, A., Andrew, A., Chapman, J., Okafor, J.C. & Robbrecht, E. (1998). The life and work of Frank White. In: C.R. Huxley, J.M. Lock and D.F. Cutler (editors). Chorology, Taxonomy and Ecology of the Floras of Africa and Madagascar. Pp. 1–16. Royal Botanic Gardens, Kew.

1. THE LIFE AND WORK OF FRANK WHITE

1.1 Introduction

SIR GHILLEAN PRANCE

At this time I feel particularly well prepared to honour Frank White as a botanist and teacher, since I have just made a short visit to South Africa and seen some of the ecosystems and species which he studied. I even saw the Ebenaceae *Diospyros mespiliformis*, the first ochlospecies (see Cronk this volume and Fig. 00), and *Euclea natalensis* in the field. Above all my trip recalled memories of Frank's storytelling about his African adventures. As I walked through the bush in the Kruger National Park, I thought of Frank "pushing his way through the bush" as he always used to say, with a short 'u' as in 'rush'. I saw many ostriches and recalled his account of being chased by an ostrich when riding on a motor-scooter (I am sure he told this story long before I taught him to drive). When I saw an elephant I thought of his story about how he climbed a tree to get away from lions and decided to have a smoke to calm his nerves. On finishing he stubbed out his cigarette on a nearby tree trunk and was startled when the angry elephant ran off in protest. Throughout his life Frank was a great humorist, especially after he had had a beer or two. He had the capacity of embroidering a grain of truth to great effect, especially in some of his tales about fellow botanists. More Frank White stories are recounted in the tributes by Angus and Chapman, but here I must introduce his life as a botanist.

Frank was born on 6th March 1927 in a small house in the town of Sunderland in the north of England. His sister Audrey Nichol describes how, in spite of growing up with no garden in an industrial town, he developed a passion for natural history at a tender age. At school, aged sixteen, he won the Hancock Essay Competition of the Natural History Society and was given Richard Perry's book, *At the turn of the tide.* In 1945 he won a scholarship to St John's College, Cambridge, where he won the Henry Humphries Prize in 1946, and achieved first class honours in both parts of his degree as well as winning the Frank Smart Prize in Botany from Cambridge University in 1948. After graduating he was exempted from military service because of the shortage of university teachers following World War II, and in August 1948 he was appointed departmental demonstrator in forest botany in what was then the Imperial Forestry Institute of Oxford University. He was to remain at Oxford in various capacities throughout his career. While at Cambridge he had participated in a trip to Lapland which resulted in a paper about peat formation (Coombe & White, 1951). However, on arrival at Oxford he was assigned the task of preparing a check list of the plants of Northern Rhodesia, and thus began his lifetime's work on the African flora.

After his first field trip to Africa, Frank began to study the taxonomy of the family Ebenaceae. He was keen to link the distribution patterns he encountered with the local ecological conditions, and became interested in chorology. He always emphasised to his students that attempting taxonomy without reference to ecology and evolution made no sense. Through the work of his students and associates he also became interested in, and himself worked on, the taxonomy of the Chrysobalanaceae and Meliaceae. In all, his taxonomic work resulted in the description of one new genus, 59 new species, 27 new subspecies and 91 new nomenclatural combinations (*fide* Angus & Chapman, 1996).

One of his most significant contributions, developed from his knowledge of distribution patterns of endemic species, was a system of phytochoria for the whole of Africa (White, 1983; 1993). A history of African chorology is given in this volume by

Friis, while Linder develops methods of analysis, and Dowsett-Lemaire and Dowsett compare plant chorology with that of birds. Possible adjustments of the phytochoria in areas of complexity are proposed by Clarke.

In 1963, Frank White was appointed secretary of a seven man committee of AETFAT (Association pour l'Étude Taxonomique de la Flore d'Afrique Tropicale) charged with the task of preparing a vegetation map of tropical Africa. The job took much longer than originally conceived by AETFAT because of Frank's meticulous attention to detail. His team gathered field information from all over Africa, and the map is the best continental vegetation map ever produced, while the 365 page Memoir is the classic treatment of African vegetation (White, 1983). In the course of these vegetation studies he became particularly interested in the Afromontane biome, and so it is appropriate in this volume to have papers by Grimshaw and the Hedbergs.

Another significant contribution which Frank made to botany was the adaptation of Edgar Anderson's (1949) scatter diagrams into a pictorialised format, and also of illustrated distribution maps, which combine distributions with illustrations of variation in taxonomic characters throughout their range. Many of his students and other workers have adopted these techniques because of their usefulness in highly variable taxa. An example is given by Pannell to illustrate her account of Frank's scientific achievements. Intrigued by the intractable variation of certain species, Frank coined the term 'ochlospecies' for polymorphic taxa which do not easily divide into subspecific categories (White, 1962a). These are discussed in this volume by Cronk.

Frank contributed accounts of several different plant families to the regional floras of Africa and elsewhere, including New Caledonia. He also initiated his own flora projects in Zambia and Malawi. His first assignment, to produce a check list of the flora of Northern Rhodesia (now Zambia), turned into a full-scale forest flora (White, 1962b). Together with Jim Chapman, Frank wrote on the evergreen forests of Malawi (Chapman & White, 1970) and his interest in the vegetation of Malawi continued until his death. During his last few years, as he struggled with ill health, he was assisted by Françoise Dowsett-Lemaire to continue with the account of the forests of Malawi. Fortunately this work will be completed because of funds left in his will for Dowsett-Lemaire to finish the book.

Frank never used taxonomic characters without thinking about both their function and their possible evolutionary significance; this led him to an interest in plant/animal interactions. He encouraged his students and research associates to think likewise, so it is not surprising that several of them were involved in studies of pollination (e.g. Hopkins, 1983), dispersal (e.g. Pannell & White, 1988) and plant/ant symbiosis (e.g. Huxley, 1986). It is most welcome that Caroline Tutin presents a paper here about gorilla foraging to remind us of Frank's interest in that study. One of Frank's final contributions was to initiate an overview of his integrated approach to taxonomy. He brought together a group of his former students and colleagues (Hopkins, Huxley, Pannell, Prance & White, 1998) to discuss functional syndromes in plant taxonomy, and he proposed that we collaborate on a book on the subject. Unfortunately his illness delayed the work, but it was finally published in 1998.

As the first D.Phil. student of Frank, I most remember him as a teacher. My initial meetings with him were when I took a summer vacation job in the Forest Herbarium dissecting and drawing *Brachystegia* flowers for A.C. Hoyle. Since Hoyle was seldom in the herbarium and offered little guidance, I was grateful when a man with extraordinarily large sideboards appeared and began to explain what I should do, and more interestingly, to describe the taxonomy and ecology of *Brachystegia*. This summer work eventually led to my becoming his research student in the department. Frank's former students are scattered through the UK and around the world, but especially Africa, from where Jonathan Okafor has sent a tribute published here. Sessions with Frank as a supervisor were rewarding and stimulating. He was also a skilful classroom teacher and made taxonomy come alive by linking it to pollination, dispersal, evolution, ecology and biogeography. Besides his formal teaching he was keen to encourage others, for instance his enthusiastic help to Rosemary Wise, the herbarium artist, in working on her *Endemic Plants of the Seychelles.*

Frank was a founder member and great supporter of AETFAT, the meetings of which he always attended. He encouraged his students to join and he participated in all the activities of the Association. Doubtless it was through AETFAT that he established his fruitful relationship with the Jardin Botanique National de Belgique in Brussels, described by Elmar Robbrecht in this section. He was also a founder Fellow of Wolfson College, Oxford, and contributed to college affairs, from initially acting chef to a lengthy period in charge of the wine cellars. In addition to the prizes already mentioned, Frank received the De Wildeman Prize of the Société Royale de Botanique of Belgium in 1988. One of the honours that most pleased him in later years, when he was so sick, was a D.Sc. from his alma mater, Cambridge University, in 1991. He was appointed Distinguished Research Curator at the Department of Plant Sciences of the University of Oxford in 1992. Details of Frank's life have been published by Angus and Chapman (1996), Léonard and Bamps (1995) and Prance (1994, 1995).

In addition to being the life and soul of the party and a storyteller *par excellence,* Frank was also a most genial host, a connoisseur of good food and drink, and a gourmet cook. Those who did not know Frank until his latter years might not have seen this side of the man. He was generous and ready to include everyone in a party. He was popular in the local pubs near where he lived and always chatted with the locals.

These are just a few reminiscences about Frank White. I am sure that the best way in which we can commemorate him is through the science to which he was so dedicated. What he would most wish is that we use the base which he has given us in African taxonomy, chorology, vegetation mapping and ecology to look to the future. Let us build on these foundations to develop African botany further.

1.2 Childhood Years

AUDREY NICHOL

Early references to Frank's interest in all things botanic are anecdotal, but have been repeated verbatim within my family for as long as I can remember. During his early childhood we lived in a house without a garden, but Frank made his own – he found some chickweed, placed it in a matchbox on the windowsill and instructed my mother not to throw it away because it was his 'garden'. Later we moved to a house with a garden and Frank took great delight in helping my father to grow flowers. When he was very small, two sisters who lived nearby used to take Frank for walks, to give my mother a little respite. These young ladies were Roman Catholics and must have taken Frank into their church. When he was about three years old Frank was told that he would soon be going to Sunday school, and his response was, "Will my church have flowers? If it hasn't I'm going to the Catholic church because they have flowers AND tulips." A photograph of Frank aged about three taken on holiday near Barnard Castle in Teesdale shows him with his hands full of wild flowers.

Sunderland is located at the mouth of the River Wear on the coast of County Durham in the north east of England. Although known for its industries of ship-building, glass-making and engineering, these were situated on a corridor along the river. The rest of the town was well endowed with parks including one which enclosed a small ravine running down to the sea. Sunderland, with the sea on its eastern border, was surrounded on its other sides by open countryside with a wide variety of habitats attractive to the young naturalist. Not many miles to the west are Weardale, Teesdale and the North Pennines. All these areas were explored by the young Frank, first roller-skating the four miles north to Marsden Bay where a large sea-stack accessible on foot at low tide provides a breeding place for seabirds. Later he extended his range using his bicycle, the bus or train. When he was seven years old Frank joined the public library preferring to borrow nature books rather than fiction (the exception being *Just William* books by Richmal Crompton). Normally the transfer from the junior to senior section of the library took place at fourteen, but by the time he was twelve Frank had persuaded the librarian to allow him a special ticket to borrow works on natural history from the senior section.

In 1938 at the age of eleven Frank passed the 'scholarship exam' to Bede Collegiate School for Boys (later renamed Bede Grammar School – referred to as 'the Bede'). The beginning of Frank's second year coincided with the start of the Second World War. It was thought that because Sunderland was such a productive shipbuilding town it would be an early target of enemy action, so the Bede was evacuated to Northallerton in Yorkshire. This presented Frank with another type of countryside to explore, undulating farmland with different varieties of birds and plants. It was here that he first saw his work in print; the following article appeared in the school magazine, *The Bedan*:

THE YORKSHIRE COUNTRYSIDE

On the 10th of September we were evacuated to Northallerton. For the naturalists there were many surprises awaiting them. A few walks around the countryside brought to view many interesting things. At first there were numerous flocks of Golden Plover which have since disappeared. Every evening I observed quite a number of Pipistrelle Bats flitting gracefully through the air. Once I was amazed to see a white bird flying over a field but I was later informed that it was a white blackbird. When the Brompton Beck overflowed which it did a few weeks ago, many surrounding fields were flooded. I saw a heron but could not approach to study it carefully. Once when out for a ramble I came across a weasel holding a rat as large as itself in its jaws. A few weeks ago I paid a visit to some woods near Brompton. As I wandered through the trees I saw an owl roosting in a pine tree; it awoke at the noise I made and flew to another part of the wood. As I ventured near the edge of a marsh a plump bird arose which I believe to be a grebe. If we have to stay here for a long time I am certain that the members of the Naturalist Society will have an interesting time.

The first months of the war were much quieter in the north of England than anticipated, so most of the evacuees were soon back, but before his return Frank was in print again, this time with a poem:

SPRING

Whene'er the Spring's begun
The flowers come to greet us –
The marshes have been brightened
With the king-cup's glorious gold,
The birds are nesting once again,
Singing their songs of old;
The furze trees look as though on fire
And make the moors aglow;
While the pale mauve of the cuckoo flower
Blooms where streamlets flow.

Two more poems along similar lines appeared in later editions.

During his teenage years Frank spent his spare time 'rambling', collecting specimens and writing an extensive nature diary. He spent every penny of his pocket money on second-hand natural history books and buying the paraphernalia necessary to every serious student of nature. He also joined the Northern Naturalists' Society based at the Hancock Museum in Newcastle-upon-Tyne. Frank was able to explore other areas of Britain at this time too. Family holidays were spent in Cardiff where he met Miss Eleanor Vachell of Caerphilly, a noted botanist who took great interest in his subsequent progress. Summer of 1943 found him with some of his school friends at a Forestry Camp near the Cheviot Hills in Northumberland, and the following year cycling to Buckdene Volunteer Agriculture Camp in East Anglia.

Frank's single-minded interest prompted his teachers to advise in a school report that he must work harder at other subjects and widen his interests to include cultural activities. His sporting activities extended to playing rugby and cross-country running for the school. Near the end of his schooldays, the teachers comments changed:- BOTANY: "This is his outstanding subject. The time he has devoted to field work has given him a maturity which is rare in a schoolboy." And later, "He has maintained very high standards of work in this subject and has passed far beyond the attainment of a school pupil. He is a student of rare distinction." In 1945 Frank was awarded a State Scholarship with his Higher School Certificate and won a scholarship to St. John's College, Cambridge. At that time, it was necessary for prospective undergraduates to pass an exam known as the 'Previous Latin'. Frank had only studied Latin for one year some time earlier. Such was his determination not to lose his place that he borrowed a complete set of Latin books and in six weeks taught himself what would normally be a five year course. Needless to say he passed.

Frank's gift of story-telling was developed and honed on me, his younger sister: verbally when I was a child and later in letters which he wrote from 1945 onwards. The most notable story is perhaps worth mentioning. One Christmas we had some holly in the house and Frank entertained my friend and me to what became an ongoing saga about a homunculus he claimed to have met, who lived entirely on holly berries and rejoiced in the name of the Ilicophagus Man. Many and varied were the exploits of this character. After Frank's death I reread my letters from him and came across a sequel. In a letter dated February 1949 he wrote "You'll never believe me I know, but on Saturday at a place called Beckley I met the cousin of an old acquaintance of yours (the Ilicophagus Man). The cousin, who calls himself the Crataegivorous Man and feeds exclusively on haws, had heard all about you. He'd had a letter from the Ilicoph saying how sad he was that you had scoffed at the tales I'd told you and treated them as idle jests." There were many similar tales which as a child I accepted as true, but as I grew older I learned to sort out the fact from the fiction. The anecdotes about his experiences in Africa were the most difficult to disentangle as somewhere within the flowery details there was usually a grain of truth.

Frank had little chance at home to develop his culinary skills since food was rationed. However, a few years ago I met a school friend of his who told me that when the Scout troop were at camp, the only days the food was edible were when Frank did the cooking! Where he learned about wine and beer remains a mystery, as our parents did not drink at all. Throughout my life as Frank's 'little sister' I seemed always to be half aware that he had singular abilities, but principally he was just my big brother whose attitude to me was that of big brothers the world over – teasing, cajoling, dominating and encouraging – and I loved every minute of it.

In 1984 Frank returned from Malawi bringing back not only his collection of specimens but also malaria. Fortunately his convalescence coincided with my summer holiday and 'little sister' was called upon to spend a month at his cottage, 'Firkins', to act as nurse, chauffeur, housekeeper and cook – although my cooking skill in no way matched his. This visit culminated in a 'garden lunch' to celebrate his recovery and it was then that I met many of his friends and colleagues. Ten years later in July 1994, I was called again to his hospital bedside with a request to look after him when he came out of hospital. Little did we know that he would never return to his beloved cottage. Until this time I had not realised the extent of the world-wide respect given to Frank's work. Within the family he was always modest about his achievements and information about his activities had to be gained by judicious questioning.

The Frank White Memorial Symposium is, to me, the supreme tribute to his life and work. I looked up the dictionary definition of the word 'symposium' and found that one of the meanings is 'a drinking party'. I am sure Frank would have approved of that!

1.3 Early Life at Oxford

ANDREW (JOCK) ANGUS

I first met Frank in December 1950 when I came to Oxford to be interviewed for the post of assistant, to work with him on the compilation of a check list of the woody plants of Northern Rhodesia. I was offered the job, accepted it there and then, and was whisked away by Frank to be shown round. Almost his first act was to change my name. That 'perception' which he wrote about later was obviously already at work, he looked me over and decided – wrong genus! "I'll call you Jock." he said. And that was that, my name reduced to synonymy!

I stayed at Oxford for five years, and momentous years they were for both of us. As well as the day-to-day work of the Forest Herbarium we made numerous trips to Kew, the British Museum and, occasionally to the herbaria at Brussels and Paris. We did a little teaching for the forestry undergraduates, and to earn some pocket money we taught courses run by the Workers Educational Association. We attended two AETFAT conferences – at Brussels in 1951 and Oxford in 1953. Between us we spent all of 1952 and January 1953 in the field in Northern Rhodesia collecting plants, living most of the time under canvas. In addition I found time to get married and become a father – which resulted in Frank becoming a godfather.

The old herbarium was in Parks Road and one had to go down a dive to get to it. Many of the specimens were kept in old ammunition boxes (Plate 1). I was soon involved in the transfer to new, purpose-built premises in South Parks Road. Life was jolly with lots of laughter, and sometimes noise, partly as a result of my predilection for whistling and breaking into song. Janet Chandler was our resident artist, and herself not exactly a quiet little mouse. Mr Hoyle the curator and Mr Pearson, retired schoolmaster and herbarium assistant, were both humorists, and Mrs Woodley, in charge of mounting the specimens, was a cheerful person much given to mirth. And of course we had Frank with his special brand of humour and his infectious laughter. So the atmosphere was very different to that of Kew which could be a bit gloomy and where the only sound beyond the shuffling of folders and hushed earnest conversation was the occasional muffled cough, which Margaret Stones, the Kew artist, likened to "a sheep hoasting in the fog." We visited Kew frequently, sometimes by train or with Mr Hoyle in his car, but mostly we went by motorbike. I had acquired a large Ariel 600cc side-valve machine. Frank rode pillion, briefcase positioned between himself and me and tied to his waist. I did get a sidecar later, and then Frank travelled in more comfort. He enjoyed motorcycling and did in fact buy an old 350cc BSA from one of the students, but he just could not get the knack of kick-starting it, and sold it on.

Frank lived at 9 Longwall Street, in lodgings formerly occupied by his predecessor, Pat Brenan. The rooms upstairs were inhabited by the poet Alvarez. Their elderly landlady, Elsie, mothered them, producing excellent meals in a small and primitive kitchen. Lunch was regularly at the 'Lamb and Flag', one of the pioneers of 'pub grub'. We strode out in our nailed veldshoen making a great noise on the pavement and usually had a large roll with a liberal filling of liver sausage or cheese, helped down by a pint of best bitter. I have never had a strong head for alcohol and if, as occasionally

PLATE 1. The Herbarium at the Imperial Forestry Institute in Parks Road, the curator A.C. Hoyle looking on. Note the ammunition boxes being used to house specimens. Photo. A. Angus, 1950.

happened, we drank an extra half pint, I used to feel as though I were floating back to the herbarium. Now the 'Lamb and Flag' was also noted for the variety and originality of the graffiti in the men's lavatory, which Frank always called 'the bog'. We used to read it. It was high quality stuff as, you can imagine in a university town. You know the kind of thing – "Philosophy gives unintelligible answers to insoluble questions", "I think I think, therefore I think", "I think I exist, therefore I exist, I think", "If you think the cost of living in Oxford is high, remember, it does include a free trip round the sun." One cold day in mid-December Frank and I laughed heartily at a message by the entrance – "A Merry Christmas to all our readers."

We cannot talk about the herbarium of those days without mentioning *Brachystegia*. Mr Hoyle, the curator, was the acknowledged authority on that genus. He had to be; he had most of the material from all the major herbaria! At the first AETFAT conference he was dubbed 'Monsieur le *Brachystegia*". He was always agonising over putative hybrids, of which there seemed to be many. He would have huge spreads down the middle of the herbarium for days and examine them endlessly with his large hand-lens. I think it used to irritate Frank a little, for one morning Frank cast a despairing eye over another row of *Brachystegia*, and said "Jock they've probably been copulating in the night and there'll be a new crop of hybrids this morning." Mr Hoyle was philosophical about his problems, I can still hear him consoling himself with the lines from the *Rubaiyat of Omar Khayam* (E. Fitzgerald's translation):

> Myself when young did eagerly frequent
> Doctor and saint and heard great Argument
> About it and about; but evermore
> Came out by the same door as in I went.

I don't remember holidays as such, but we did take a break now and again when we felt we needed it – a walk in Bagley or Wytham Woods, or punting on the Cherwell. Frank once came to Exeter to stay with my wife's parents, and I took him for a hike across Dartmoor. We walked from Moretonhampsted to Two Bridges and tramped over the moor to visit Wistman's Wood, where we camped for the night. He had always wanted to study this isolated ancient oak wood and he gave it a good going over. On the way to Okehampton we tried, but failed, to find the famous letterbox in the middle of the moor. That was a hike of 25 miles, but we had to take a bus because Frank developed an enormous blister covering all of one heel. It put him out of action for a couple of days.

Frank liked spinning yarns, one of his favourites from those days concerned the use of 'pidgin English', the lingua franca of the West coast of Africa. He used to tell of a West African from upcountry who had been down to the coast for the first time and had seen a large ocean-going steamship. Back home in his village, and trying to convey what he had seen to friends and relatives, he described the ship as "One piecey bamboo and two piecey puff-puff, and a man can walk about for himself and see bloody well." Frank would do the actions and round the story off with his usual guffaw. There was one episode during our expedition to Northern Rhodesia which I suspect Frank may have later elaborated. We were botanizing on the border with the Congo north of Mwinilunga (Plate 2). The border was a cleared line through the bush, dead straight, with piles of stones at regular intervals. It was used as a path by the Africans. One day our vehicle was parked on the border. My gun was leaning against a wheel, and I, with bushy beard, was just sharpening the axe which I used to collect bark and wood specimens. There came into view in the distance an African on a bicycle laden with merchandise. As soon as he spotted me he fell off the bike and disappeared into the bush. We sent some of our men to fetch him back . He was clearly very frightened and we asked why he had fled. Our interpreter told us – "Bwana, he thinks you cannibal." Another incident, this time in the Southern Province, was the bogging down of our vehicle in the middle of a river crossing. You should know that Frank never took much

PLATE 2. Frank White and O. Kerfoot at camp by Lake Tanganika in 1952. Photo A. Angus. Note the *Vellozia* in the foreground.

interest in the mundane practicalities of travelling and camping; he was content to let others do the chores and he got on with his work. So, while we were all up to our knees in muddy water, pushing and shoving and becoming filthy, Frank stood up on the bank, well clear of any flying mud, beckoning us on with secateurs in hand, shouting "Push, boys, push!" ('push' pronounced like 'rush'). That pronunciation of the vowel 'u' was one feature of his speech which betrayed his north of England origin. He had cultivated an Oxford accent during his school days, I believe, but he would say 'sugar' as 'shuggar' rather than shoogar, and 'put' like golf putting. It was Bill Bainbridge who reminded me that the forest flora was first planned to be a volume that one could take into the field and be small enough "to push into your bushjacket pocket." The outcome was very different – an unwieldy tome.

After I had left Oxford to work in Northern Rhodesia, I always visited him when back on home leave. I remember staying at Pin Farm, South Hinksey, where Frank lived for a time. I shared a room with a fellow guest and slept on the floor, or tried to – it was very cold! Frank used to cycle into work humping his rusty old bike up the footbridge over the railway, but when I stayed we walked. By force of habit from riding the bicycle he still tucked his right trouser leg into his sock! Later when he was at Taston (Plate 3), he insisted that I bring all my family to stay. He gallantly made room for all six of us, converting his upstairs room into a dormitory with inflatable mattresses arranged in parallel. Frank was truly a kind and generous host.

In conclusion I want to tell you about a word that Frank used a lot. It is the Bantu word 'fundi', meaning 'an expert, one who has great knowledge'. Frank recognised great fundis and lesser fundis. Pat Brenan, for example, was a great fundi because he had great knowledge of the African flora as a whole and specialist knowledge of parts of it. A lesser fundi might be an authority on just a single family or genus. But he respected them all and he consulted them. Long before the end of his day, Frank himself had become a great fundi. I think the label suits him; he would have liked it.

PLATE 3. Frank White at his house 'Firkins' in Taston, Oxfordshire, 1971. Photo by A. Angus.

1.4 African Experiences

JIM CHAPMAN

Not far from Chisenga, on the border between Tanzania and Malawi, are the Misuku Hills, where long high ridges are crowned by impressive forests, whence come the streams which supply the farming community below. In 1952 one of my first jobs with the Forestry Department of northern Nyasaland, was to demarcate the boundaries of these forests – 30 miles of ups and downs, with compass, chain and Abney level, and through seemingly endless discussions with the people, to obtain their consent. Having been taught at Bangor by Paul Richards, newly come from Cambridge, where Frank had been one of his students, I could not but look forward to the botany. We were fortunate in our Chief Conservator, Richard Willan, who encouraged botanical exploration; new arrivals were given a plant press. This meant official blessing, important because not all senior officers shared his enthusiasm. The *Flora Zambesiaca* was still only a concept and published information almost nil, other than a preliminary check list and the Brass-Vernay Expedition report, published by the New York Botanical Garden. This did not, however, include the Misuku forests.

Although for us botany had always to take second place, we soon had several very full plant presses. One day on our return to the little rest house at Chisenga, where until then we had been the sole occupants, it was to find two Northern Rhodesian Forestry Department vehicles and a line of plant presses on the veranda. We were soon aware that the two new arrivals were no ordinary people, Frank with his twinkling eyes and debonair presence, and Jock, a romantic figure with his abundant red locks and beard to match, both full of enthusiasm and fun. That evening, over beer, Frank suggested he look through our collections – he named them all, even the more unusual trees such as *Cylicomorpha* and *Mitragyna rubrostipulata*. He then remarked that our collections would have come to him eventually. The next day, while Frank worked on his specimens and notes, Jock and I followed a forest-fringed stream to the top of the Mafinga ridge. Some of the collections we made are cited in the Forest Flora. The following day they left. So began a lifetime's friendship.

Many other African foresters of this time came to know Frank when doing 'the Oxford year'. Recently I had a letter from Bob Fishwick, one-time conservator of the Sudan Zone of northern Nigeria. Frank had told me how Bob smoothed his progress when he was in the country in connection with the book *Nigerian Trees*. Bob wrote "Frank was indeed a remarkable person – I vividly remember him drinking warm Guinness cooled a little by tonic water in a village in the heat of April. When we were on leave we saw Frank in Oxford and remember trawling round 'Campaign for Real Ale' pubs where they did not play piped music".

It was while I was at Oxford that Frank, with Janet Chandler and Mrs Woodley, organised a Herbarium party to commemorate the mounting of the 50,000th specimen. Pat Brenan was guest of honour. The arrangements included a creche for our children. On leaving Malawi in 1965, I spent part of the summer at Taston, where Frank had just come to live, writing the first draft of Part II of *Evergreen Forests of Malawi*. The previous owners of Frank's Cotswold stone cottage had gone a long way to destroying its character. I remember the satisfaction of demolishing a hideous porch with a sledgehammer, and the hard graft of barrowing loads and loads of soil. A professional dry stone waller, Ralph Stowe, who became a close friend of Frank, was working there too. The three of us would often adjourn on those summer evenings to 'The Bull' at Charlbury where Frank was already an established figure. At weekends there were work parties, no shortage of volunteers, with an abundance of delicious food and a firkin of Garne's Burford ale always on tap. Frank also gave two full-scale parties that summer. I remember Arthur Exell coming to one of these memorable occasions, with the cottage packed out, and going on well into the small hours.

Thanks to Dick Brummitt, 1981 found us back in Malawi at Zomba where I stood in for James Seyani while he undertook his *Dombeya* study at Oxford. It was the perfect opportunity for Frank to make an extended visit to the country after 28 years absence. This was in connection with the *Evergreen Forest Flora of Malawi*, or as it was called then, 'The Evergreen Forest Trees and Shrubs of Malawi', on which we had begun to collaborate in 1963. This visit was made possible through the ready support of David May, Chief Forest Officer. That it was such a success was largely due to Rodney Nkaonja, Senior Forest Research Officer. Frank's arrival in March coincided with a countrywide petrol shortage which threatened to jeopardise much of the itinerary. Rodney saved the situation by making available the Department's only diesel Landrover. In what was inevitably a short visit to the Misuku forests in the far north Frank nonetheless spotted *Alangium chinense*, the first record for Malawi. Thence to the Nyika, and his meeting with Françoise Dowsett-Lemaire and Robert Dowsett; this was the start of a most fruitful botanical collaboration and a great friendship, which saw Françoise working with Frank in the hospital at Oxford until shortly before his death. We well remember Frank's delight over this meeting, especially his appreciation of Françoise's cooking after a diet mostly of nsima, stodgy maize porridge, for which his travelling companion, Bruce Hargreaves, had an inordinate liking. After Frank's return to Zomba, we visited the Nkhulubvi Thicket, shrine of Mbona, the rain-maker. We had first to present a sizeable roll of calico to the Guardian. No footwear was permitted, the answer was a wide roll of Elastoplast, with which we bound up our feet. I can see Frank now, 'treading delicately' among the thorns below the Acacia trees.

The main field objective was a 17 day stay on Mt. Mulanje, which we approached with some trepidation. Frank's breathing was already a problem, so to prepare him for the exertion of getting up to plateau level, a climb of over 3,500 ft, my wife Betty endeavoured to persuade him to walk a little each day, if only the mile of contour road from our house to the Herbarium, but to no avail. Reaching Likabula, at the foot of the mountain, we were really concerned, the local forest officer, meeting Frank, refused to believe he would be climbing the mountain the following day. But we need not have worried. The attraction of Mulanje was such that I believe Frank would have gone on hands and knees, rather than fail. A suitable person was found to accompany him throughout the trip, keeping two meters behind Frank, and to have open a big plastic bag for the ecological collections he was continually making. The two of them set off at first light. When the rest of us reached the Chambe hut in the evening, Frank was approaching the steps. Every day saw him covering more ground, as we moved from plateau to plateau. Being May, the weather was at its best. The two would get back to camp early in the afternoon. Thereafter, until sundowner time, Frank would sit surrounded by his collections, writing up his notes, happy as a king. In the evenings, we had lots of laughs and Frank was consistently lightsome and well. At Sombani, he spotted *Olea europaea* subsp. *africana*, the sole record for Malawi south of the Viphya Plateau. Besides myself, Len Brass and other visiting botanists had passed that way and missed it. Frank's eye for detail was specially apparent to me when he and Françoise compared the bark and leaf characters of *Rawsonia lucida* and Mt. Mulanje's *R. burtt-davyi*.

The descent from Mt. Mulanje was very different from the climb up. Some botanical friends, including Pat Jenkins, had come up specially for the weekend, to "help you get

Plate 4. The Mt. Mulanje team, May 1981. Standing, left to right: Frank White, Anne Killick, Hassam Patel, John Killick, Pat Jenkins. In front: Hazel Meredith with Fisher and Betty Chapman with Skinner. Photo J. Chapman.

Frank down safely"(Plate 4). From Madzeka we went down the Little Ruo path, steep and slippery rocks with difficult stretches involving makeshift ladders. When we looked for him after breakfast, he and his retainer were not to be seen. We eventually caught up with them 2,000 ft lower down, Frank busily snipping specimens. Back at Zomba, Frank said that he felt in better health than for more years than he could remember. He could not have paid Betty a bigger compliment than when he said that he would never have dreamt it possible to go to bush and yet not miss a single creature comfort.

The evening before Frank's departure to Oxford, we were having dinner with some Chancellor College friends. Frank excelled himself in the graphic account of his South African ostrich adventure. He was heading for the Karroo on his not-so speedy moped. Beyond Oudtshoorn, where there was an ostrich farm, chugging up a long incline on the seemingly deserted road, he had a sudden premonition of danger. Stalking up the hill behind him, still a quarter of a mile distant, but gaining fast, was an ostrich. Coaxing his little machine as hard as he could it became all too obvious that it would be a very close-run thing. Indeed just as he gained the crest and the moped began to pull away, the huge bird drew alongside and delivered a vicious peck at Frank with such force that his helmet was permanently dented. Fortunately, the little machine began to draw ahead. However the angry bird did not abandon the chase until Frank, at full throttle, was already halfway down the slope towards Miering's Poort. As our host opened the car door for Frank when we were leaving, he asked, in awe, if the ostrich had *really* left a dent in his hard hat and *did he still have it?*

Frank's last visit to Africa came several years later, again at Mulanje. But with his breathing now making it very difficult for him, we spent our time on the lower slopes round the foot of the mountain. One day Frank and ourselves, with others including several botanically inclined ladies, went to Lichenya Forest Reserve, a fascinating graveyard forest, then still more or less intact, in an oxbow on the river south of the mountain. Much as the ladies admired Frank's red straw hat, it was clearly he himself who was the attraction, for they hung on every word he spoke.

Our friendship with Frank, extending over 40 years, has meant that our African experience has never lacked excitement and interest. Add to this his warmth to us as a family, we inevitably have a continuous sense of loss at his death. When Rosemary Wise, the current Oxford botanical artist, in a letter written not long after his death, said "When I returned from the Seychelles and had no Frank to talk to .... we always had this long chat on my returns, to discuss any progress that had been made." We understood exactly what she meant.

1.5 Frank White the Teacher

JONATHAN C. OKAFOR

I was one of Frank's students from Nigeria and became a member of Ronald Keay's staff when he was Director of Forest Research. As a student of Frank White from 1964 to 1965, at the Forest Herbarium, Oxford, I studied the *Combretum collinum* group of species. I found him to be a patient, diligent, painstaking, hardworking and skilful taxonomic teacher who was able to strike a delicate balance between hard work and pleasure. He arranged the loan of specimens and visits to the herbaria at Kew, Paris and Brussels and took me in his Morris Minor car to these places. At weekends, Frank held 'working' parties in the 'White Cottage' ('Firkins') at Taston, when we roasted meat, and kegs of beer were readily available He was kind and humane. He was also of good humour and an impressive personality. His handwriting was impeccable and his English was instructive, crystal clear and explicit.

His name is immortalised through numerous students and friends, as well as many and varied texts, publications and maps. The world has lost a rare and talented taxonomist, ecologist and vegetation analyst of the highest order.

PLATE 5. Frank White with students in the morphology lab. of the Botany School (now Plant Sciences Department) of Oxford University.

1.6 The European Colleague

ELMAR ROBBRECHT

White frequently stayed a few weeks or so in our National Botanic Garden at Meise. He used to join us for finishing many of the large papers which he published in the institute's serials, away from the everyday disturbances of Oxford, and taking profit from direct contact with the editors of his work, Mr Bamps or Dr Robyns, as well as the support of a large African herbarium and vast library. Of course the epicurean White was also fond of coming to Belgium for its traditions of beer brewing and wine drinking.

I was an inexperienced taxonomist when I first learned the typical sign of his arrival: an impressive metallic step in the corridors of our building. I was then struggling with the very large genus *Tricalysia* in the Rubiaceae. White was amused and interested to learn that his *Coffea* species 1 in the *Forest Flora of Northern Rhodesia* was in fact a *Tricalysia* disguised by having one-seeded fruit locules (Robbrecht, 1987: 141); it is not surprising that it was taken for a coffee in the early sixties considering the state of generic delimitation of the African Rubiaceae then. This is only one incident of course, much more important were White's ideas and the discussions I had with him on geoxylic speciation, complete or incomplete. Without the backing of his publication *The Underground Forests of Africa* (White, 1976) it would have been a much harder job to delimit and understand the numerous geoxylic representatives in Zambezian Tricalysias.

I have always been impressed by his treatment of the Rubiaceae in the *Forest Flora of Northern Rhodesia* (White, 1962b); it is so much more than a classical flora compilation. It classifies the Zambian representatives in then up-to-date subfamilies and tribes (partly personally delimited) and includes a section on pollination mechanisms resulting from keen field observations – one of the first overviews of this aspect of the Rubiaceae. His introduction to the flora records that Mr Angus had done much of the preparatory work for the large and difficult Rubiaceae, but by talking with White I learnt that he had seen most of the rubiaceous pollination evidence in the field, himself.

But of course White's most important work for the practising taxonomist has been his immense phytogeographical contribution, so useful for understanding speciation and relationships between species; I always counter-checked with it when delimiting and mapping a species. But he warned me, in a letter of February 9th 1979, that "the final answer must be taxonomic and not chorological," so I decided that the well distinguished *Tricalysia ovalifolia* and *T. sonderiana* are a pair of related species with an unusual vicariant distribution (*T. sonderiana* being a Tongoland-Pondoland endemic crossing the Limpopo River for more than 1,000 km northwards (Robbrecht, 1979: 258)). More recently I presented a phytogeographical survey of the African Rubiaceae (Robbrecht, 1996); needless to say how the discussion therein had been fertilised by White's work. I dedicated this paper to his memory – 'father of modern African phytogeography.' My survey was made at the generic level, because of the incomplete state of knowledge of the Rubiaceae. Interestingly its results fully corroborate White's chorological conclusions which were based on analysis of specific endemism. Léonard told me that Frank White would have cried "Hurrah!" for this result.

The National Botanic Garden of Belgium is proud to have contributed to the dissemination of White's phytogeographical contributions, namely by publishing several of these papers, including a very long one at the end of his life, sketching the history and methodology of his chorological classification (White, 1993), as well as by Bamps' French translation of White's Memoir on the vegetation of Africa accompanying the AETFAT map (White, 1986), which has been of so much use to French-speaking African botanists. Our institute is also associated with the extension of White's chorological system for Africa to south-west Asia, which he proposed with his friend Léonard (White & Léonard, 1991). I am sure that the new *Flora of the Arabian Peninsula and Socotra* (Miller & Cope, 1996) will be the first of many to quote White and Léonard's phytogeographical division of south-west Asia. More details of White's relationship with our institute can be found in a paper in our Bulletin last year by his friends Léonard and Bamps (1995). Here I have mentioned only the most significant aspects of White's collaboration with our institute and my own debt to the great scientist whom we lost too early.

References

Anderson, E. (1949). Introgressive hybridization. Wiley, London, New York.

Angus, A. and Chapman, J. D. (1996). A tribute to Frank White (5th March 1927 to 12th September 1994). *Bothalia* **26**: 69–76.

Chapman, J. D. and White, F. (1970). The evergreen forests of Malawi. Oxford, Commonwealth Forestry Institute, Oxford.

Coombe, D.E. and White, F. (1951). Notes on calcicolous communities and peat formation in Norwegian Lappland. *J. Ecol.* **39**: 33–62.

Hopkins, H.C. (1983). The taxonomy, reproductive biology and economic potential of *Parkia* (Leguminosae: Mimosoideae) in Africa and Madagascar. *J. Linn. Soc. Bot.* **37**: 135–167.

Hopkins, H.C., Huxley, C.R., Pannell, C.M., Prance G.T. and White, F. (1998) The Biological Monograph: The importance of field studies and functional syndromes for taxonomy and evolution of tropical plants. Royal Botanical Gardens, Kew.

Huxley, C. R. (1986). Evolution of Benevolent Ant-plant Relationships. In: B. E. Juniper and T. R. E. Southwood (editors). Insects and the Plant Surface. Pp. 257–283. Edward Arnold, London.

Léonard, J. and Bamps, P. (1995). Le Dr. Frank White (1927–1994) et le Jardin botanique national de Belgique. *Bull. Jard. Bot. Nat. Belg.* **64**: 3–11.

Miller, A.G. and Cope, T.A. (1996). Flora of the Arabian Peninsula and Socotra. 1: xxi + 585 pp. Edinburgh University Press, Edinburgh.

Pannell, C.M. and White, F. (1988). Patterns of speciation in Africa, Madagascar, and the tropical Far East: regional faunas and cryptic evolution in vertebrate- dispersed plants. *Monogr. Syst. Bot.* **25**: 639–659.

Prance, G.T. (1994). Frank White. *Oxford Plant Systematics* 2: 16.

Prance, G.T. (1995). Frank White (1927–1994). *Taxon* **44**: 462–468.

Robbrecht, E. (1979). The African genus *Tricalysia* A. Rich. (Rubiaceae - Coffeeae): 1. A revision of the species of subgenus *Empogona. Bull. Jard. Bot. Nat. Belg.* **49**: 239–360.

Robbrecht, E. (1987). The African genus *Tricalysia* A. Rich. (Rubiaceae). 4. A revision of the species of sectio *Tricalysia* and sectio *Rosea. Bull. Jard. Bot. Nat. Belg.* **57**: 39–208.

Robbrecht, E. (1996). Generic distribution patterns in subsaharan African Rubiaceae (Angiospermae). *J. Biogeogr.* **23**: 311–328.

White, F. (1962a). Geographic Variation and Speciation in Allica with Particular Reference to *Diospyros. Syst. Assoc. Publ.* **4**: 71–103.

White, F. (1962b). Forest Flora of Northern Rhodesia: xxvi + 455 pp. Oxford University Press, Oxford.

White, F. (1976). The underground forests of Africa: a preliminary view. *Gard. Bull. Singapore* **29**: 51–71.

White, F. (1983). The Vegetation of Africa: a descriptive memoir to accompany the UNESCO/AETFAT/UNSO vegetation map of Africa. *Natural Resources Research* **20**. UNESCO, Paris.

White, F. (1986). La végétation de l'Afrique. Mémoire accompagnant la carte de la végétation de l'Afrique Unesco/AETFAT/UNSO (French translation by P. Bamps). *Recherches sur les resources naturelles* **20**: 384 pp. OSTROM/UNESCO.

White, F. (1993). The AETFAT chorological classification of Africa: history, methods and application. *Bull. Jard. Bot. Nat. Belg.* **62**: 225–281.

White, F. and Léonard, J. (1991). Phytogeographical links between Africa and Southwest Asia. *Fl. Veg. Mundi* **9**: 229–246.

Pannell, C.M. (1998). The scientific achievements of Frank White. In: C.R. Huxley, J.M. Lock and D.F. Cutler (editors). Chorology, Taxonomy and Ecology of the Floras of Africa and Madagascar. Pp. 17–23. Royal Botanic Gardens, Kew.

2. THE SCIENTIFIC ACHIEVEMENTS OF FRANK WHITE

C.M. PANNELL

**Department of Plant Sciences, University of Oxford,
South Parks Road, Oxford, OX1 3RA, UK**

Abstract

The scientific methods and achievements of Frank White are briefly reviewed. He recognized different kinds of species in his taxonomic treatments of woody African families, ranging from monotypic to ochlospecies. His use of pictorialized distribution maps and scatter diagrams was instrumental in his successful resolution of variation in some species. His greatest achievement was the AETFAT/UNESCO vegetation map of Africa and delimitation of the phytochorological regions into which it is divided. A summary of the way in which this was achieved is given. Some recent projects which have been carried out in the context of Frank's classification of African vegetation are referred to, as is his interest in seed-dispersal by vertebrates.

Introduction

Frank White was one of the great scholars of African plants and vegetation. He is best known for his Vegetation Map of Africa (1983a) and the descriptive memoir which accompanies it (1983b) and many students of the African flora have had some experience of the vast knowledge and understanding of facts which underpinned this synthesis. Three words: 'taxonomy', 'ecology' and 'chorology', frequently used by Frank in his publications, sum up his approach to his research.

History and taxonomic base

From the time that he first arrived in Oxford, newly graduated from Cambridge, in 1948, it was Frank's goal to study and describe the vegetation of Africa. To do this, he set about developing a sound and detailed empirical base, focusing on, but by no means confining his attention to, the woody flora. He soon discovered that many of the woody plants he encountered were in need of taxonomic revision. Among these, he chose to concentrate on the ebony family (Ebenaceae) and, in collaboration with others, the mahogany family (Meliaceae) and the Chrysobalanaceae. Throughout his working life, he emphasized the importance of monographic study of plant families (and communities) and whenever he wrote flora accounts of these families it was within the framework of his monographic studies.

Altogether, he visited Africa 12 times and spent a total of four years there. During those visits he got to know all but one of the major regional phytochoria and most of the major vegetation types.

Pictorial presentation

Frank frequently said that his mind worked visually and, from the beginning, he illustrated his taxonomic accounts with distribution maps. From at least 1962 onwards, he also used pictorialized distribution maps and scatter diagrams (White, 1962). He did this to analyse and illustrate the variation of the most difficult of the species he was studying. Silhouettes of leaves were a distinctive feature of these diagrams (Fig. 1). Several outstanding botanical illustrators worked with Frank over the years and he was proud of the fact that most species could readily be identified using accurate and detailed illustrations, with a minimum of accompanying text.

Species delimitation and concepts

Relatively early in his work on the family Ebenaceae, it emerged that the majority of species of *Diospyros* are clearly defined (White, 1962). Even when two or more species are closely related, they can be readily distinguished by a series of correlated characters. This is usually backed up by differences in their distribution and/or ecology. A few species, however, show complex or even chaotic variation. If Frank was able to extract a pattern related to the ecology and distribution of different genetic ecotypes, which he called 'phenotypes' (this term is usually applied to environmentally-, not genetically-determined variants), he would recognise subspecies, and treat the species as being polytypic. He believed that, in such cases, subspecies 'are necessary if the patterns are to be succinctly described and discussed in evolutionary terms' (Hopkins *et al.*, 1998). This approach revealed, for example, that *Euclea natalensis* (also Ebenaceae) is a ring species similar to those found in some bird species (see Mayr, 1942, 1963; Cain, 1954).

This left a few widespread species, such as *Disopyros natalensis, D. ferrea* and *D. mespiliformis*, which show such chaotic variation that there is insufficient pattern to describe it formally. For these, Frank coined the term, *ochlospecies* (see also, Cronk, this volume). He advocated that only informal names should be used for variants within these complexes, thereby avoiding giving a false impression of order (Hopkins *et al.*, 1998). He believed that the major phenotypes in these ochlospecies, such as the rheophytes in *D. natalensis*, have arisen more that once and are therefore polytopic in origin. In discussing the history of the taxonomy of *D. ferrea* in the Far East, he says, with characteristically colourful language:

> 'by [1986] well over a hundred infraspecific names had been published in the *D. ferrea, D. elliptica* complex, resulting in one of the greatest nomenclatural and taxonomic jungles ever created by man.' (1993b: 216)

His own analysis of this complex for the *Flora of New Caledonia* (1993b, 1993c), however, resulted in the recognition of 23 discrete species, nearly all of which he found to have:

> 'distinct, sometimes unique, ecogeographical distributions closely linked with climate and the geological substrate', (1993b: 181)

Yet, even after recognising these distinct species, there remained one ochlospecies, *D. parviflora*, which he was unable to subdivide (Fig. 1).

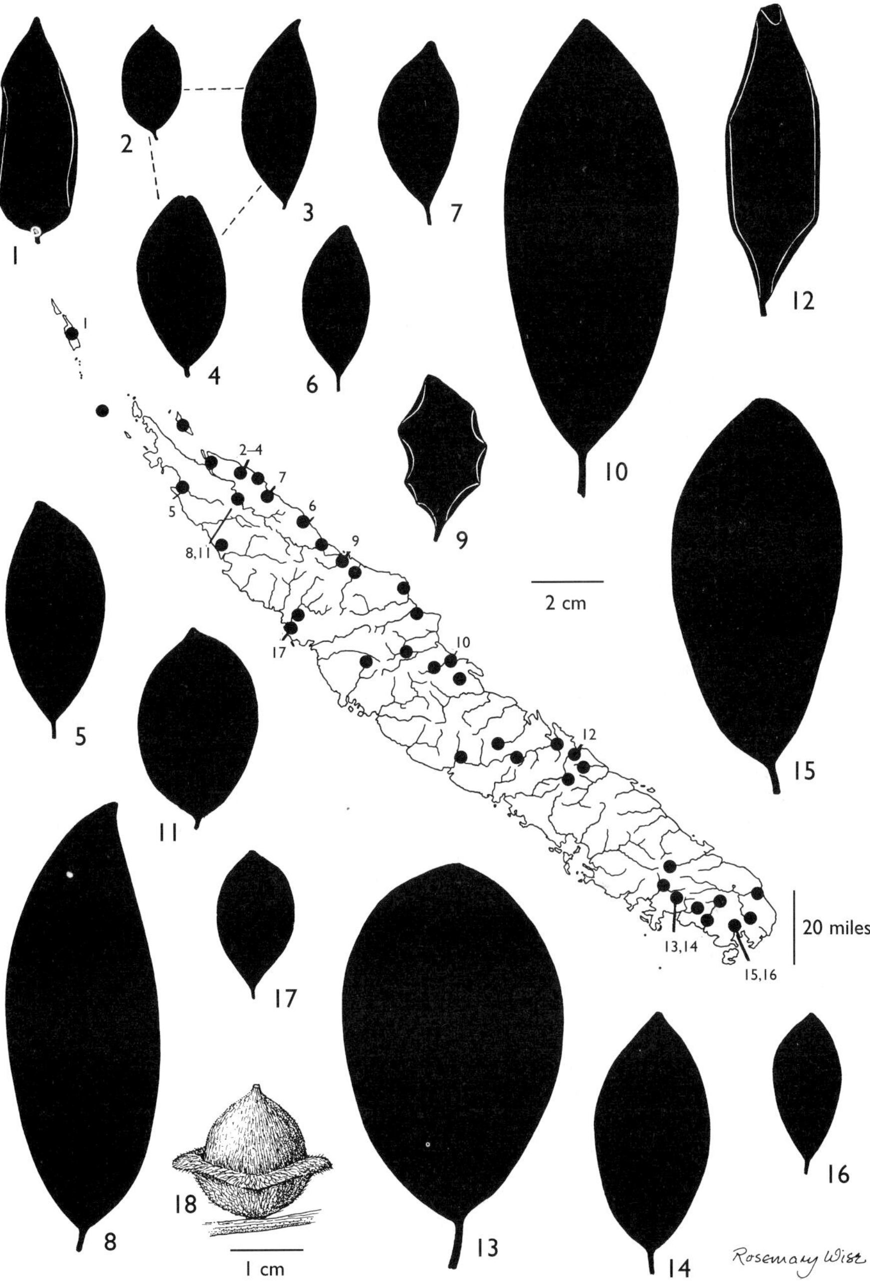

FIG. 1. *Diospyros parviflora* in New Caledonia. **1–17**, leaves selected to show the range of variation in shape and size of the species; **18**, fruit and fruiting calyx (1, *Veillon 2703*; 2, *Webster & Hildreth14810*; 3, *Thorne 28157*; 4, *MacKee 24636*; 5, *MacKee 36424*; 6, *Schlechter 15533*; 7, *Schmid 884*; 8, *MacKee 33473*; 9, *Le Rat s.n.*; 10, *MacKee 36837*; 11, *Mackee 28998*; 12, *MacKee 13970*; 13, *McPherson 1537*; 14, *Brinon 865*; 15, *Cribs 1388*; 16, *Franc 1696 A*; 17, *MacKee 15499*; 18, *MacKee 33473*). Reproduced with permission from Bull. Mus. Natl. Hist. Nat., B,

The Vegetation Map of Africa

By 1970, Frank had completed the AETFAT/UNESCO Vegetation Map of Africa, which had been commissioned seven years earlier. It was an extension and refinement of maps which had preceded it, notably the AETFAT map of Ronald Keay (1959), and was based on physiognomic vegetation type. For various reasons, the map was not published at that time. Frank then embarked on extensive new fieldwork in Africa and gathered information from at least 100 other members of AETFAT who were working in Africa and Madagascar. As a result, he proposed that the vegetation map be divided into phytochorological regions. These were based on the floristic distinctness of different parts of the continent and the distribution of the constituent plant species, but he was still not completely satisfied with the amount of information he had and admitted that the boundaries were drawn partly intuitively. They have, for the most part, been confirmed by later work and especially by the many standardized distribution maps published since 1969 by the Jardin Botanique National de Belgique, Meise (*Distributiones Plantarum Africarum*).

Frank defined 18 phytochorological regions, which were based on the number of endemic species, rather than families or genera as had previously been attempted in various parts of the world. The recognition of these regions represented a breakthrough for the description of the phytogeography of Africa. By combining the names of his chorological regions with descriptive terms for the physiognomic vegetation types he produced a terminology which was less cumbersome and more informative than earlier attempts to describe vegetation by physiognomy alone. Hence, Guineo-Congolian forest, Afromontane forest and Mediterranean shrubland each succinctly provides information on geography, flora and vegetation type.

The chorological regions themselves fall into five main types: there are seven Regional Centres of Endemism (each of which has more than 50% of its phanerogamic species confined to it and a total of more than 1000 endemic or near-endemic species of phanerogam), one Archipelago-like Regional Centre of Endemism, one Archipelago-like Region of Extreme Floristic Impoverishment, six Regional Transition Zones and three Regional Mosaics. The use of transition zones overcame the problem of precise delimitation of the regional centres of endemism. He always emphasized, however, that 'The boundaries of all the regional centres of endemism would repay detailed study. This is particularly true for regions of physiographic and climatic complexity, such as the Cape and Afromontane regions.' For these, he said that 'Their boundaries are shown only crudely on the Vegetation Map, not always because the information was not available, but sometimes, as for the Cape region, because of cartographic difficulties'. For the Cape and Afromontane regions, he advocated that the Vegetation Map 'be supplemented by other smaller maps showing more detail' (White 1993a: 258, see *Nymania capensis*, p. 259).

He assigned species with wide distributions, and which significantly transgressed the borders of his chorological regions, to various categories, the most well-known of which is 'ecological and chorological transgressor' (the ochlospecies, *D. natalensis*, for example, occurs in six regional phytochoria). As with the taxonomically complex species, he found a great deal of interest in these misfits. In his view, ecological and chorological transgressors provided evidence for direct migration across the African lowlands of plants from one mountain system to another, far distant, one. The distribution and ecology of another group, called 'nomads', suggested the existence of a 'Southern Migratory Track' by which Frank suggested that some species have migrated between the mountains of East and West Africa (White, 1981, 1983c).

Perception

His remarkable achievement in producing a descriptive framework for all physiognomic vegetation types and phytochoria in Africa is a testimony to Frank's extraordinary memory for facts, coupled with an ability to collaborate with people with a wide range of interests, and his remarkable capacity for synthesis. To these qualities, we must add 'perception'. In Frank's opinion, 'most taxonomy (and also much chorology and ecology) would be both prohibitively time-consuming and deadly dull, if it were not possible to detect patterns by using what Adanson and Burtt called perception' (Burtt, 1965: 428; White, 1971: 93).

> 'In this sense, perception is based on the capacity of the human eye and mind to detect pattern using large amounts of visual and factual data before they have been consciously analyzed. ..... Without its use, chorology would never have emerged as a serious intellectual discipline, but without subsequent rigorous analysis, perception is of limited value.' (White 1993a: 253).

New work arising from the Vegetation Map

It was always Frank's hope that his vegetational analysis of Africa would be continually confirmed, improved and refined by new empirical studies. For 11 years after the publication of the map he saw much new work which did these things. He was particularly pleased to extend his classification to Southwest Asia where a phytochorological analysis by Professor Léonard concurred with and extended the classification that Frank had established for north and west Africa (White & Léonard, 1991).

In collaboration with Françoise Dowsett-Lemaire and Jim Chapman, Frank once again focused on the vegetation of Malawi, and Françoise has now completed their book. During the last few years, John Grimshaw has concentrated on the vegetation of Mt Kilimanjaro in Tanzania and has found that his detailed analysis of the vegetation confirms and extends Frank's treatment of the Afromontane region. John Grimshaw belonged to one of the last undergraduates classes which Frank took such pleasure in teaching.

Seed dispersal by vertebrates

A summary of Frank's scientific achievements would not be complete without reference to his interest in the interactions between plants and animals. Increasingly through his career, he looked at the morphology of flowers and fruits and the vegetative parts of plants in the context of their function and encouraged others to do the same. During the 1980s, he expressed great excitement at the results of several of us who were studying pollination, dispersal and ant-plant interactions in the field. These included Helen Fortune-Hopkins' study of pollination of *Parkia* by bats (Hopkins *et al.*, 1998), Françoise Dowsett-Lemaire on seed-dispersal by birds on the Nyika plateau in Malawi (Dowsett-Lemaire, 1988), Caroline Tutin and her co-workers studying gorillas in Gabon (e.g. Tutin *et al.*, 1986), Anat Barnea (see Barnea *et al.*, 1993) on toxins in fleshy fruits eaten by birds in Oxfordshire, and myself working on dispersal by birds and primates in the South East Asian mahogany genus, *Aglaia* (e.g. Pannell & Kozioł, 1987). Partly as a result of the latter, the research on the rubiaceous ant-plants by Camilla Huxley & Matthew Jebb (Hopkins *et al.*, 1998) and his own correspondence with botanists working in New Caledonia, Frank's interest in South East Asia and the tropical Far East grew and he applied this to plant speciation on either side of faunal boundaries (Pannell & White, 1988). The breadth of his own knowledge meant that he was able to interpret the results of detailed studies in one site or on one genus and place them in a much wider context. This often suggested an answer to a question, which he may have been pondering for years, about structure, function or distribution of plants he knew in the field. He confessed that he found this period one of the most exciting of his entire scientific career. In the last few years of his life, he resurrected his boyhood interest in British birds by making observations, from his kitchen window at Taston, of birds taking food which he put out for them. He paid particular attention to what they ate, the way they handled food and the interactions between birds of the same and different species. So, even when he was no longer able to carry out field work in Africa, he maintained an intense interest in British plants and animals and in the field work of others. We and many others are indebted to Frank for his inspiration and encouragement and for sharing his immense knowledge with us.

Acknowledgements

I am grateful to those in the Botany Department, the Plant Resources Center and the Life Sciences Library at the University of Texas at Austin, and especially to Professor Beryl Simpson, who welcomed me and generously provided facilities during the period that this paper was written. The Linnean Society paid my airfare to return to England for the Symposium.

References

Barnea, A., Harborne, J.B. and Pannell, C. (1993). What parts of fleshy fruits contain secondary compounds toxic to birds and why? *Biochem. Syst. & Ecol.* **21**: 421–429.

Burtt, B.L. (1965). Adanson and modern taxonomy. *Notes Roy. Bot. Gard. Edinburgh* **26**: 427–431.

Cain, A.J. (1954). Animal species and their evolution. Hutchinson's University Library, London.

Dowsett-Lemaire, F. (1988). Fruit choice and seed dissemination by birds and mammals in the evergreen forests of upland Malawi. *Rev. Écol.* **43**: 251–285.

Hopkins, H.C.F., Huxley, C.R., Pannell, C.M., Prance, G.T. and White, F. (1998). The biological monograph: the importance of field studies and functional syndromes for taxonomy and evolution of tropical plants. Royal Botanic Gardens, Kew.

Keay, R.W.J. (1959). Vegetation map of Africa south of the tropic of Cancer. Explanatory notes. With French translation by A. Aubreville: 24 p., with coloured map 1: 10 000 000. Oxford University Press.

Mayr, E. (1942). Systematics and the origin of species. Columbia University Press, New York.

Mayr, E. (1963). Animal species and evolution. Belknap Press of Harvard University Press, Cambridge, Massachusetts.

Pannell, C.M. and Kozioł, M.J. (1987). Ecological and phytochemical diversity of arillate seeds in *Aglaia* (Meliaceae): a study of vertebrate dispersal in tropical trees. *Philos. Trans., Ser. B.* **316**: 303–313.

Pannell, C.M. and White, F. (1988). Patterns of speciation in Africa, Madagascar, and the tropical Far East: regional faunas and cryptic evolution in vertebrate-dispersed plants. *Monogr. Syst. Bot. Missouri Bot. Gard.* **25**: 639–659.

Tutin, C.E.G., Parnell, R.J. and White, F. (1996). Protecting seeds from primates: examples from *Diospyros* spp. in the Lopé Reserve, Gabon. *J. Trop. Ecol.* **12**: 371–384.

White, F. (1962). Geographic variation and speciation in Africa with particular reference to *Diospyros*. *Publ. Syst. Assoc.* **4**: 71–103.

White, F. (1971). The taxonomic and ecological basis of chorology. *Mitt. Bot. Staatssamml. München* **10**: 91–112.

White, F. (1981). The history of the Afromontane archipelago and the scientific need for its conservation. *African J. Ecol.* **19**: 33–54.

White, F. (1983a). The Vegetation of Africa: a descriptive memoir to accompany the UNESCO/AETFAT/UNSO vegetation map of Africa. *Natural Resources Research* **20**. UNESCO, Paris.

White, F. (1983b). UNESCO/AETFAT/UNSO Vegetation Map of Africa. Scale 1: 5 000 000. UNESCO, Paris. (French translation by P. Bamps, 1986).
White, F. (1983c). Long-distance dispersal and the origins of the Afromontane flora. *Sonderb. Naturwiss. Vereins Hamburg* **7**: 87–116.
White, F. (1993a). The AETFAT chorological classification of Africa: history, methods and applications. *Bull. Jard. Bot. Belg.* **62**: 225–281.
White, F. (1993b). Twenty two little-known species of *Diospyros (Ebenaceae)* from New Caledonia with comments on section *Maba. Bull. Mus. Natl. Hist. Nat.*, B, *Adansonia* **14(2)**: 179–222.
White, F. (1993c). *Ebenaceae.* In: Ph. Morat and H.S. Mackee (editors). Flore de la Nouvelle-Calédonie 19: 3–89. Association de Botanique tropicale, Paris.
White, F. and Léonard, J. (1991). Phytogeographical links between Africa and Southwest Asia. *Fl. Veg. Mundi* **9**: 229–246.

Friis, I. (1998). Frank White and the development of African Chorology. In: C.R. Huxley, J.M. Lock and D.F. Cutler (editors). Chorology, Taxonomy and Ecology of the Floras of Africa and Madagascar. Pp. 25–51. Royal Botanic Gardens, Kew.

3. FRANK WHITE AND THE DEVELOPMENT OF AFRICAN CHOROLOGY

IB FRIIS

Centre for Tropical Biodiversity, Botanical Museum and Library, University of Copenhagen, 130, Gothersgade, DK-1123 Copenhagen K, Denmark

Abstract

A historical review of some general ideas of chorology and the chorological classification of Africa is given. Early ideas about chorology developed in the immediate pre- and post-Linnaean era, but the first reasonably complete chorological analysis of Africa, with a map of the phytochoria, was produced by Engler only in 1882, the year Darwin died. The permanence of some of Engler's African phytochoria in subsequent phytogeographical classifications is indeed striking; this is thought to illustrate very basic differences in the African flora, especially those between forest and woodland floras. The various chorological concepts and classifications of Africa proposed by Frank White are seen in the light of the development of ideas, and as a basis for future work. The new opportunities to test the broad chorological divisions by, for example, data provided by remote sensing are mentioned, as well as the possibilities opened up by new methods and concepts, for example the opportunity of mapping endemic (restricted range) species by computer technology, and new methods (such as DNA studies) to distinguish between palaeo- and neo-endemics.

Introduction

The first lecture by Frank White that I attended, at the age of 25, was an analysis of the inter-relationship between chorology, ecology and taxonomy, presented at the 7th AETFAT meeting in Munich (White, 1971). This lecture was, as we shall see, one of the focal points in the development of Frank White's ideas of chorology. I come from Scandinavia, where there are strong traditions of phytogeography, and of paying as much attention to biological classification of plants according to their ecological adaptations as to their phylogenetic relationship (Schouw, 1822, 1823a; Warming, 1892, 1908a, b; Raunkiær, 1904, 1905; Du Rietz, 1931; Weimarck, 1941; Fries & Fries, 1948; Böcher, 1977; and Hedberg, 1951, 1961, 1964, 1965; authors often referred to by White). The lecture in Munich prompted my long-lasting interest in White's ideas on plant geography which lead me to an analysis of the forests of the Horn of Africa (Friis, 1992) and of the entire flora of the southern Sudan east of the Nile (Friis & Vollesen, 1998, in prep.).

The last discussions I had with Frank White on chorology took place in the Forest Herbarium (FHO) in 1993, when he was concluding his review of the work with the AETFAT vegetation map and the chorological classification expressed in the "memoir" (White, 1993a). In this large review of finished and unfinished work he briefly outlined the history of some ideas of chorology in his Chapter 2, 'Chorological background.' During our discussions, I mentioned the similarity between the criteria for defining chorological units used in the manual of plant geography by Schouw (1822) and the criteria used by White (1983b), especially the similarity between Schouw's definition of

'Phytogeographical Kingdoms' and White's of 'Regional centres of endemism'. White mentioned the similarity in the published version of the review, and dedicated it to Schouw with the words "one of the more percipient of the founding fathers of chorology." In connection with our discussion, we talked about the need for a historical review in order to enable new generations of botanists to grasp the ideas of chorology and to boost their interest in the subject. White must have approved of the idea because he (White, 1993a) mentioned the intention to write a paper to be called "Some old and new concepts in chorology: a historical review (White in preparation)", but unfortunately the paper never appeared.

The present paper is, of course, very different from anything Frank White would have written, but it deals with at least some of the themes we talked about in 1993. I will trace the history of some of the ideas of chorology which are especially relevant in relation to African chorology. I will also try to see White's work in this light, and trace a few of the themes which we will have to deal with in the future. But first, the term chorology itself must be considered. According to Monod (1957) it is derived from Greek "choros" or "chora", a country or a region, which again is part of a whole range of Greek words, some meaning splitting, division or fragmentation. The dictionary definition of chorology is: "the science of the geographical distribution of anything", but here we will restrict ourselves to plants and animals, which was the original meaning of the word when coined by Häckel (1866). I have earlier dealt with some aspects of this subject in a review of the chorological division of Ethiopia (Friis, 1986). Schnell (1976–1977) has reviewed the major chorological classifications of Africa, and Werger (1978) has summarised the proposed chorologies for southern Africa; here there will of necessity be a certain overlap with these papers.

Early development of plant geography and the beginning of chorology

The roots of plant geography are in the 18th and the first half of the 19th century, largely in the period between the early works of Linnaeus and the late works of Darwin. Some problems from this period are still with us, so I will outline the early development of plant geography as it is relevant to African chorology.

Linnaeus, Willdenow and the centres of origin

It has often been said, for example by Drude (1890), that chorology, and plant geography in general, had its origin in studies of the distribution of arctic-alpine flora in relation to the floras of warmer climates, and it is notable that Linnaeus was aware, from the travels of Tournefort (1717), that northern plants occurred at high altitudes on the mountains of Asia Minor. I have not searched through all original works of Linnaeus for statements on plant geography. However, it is clear from aphorism no. 132 of *Fundamenta botanica* (Linnaeus, 1736) that at that time he believed all species originated from individuals which had been created at the same time, but that he also carefully noted the current habitats and the general area of the species, and that he was aware of the existence of particular floras specific to certain regions. He wrote little about how to reconcile these notions, but in a speech on an academic occasion, *Oratio de telluris habitabilis incremento* (Linnaeus, 1744), he discussed in numbered aphorisms the changing shorelines of the Baltic Sea, and what could be concluded from these observations. In 1744 he went as far as to speculate on a single centre of origin of the world's floras: [text quoted beginning with aphorism no. 16:] "All the continents of the earth were in the first ages immersed under a vast ocean, except a single island, where all animals lived ... and all plants grew. ... [In order to allow for the existence of tropical plants and animals this island, presumably identical with the Biblical Paradise, must, according to aphorism no. 46, have been] situated under the Equator. [And in order to allow also for plants and animals from a cold climate it is according to aphorism no. 47 necessary to assume] a high mountain in the middle of this region where the climate would be cool." He did not suggest how the very different floras of the world developed through dispersal from the primordial island. A. De Candolle (1855: vi) called the whole idea "bizarre and erroneous", and it would not have been worth mentioning, had it not been for the almost contemporary statement by Buffon (1761: 96) on mammals that leads to conflict with the Linnaean idea of one common centre of origin for all created beings; according to Buffon it is "a general fact, which at first appeared very strange, and which no one before us even suspected, that no species of the torrid zone of one continent is found in the other." Buffon's idea was later developed by A.-P. De Candolle (1821) in connection with the establishment of the first system of phytochoria which were precisely designed to demonstrate such floristic differences under conditions which should otherwise have been assumed to be identical. Moreover, the Linnaean interest in "primordial mountains" was shared by Willdenow, one of the first founding fathers of chorology.

Willdenow's general handbook of botany (Willdenow, 1792) contains the first outlines of a general presentation of plant geography, called *The History of Plants*. However, I will quote from the second edition in which Willdenow (1798) expanded the "history of plants" considerably. At the beginning of his account Willdenow directly stated one of his main themes, that the physical conditions of the locality determines the plants which grow there (section no. 348): "Plants grow where it has been determined that they can grow." He restated that one can find plants from the polar areas on the peaks of high mountains closer to the equator, and he was not surprised that different species grow at the same latitude in Asia, Africa and America. He described the influence of climate on the physiognomy of plants growing in the polar regions, in Europe, Asia, Africa, America, Australia, the Mediterranean region, Arabia and other areas. He also gave a list of the vicariant species which occur in north-east Asia and North America, and described similarities between the Cape flora of South Africa and parts of the Australian flora (section no. 351).

Willdenow's idea was that mountains were stable places in the great revolutions the earth has undergone (section no. 353), although (section no. 365–366) he found it difficult to point to specific places from where plants have originated and begun their dispersal over the globe. However, he attempted to do so for Europe and subdivided the entire European flora into five main floras [here corresponding to floristic regions], each with a distinct area of origin: 1) The Nordic flora, distributed from Scandinavia and part of Great Britain to Russia; 2) the Helvetian flora, distributed from the mountains of Switzerland, Bayern and Tyrol, through part of France to southern England, 3) the Austrian flora, distributed from Poland through the Czech Republic, Slovakia, Hungary, Austria, to the Balkan Peninsula; 4) the flora of the Pyrenees, distributed through Spain, Portugal and the Balearic Islands; and 5) the flora of the Apennines (Italy with the large Mediterranean islands). This is the first time the idea of specific flora areas (phytochoria) is expressed with a reasonably clear indication of their extent. Willdenow points out that the plants must have been distributed from mountain centres to more peripheral places, a subject which is further illustrated (section no. 366) in his discussion of the mixing of floras in places where several floras meet.

Willdenow had no theory about the origin of species, and his floristic data were very far from complete. In an early work Schouw (1816) reacted against Linnaeus' and Willdenow's theories about the origins of floras, which Schouw, like Willdenow, called "the history of plants." Schouw clearly stated that nothing certain could be deduced about this, and that the speculative subject called the "history of plants" (the distribution of plants in the past; historical plant geography) should be strictly kept apart from the exact science of the current distribution of plants (chorology).

This sharp distinction between historical and analytical plant geography was also White's. He referred (White, 1979) to the special associations which might wrongly be attached to his own chorological term "Centre of endemism" (that they sounded like "centre of origin" in the Linnaean and Wildenowian sense), pointing out that his term was entirely descriptive and analytical.

Humboldt, Brown, Wahlenberg and Hornemann, the physical environment, life forms and 'arithmetica botanica'

Humboldt is often considered the founder of plant geography with his "Essai sur la géographie des plantes" (Humboldt, 1806; in German, Humboldt, 1807). This work summarised many subjects which had been touched upon by earlier authors, but new emphasis was placed on the importance of physical factors on the physiognomy of vegetation and plants; the work is therefore also an important starting point in physical geography. Humboldt was the first to introduce physiognomic life forms as well-defined concepts in plant geography. Humboldt's life forms of vascular plants are: the banana form, the palm form, the form of the tree ferns, the *Aloë* form, the *Pothos* form, the form of coniferous trees, the form of the orchid, the form of the mimosaceous tree, the form of the malvaceous tree, the form of the woody liana, the form of the liliaceous plant, the cactaceous form, the form of the casuarinaceous plants, and the grass and sedge form. The concepts of life forms now used in the classification of vegetation formations were developed by Warming (1892, 1908a & b, 1923), Raunkiær (1904, 1905), Du Rietz (1931) and others.

Brown (1814), in Flinder's *Voyage to Terra Australis*, analysed the frequency of the families of flowering plants in remote areas. His studies continued with analyses of plants from the Congo River collected by Christian Smith on his journey with Captain Tucky (Brown, 1818). The calculations used by Brown, sometimes called "arithmetica botanica", were initially used by A.-P. De Candolle (1813) for statistical studies of the relative importance of plant families.

Humboldt's last phytogeographical work appeared as the introduction to the large taxonomic work *Nova Genera et Species Plantarum* (Humboldt, Bonpland & Kunth, 1816), and independently in a slightly expanded version (Humboldt, 1817). The work took up the statistical analyses of A.-P. De Candolle (1813) and Brown (1814) in order to establish the proportions of the larger families in various major geographical areas and thus characterise these regions floristically. A prerequisite for this was a natural system, and Humboldt adopted the system of Jussieu (1789). The work intended to compare floras of large areas at the same latitude, such as geographically well-defined areas in the Old and the New World, and large areas at different latitudes; it documented that the distribution of families and genera was far from uniform. Also in other respects Humboldt was a pioneer — with regard to zonation of mountain floras and the relationship between vegetation physiognomy and climate, Humboldt (1817) has long been an inspiration for plant geographers.

In Scandinavia Humboldt's ideas were well received, the ground having been prepared by a number of almost contemporary works by Wahlenberg on zonation of arctic-alpine flora and vegetation (Wahlenberg, 1812, 1813, 1814). Together, Wahlenberg's works formed an excellent comparison of the flora and zonation of vegetation in northern Sweden, the Alps and the Carpathian mountains. In the introduction to the first work the Scandinavian flora is compared with the flora of Siberia, as described by Gmelin (1747–1769). Wahlenberg's studies of the correlation between average temperature and plant distribution has been another starting point for numerous phytogeographical studies.

An early attempt to compare the floras of Guinea and Congo was made by Hornemann (1819). He based his comparison on Smith's collections from the Congo River and Thonning's collections from Ghana, but his floristic data from only two sources were of course very incomplete. Hornemann agreed with Brown about the importance of the Cyperaceae in the tropics; on this point they both disagreed with Humboldt who had suggested that the family had the lowest number of species near the equator. Hornemann and Brown also agreed that within the Poaceae the Paniceae is the most prominent group in the tropical lowlands, while the Poeae are more prominent in the uplands and in the extratropical regions. White (1983a: 168) also discussed the importance of the various grass tribes in Afromontane habitats. Hornemann pointed out that palms are very poorly represented in Africa, while Leguminosae are very prolific; even today these observations are recognised as striking features of the tropical African flora. Although White did not apply statistics to family prominence, he has in the floristic characterisation of his phytochoria pointed out prominent families and genera (White, 1983b: 74, 87, 111, 147, etc.)

A.-P. De Candolle, Schouw and Meyen, the beginning of chorology and vegetation geography

Chorology, in the sense of the word used here, appeared when botanists began to divide the globe into areas with more or less well defined floras, or combined this with other factors which also characterise the areas. The first system which classified the entire world into units on such criteria was proposed by A.-P. De Candolle (1821). His 20 phytochoria were so defined that a sufficiently large number of species were specific to each area [endemic], "... un certain nombre de plantes particulière, qu'on pourrait nommer véritablement aborigénes."

Soon after, Schouw (1822, 1823a) presented in his general textbook of plant geography a rather similar classification of 22 phytochoria, based on more detailed criteria. Schouw was the first to use maps to depict the distribution of plants, mapping both the distribution of species (*Fagus*) and whole families, e.g. Cactaceae, and showing the precise position and delimitation of his phytochoria (Schouw, 1823b, 1824). One of the points which interested White (1993a) was that Schouw's criteria for defining phytochoria are surprisingly detailed. He used presence or absence of characteristic endemic species, genera and families (called taxa "specific to the area" by Schouw). Another of Schouw's ideas about chorology was that the phytochoria should be defined on criteria which could be observed directly, without making any assumptions about the historical development of the flora (in opposition to Willdenow). White agreed completely with this (see, for example, White, 1993a: 229).

Schouw's criteria for establishing phytochoria of the highest rank, "kingdoms", were: (1) at least 50% of the species should be endemic, (2) at least 25% of the genera should be endemic, and (3) at least one or more families should be endemic or at least mainly distributed within the floristic area. White (1979) required that a Regional Centre of Endemism should both have more than 50% of its species confined to it, and a total of more than 1000 endemic species. The percentage and total number of endemic genera, and the presence of endemic families were also considered important factors by White, but he felt that neither specific criteria for percentage nor limits for total numbers could be given.

Schouw also accepted less well marked phytochoria, based on species endemism, which were classified as "provinces" within the kingdoms. Schouw's system is therefore hierarchical. Schouw recommended that the definition and delimitation of phytochoria should be based on the study of congruent distributional areas for many species, but he realised the difficulties of phytogeographical comparisons between closely adjacent areas because of the many species which may just transgress the border and thus blur the pattern. He therefore suggested (Schouw, 1822: 447) that areas should be delimited primarily on the physiognomy of their vegetation, and then the floras of the central parts of the areas so defined could be compared according to his principles. He also suggested guidelines for the description of phytochoria, and emphasised that a good description should include an account of the endemic taxa, the characteristic vegetation, and the characteristic life-forms (trees, shrubs, herbs, etc.). Schouw's methods are therefore basically floristic, but with notable physiognomic elements, as is the system of White (1983b), and there is clear similarity between the way Schouw characterised his phytochoria and what White chose to include in the descriptive 'memoir' (White, 1983b).

Schouw also documented that mountain floras were sometimes discontinuously distributed. In this, he used the work of others, especially Wahlenberg. He pointed out that of 125 species observed in the mountains of Central Norway only five did not also occur in Switzerland (Schouw, 1822: 446). This is also a rejection of Willdenow's distinction between Scandinavian and Alpine flora areas. According to Schouw, areas within the same mountain system could sometimes belong to two different phytochoria because the floras of the higher regions and the foothills could be so different that the conditions for the establishment of two phytochoria of high rank were present (Schouw, 1822: 448). As can be seen from the phytogeographical map (Schouw, 1823b, 1824), phytochoria could be disjunct, for example his arctic-alpine phytochorion. The parallel with Hauman's discontinuous Afroalpine (Hauman, 1955) and White's Afromontane Region is clear.

For the floristically better known parts of the world, Schouw's chorological "kingdoms", which were named after prominent plant families, agree fairly well with the phytochoria known today, but tropical Africa, especially the inland regions, was poorly known, and Schouw decided not to classify the area (Schouw, 1822: 461). On his lithographed phytogeographical map Schouw (1823b, 1824) has left the inland areas of tropical Africa blank, as can be seen from the redrawn version shown in Fig. 1. Schouw (1833, 1834) later proposed a system with 25 phytochoria based on prominent genera, but both the system itself and the reasoning behind it are inferior to what had been proposed in 1822–23. In the German translation (Schouw, 1834), he even proposed a new terminology for the phytochoria, based on the names of prominent botanists; an idea which was ridiculed by A. De Candolle (1855). The phytochoria of White and most other modern phytogeographers are named from geographical features.

The work of Meyen (1836) is an approach to vegetation geography, based on the physiognomy of plants and vegetation and their relation to the climate. The first part of his book deals with the climate of the earth, the second part with plants of various physiognomic formations and factors which may limit plant distribution, the third part is an elaboration of Humboldt's plant forms, a physiognomic description of vegetation types, and the final section deals with "arithmetica botanica." His division of the world into phytogeographical areas is based on climate and physiognomy. Many plant geographers followed these lines in the 19th and early parts of the 20th century, e.g. Schimper (1898) and Gräbner (1910). The works of Schnell (1970–1971) and Knapp (1973) are also in that tradition.

A. De Candolle, the great critical review

A. De Candolle's large work on plant geography (De Candolle, 1855), with the subtitle *The geographical distribution of plants in the present period*, begins with discussions of temperature, light and humidity, followed by sections on distributions of species, genera and families, physiognomy of vegetation, differences in prominence of species, and finally in chapter 25, a discussion of chorology, "De la division des surfaces terrestres en région naturelles." He rejected Willdenow's idea of the phytochorion representing an area the flora of which had single origin, the mountains of the region, "n'offrent aujourd'hui qu'un seul mérite, celui de prouver le changement qui s'etait fait dans la science, depuis l'epoque où Linné avait accrédité la théorie d'un centre unique pour toutes les espèces du règne végétal." He also criticised the 20 phytochoria proposed by his father (A.-P. De Candolle), who had defined areas so that a large number of species were specific to them, by not applying strict criteria. According to A. De Candolle, Schouw had followed a better line by defining clearly a set of criteria; however, Schouw's set of criteria was criticised for being arbitrary; why not demand

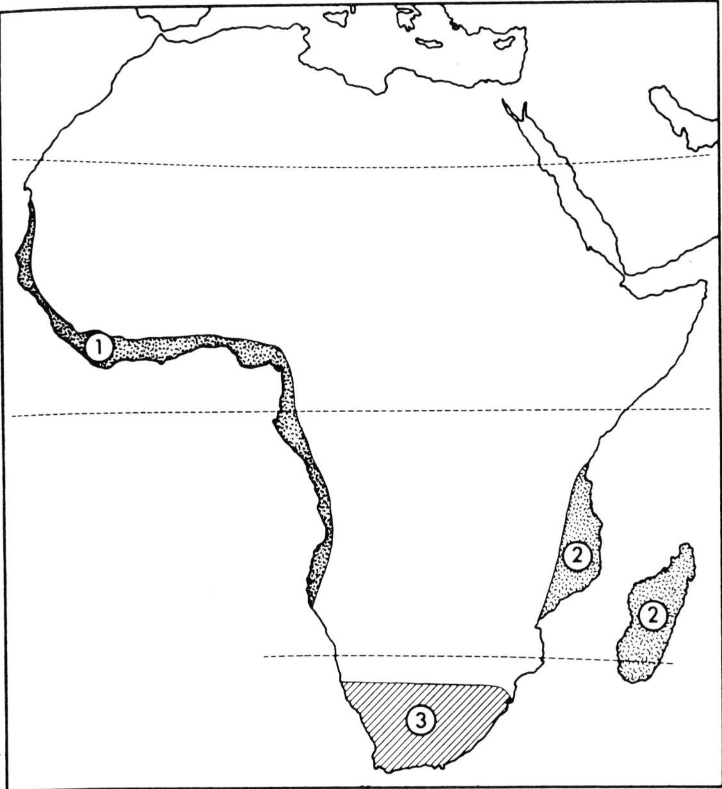

FIG. 1. The phytochoria of Africa and Madagascar, according to Schouw (1822). Reproduced from a redrawn map in Werger (1978). 1. Regum Africae occidentalis. 2. Regnum Africae orientalis. 3. Regnum Mesembryanthemorum et Stapeliarum.

that $^2/_3$ of the species should be endemic, and why not require a higher fraction of endemic genera? De Candolle further criticised Schouw's later set of chorological criteria (Schouw, 1833, 1834) for being more arbitrary than the criteria of Schouw's early works. White (1979) addressed these problems and stated: "In the author's opinion lack of agreement has partly stemmed from ignorance, partly from lack of clearly stated objectives, and partly from insufficiently refined methods of analysis." White's criteria for his Regional Centre of Endemism (that it should have both more than 50% of its species confined to it and a total of more than 1000 endemic species) were admittedly arbitrarily chosen, but continued testing showed that they gave the "best fit" with reality.

A. De Candolle completely rejected all previous attempts at producing a chorological classification of the world, claiming that the results had so far been too different, and that no two scientists had been able to produce identical classifications. A natural classification of the world into phytochoria seemed impossible because: 1) The plant world was too poorly known, and 2) scientists did not apply sufficiently logical criteria. White was completely aware of these risks, and saw chorology as a continually tested approximation (White, 1993a).

The first development of African chorology

Grisebach (1871) was the first after A. De Candolle's synthesis to summarise the steadily increasing information on the African flora. He divided Africa into four phytochoria: 1) the Mediterranean north of the Sahara; 2) the Sudan region, covering most of tropical Africa; 3) the Kalahari region, including Namibia and the Karoo; and 4) the Cape region, largely restricted to south of the Orange River. This classification is based on a combination of physiognomy of the vegetation and "natural floras" ("natürlichen Floren"), with no specially defined criteria of species richness or endemism. Grisebach's definitions of the Mediterranean and Cape floristic regions are elaborations of the phytochoria defined by A.-P. De Candolle and Schouw, and their distinction within African chorology is still upheld. Grisebach's other African phytochoria have been much modified.

Engler was a major player in the development of ideas for African chorology after the general acceptance of Darwin's *Origin of species*. Engler elaborated on the phytogeographical division of the world in several major works. In the first of these (Engler, 1882), which actually appeared the year Darwin died, Engler drafted 36 phytogeographical aphorisms ("Leitende Ideen") most of which are more related to historical plant geography than to chorology. Some of the aphorisms are well known and fairly widely accepted, for example: "21. The difference between old and new flora areas with a large degree of endemism normally consists in a low number of species in the genera of the former." "29. The presence of monotypic genera in a flora area is often somewhat accidental, and is only relevant in characterisation of the area in so far that it shows that the former conditions have lasted longer in this area; monotypic genera are hence only of relevance in the characterisation of larger areas in which they are generally distributed." "31. Sharp boundaries between the flora areas do not exist; elements of one area always transgress the boundaries and indeed to a varying degree in various periods of the history of the earth." Other of Engler's aphorisms are now highly controversial, for example: "7. Observations have shown that closely related forms in a species complex develop in the same place."

In his discussion of the African flora, Engler (1882) analysed the conclusions of Grisebach (1872), and argued for placing the Cape flora in a region ("Reich") together with extratropical Australia and New Zealand, but not the extratropical part of South America. In tropical Africa, Engler argued for the recognition of two tropical African "floras", a forest flora, and a savanna, bushland and grassland flora. These floras occupied two distinctive areas, the West African Forest region ("Westafricanisches Waldgebiet") and the African-Arabian Savanna, Bushland and Grassland region ("Africanish-arabisches Steppengebiet"). The borderline between the two regions follows the assumed border between the evergreen forest and the savanna woodlands and grasslands. Using sparse floristic data from the early volumes of the *Flora of Tropical Africa*, Engler concluded that 32% of the tropical African genera occurred also in the tropics of both America and Asia, 19% in both tropical Africa and Asia, 6% in both tropical Africa and America, 7% in both tropical and South Africa, and 4% in tropical Africa and Madagascar, while 23% were endemic to tropical Africa, and the rest had other distributions. Tropical Africa by itself did therefore not quite meet Schouw's criterion of a generic endemism of 25% in order to be recognised as a region. Engler also pointed out that both the generic and the specific endemism of the West African Forest region was higher than in the African-Arabian Savanna, Bushland and Grassland region. The classification is shown on a coloured lithographed map of the entire world, and has been redrawn here as Fig. 2. Of Engler's conclusions about African chorology, the distinction between the Guineo-Congolian forest flora and the Sudano-Zambezian savanna flora (as White (1983b) chose to call them) is the one which has best stood the test of time, and it reappears in all modern classifications.

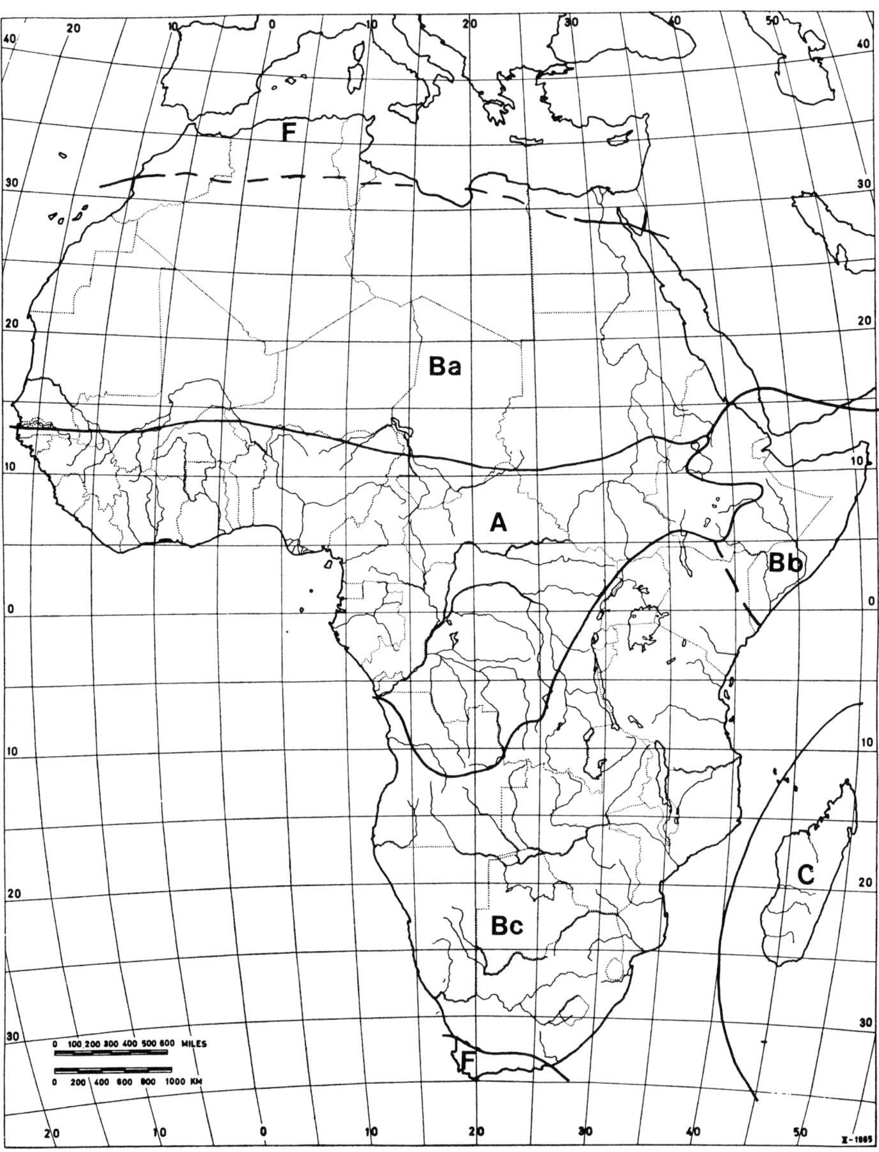

Fig. 2. The phytochoria of Africa and Madagascar, according to Engler (1882). Reproduced from a redrawn map in Friis (1986). **A** West-African Forest Region ("Westafrikanisches Waldgebiet"). **B** African-Arabian Grassland Region ("Afrikanisch-arabisches Steppengebiet"). **Ba** North Africa-Indian Grassland Province ("Nordafrikanisch-indische Steppenprovinz"), **Bb** Abyssinian Province ("Abessinische Provinz"). **Bc** South African Province ("Südafrikanische Provinz"). **C** Malagasy region ("Malagassisches Gebiet"). **F** Mediterranean Region and Cape Region ("Mittelmeergebiet"; "Gebiet des Caplandes"). Reproduced from Symb. Bot. Upsal. 26 (2): 68–85, with permission.

Later, Engler (1899) abandoned the basic distinction between a forest- and a savanna-flora, and suggested that the floras of tropical Africa were mutually more closely connected than they were with any other area. The whole of tropical Africa was therefore placed together in a phytochorion termed "African forest- and savanna region" ("Africanisches Wald- und Steppengebiet").

This idea reappeared in his last works on African phytogeography (Engler, 1908, 1910: XVIII–XX). Both contain a detailed hierarchical classification (Engler, 1910). I have traced no original maps showing this classification, which due to its details is difficult to map on a single sheet. It has been said that the Englerian chorology of Africa was mainly based on endemism in taxa of higher rank. This is only partly true: in Engler (1882) the phytochoria of Africa were largely based on species endemism. A simplified version of Engler's chorological classification of Africa was used in Engler & Gilg (1912), and representations based on this have been in use until recently. Thonner (1908) showed the classification on a chorological map of Africa. Good's chorological classification of Africa (on the phytogeographical map of the world in Good (1947) and subsequent editions) has a strong Englerian flavour. The system is almost identical with the one adopted for the latest edition of Engler's *Syllabus der Pflanzenfamilien* (Mattick, 1964). Robyns (1948) has also largely followed Engler's ideas. Takhtajan (1969) stated clearly that his chorological map is based on that of Engler, and that he based his phytochoria on the presence of endemic families and, if such are not present, then on a high frequency of endemic genera or very high occurrence of species of certain families, in the Cape Kingdom for example an abundance of Asclepiadaceae, Rubiaceae, Lamiaceae, etc. Similar principles were applied by Takhtajan (1986).

Development of modern chorological classification systems for Africa

Eig (1931a, b, 1933) made a thorough analysis of the flora of Palestine, and gave very careful definitions of the elements used in his analysis. His definitions of the chorological areas and elements have frequently been referred to by students of African chorology, and they agree well with the definitions used by White.

Chevalier (1933) mapped and discussed the West African phytochoria; his classification is very similar to the one accepted now. He called attention to what he called the Dahomey-Lagos recess ("échancrure") in the Guinean forest zone.

Lebrun (1947) resurrected the idea that there was a principal distinction between the forest and savanna floras, and that they should therefore be placed in different phytochoria of high rank. This was partly echoed in the palaeoecological views expressed by Aubreville (1949a, b, c), in which he distinguished the "forêt dense humide" and the "forêt claire et savanes boisées" according to the different histories of climatic changes, which the two systems have gone through. The conclusions about the different forest and savannah floras are independently confirmed by modern floristic evidence, as for example presented by White (1979). However, Lebrun placed the tropical African mountains in the same chorological categories as the surrounding lowlands, some mountains raising out of the forest region, others out of the savannah region (Fig. 3).

Hauman (1955) was the first to take up, for tropical Africa, Schouw's old idea of placing the higher parts of mountains into a separate phytochorion because of a distinctive flora. He proposed a discontinuous phytochorion called the Afroalpine region for the peaks of the East African mountains. This proposal agreed with the work of Hedberg (1951, 1955), who had analysed the zonation of the vegetation on the tropical African mountains, and a few years later revised the Afroalpine flora (Hedberg, 1957).

In a monumental synthesis of what had up to then been achieved in African chorology, Monod (1957) chose to adopt a hierarchical classification of African phytochoria (the

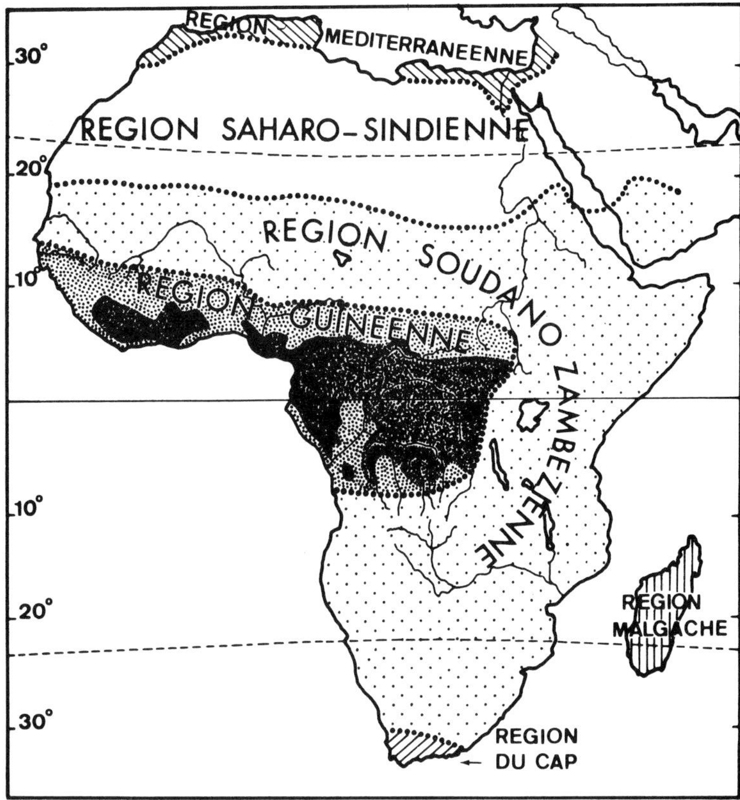

FIG. 3. The phytochoria of Africa and Madagascar, according to Lebrun (1947). Reproduced from Schnell (1976–1977).

terminology in French being "Region" and "Domaine", with subdivisions where found appropriate; Fig. 4). Monod followed Lebrun in stressing the contrast between Guineo-Congolian forest region and the Sudano-Zambezian woodland and savanna region. To accommodate the different mountain floras, he devised a "Faciès montagnard" of the various regions in which mountains occurred. The paper also gave a methodological review and emphasised the importance of large data sets through collaboration between many botanists. In the long introduction he also listed more than 40 slightly different chorological classifications which had been proposed for Africa since the early works of Engler, apparently confirming A. De Candolle's disillusioned words about chorologists never being able to reach consensus.

Frank White and African chorology

With the work of Lebrun, Hauman, Hedberg and Monod we reach the time around 1960 when Frank White began to work on African chorology. Throughout his life as a botanist it was White's intention to improve the taxonomic and ecological foundations of chorology; this broadly based interaction between three classical botanical disciplines is an important characteristic of White's work. Another characteristic is the close connection between White's theoretical works and his detailed empirical studies;

35

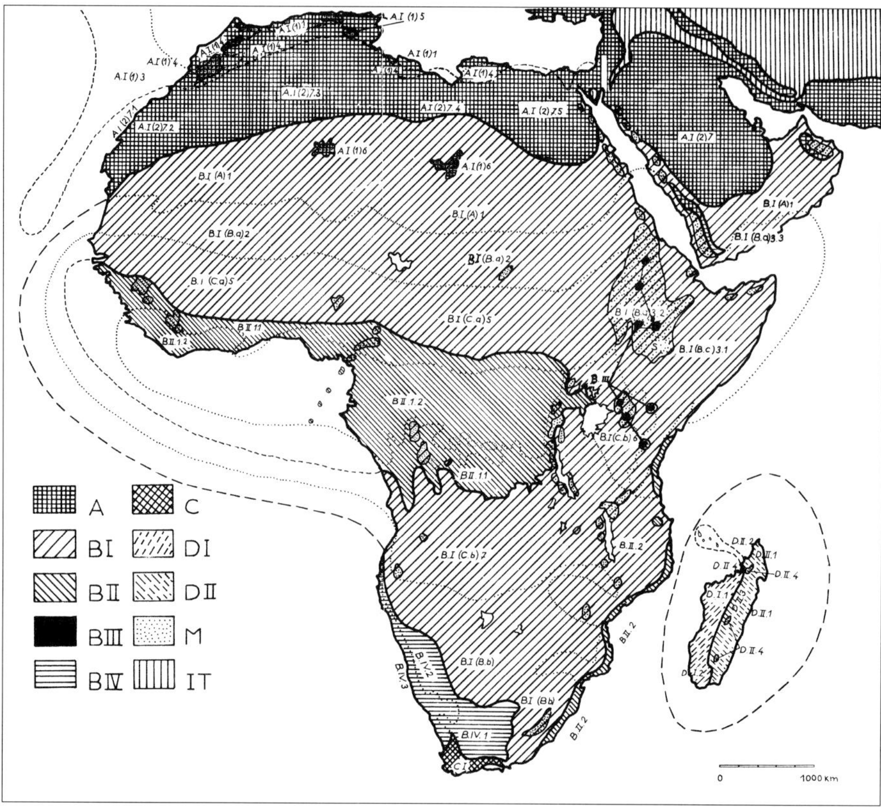

Fig. 4. The phytochoria of Africa and Madagascar, according to Monod (1957). Reproduced from Schnell (1976–1977). **A** GROUPE MÉDITERRANÉEN. **AI** Région méditerranéenne. **AI(1)** Sous-Region eu- méditerranéenne. **AI(2)** Sous-Région saharo-sindienne. **B** GROUPE TROPICO-AFRICAIN. **BI** Région sudano-angolane. **BI(A)** Types sahariens. **BI(B)** Types sahéliens. **BI(C)** Types soudaniens. **BII** Région guineo-congolaise. **BIII** Région afro-alpine. **BIV** Région Karoo-Namib. **C** GROUPE DU CAP. **CI** Région du Cap. **D** GROUPE MALAGACHE. **DI** Région occidentale. **DII** Région orientale. **IT** IRANO-TOURANIEN. **M** Faciès montagnard. The list of hierarchically arranged phytochoria is far from complete; no names of "Domaines" are listed here.

new experience was acquired, and at intervals this experience was summarised in theoretical syntheses. In the following, I will first summarise the theoretical work, and then some of the empirical studies.

The theoretical papers on chorology

The old problem of whether a chorological classification should be hierarchical or not was addressed in White's first major work with a theoretical approach (White, 1965). He concluded that in African chorology the emphasis should be on the highest level, the Region. In this work, the criteria for defining Regions were the presence of distinctive floras with high degree of endemism, distinct vegetation types, that is distinct physiognomy, and different life form spectra. Already here the importance of representative samples was stressed, as a total analysis of the entire flora would tend to blur the picture. Division of phytochoria into lower categories would be unprofitable because many species would transgress their boundaries, and the units would vary considerably as to the distinctness of their flora.

In his survey of ideas for the interaction between chorology, ecology and taxonomy, White (1971) stated as an aphorism that chorological patterns are too complex to provide an objective basis for a formal hierarchical classification except at the highest levels. The aphorism was supported by the reasons that: 1) Adjacent phytochoria may differ greatly in floristic richness and in the complexity of the patterns formed by their subordinate elements. 2) The replacement zones separating adjacent phytochoria vary greatly in their width and steepness. 3) The difference between adjacent phytochoria may be due either to both of them having distinct endemic floras, or to one of them being merely an impoverished variant of the other. As a consequence of these statements, White again argued that 'centres of endemism' (here only on the level of the local centres of Weimarck (1941)) can be more profitably recognised than the lower ranking phytochoria of formal systems, because centres of endemism need not have contiguous boundaries. Another conclusion was that it is much more important to recognise patterns of endemism than to elaborate complex and static hierarchical classifications.

The consequences of these considerations emerged in the work with the vegetation map of Africa and White's accompanying 'memoir' (White, 1983b). While waiting for the publication of these White presented an essay on the ideas behind the work (White, 1976a). In this he re-examined the old question of coincidence of boundaries between physiognomically and floristically defined phytogeographical areas. He restated that there is no *a priori* reason why the pattern lines on a vegetation map based on the physiognomy of vegetation should coincide closely with those of a chorological map based on coinciding distributional limits of species. But the results of his work with the vegetation map of Africa showed that if the chorological map of Africa was based on chorological data alone, rather than on transferring pattern lines from a detailed vegetation map, the pattern lines would not have been significantly different. Phytogeographical maps showing phytochoria of high rank can therefore be based on vegetation maps. Only phytochoria of high rank were to be considered, and for the reasons quoted from White (1971), the vegetation map would be divided into a system of Regional centres of endemism, Transition zones, and Regional mosaics also used in the 'memoir'. Although the complete reasoning behind it was only presented in the next paper (White, 1979), the idea of using Centres of endemism at the rank of the traditional chorological unit, the 'Region', was already put to use (White, 1976a).

The paper which analysed the Guineo-Congolian Region (White, 1979) also contained some theoretical considerations of wide-ranging importance. White here pointed out that a system based on species endemism would serve the chorological "middle-level" — the Region — best, rather than placing too much emphasis on endemic genera and families. White also pointed out that it is possible to reach a reasonable consensus about "Regional centres of endemism" because of the objective criteria on which they are based. It would be much more difficult to reach consensus if floristically poorer phytochoria were to be given rank in a hierarchical system; the position and circumscription of the floristically poor phytochoria was one of the main causes for differences in opinion among plant geographers interested in chorology. White (1979) first stipulated the criteria of a Regional Centre of Endemism: it should both have more than 50% of its species confined to it, and a total of more than 1000 endemic species. It was also established that Regional centres of Endemism always have Transition zones between them, but these may be so narrow that they are difficult to map, e.g. the Transition zone between the Cape and the Karoo-Namib Regions, and the horizontal projection of the transition zone between the Afromontane Region and the surrounding phytochoria. One of the criteria for Transition zones was that they have less than 50% endemic species.

While the Transition zones were simple replacement zones, more complex situations involving the meeting of several Regional Centres of Endemism needed a new term, the Regional Mosaic. This concept seems to be entirely White's own idea, and one of the more controversial ones. The phytogeographical area around Lake Victoria contains a mosaic of vegetation types and floras, is much influenced by human activity, and holds relatively few endemic species. The Indian Ocean Coastal Belt has a similar complexity and history of human influence, but has a higher degree of endemism (see Clarke this volume).

The chorology of African savanna woodland

The African woodlands were originally an exception to White's disapproval of hierarchical classification of phytochoria (Fig. 5); according to White (1965) the entire non-forested lowland region of tropical Africa, the Sudano-Zambezian Region, could readily be divided into Domains: a northern Sudanian, a north-eastern Oriental (covering the lowlands of Ethiopia, Djibouti and Somalia, as well as parts of the Sudan, lowland Kenya and Northern Tanzania), and a southern Zambezian Domain. Areas with a specially high degree of endemism within regions could be considered 'centres' or 'subcentres of endemism and such subcentres might or might not be present in each region. The idea of centres was at this point, at least in part, inspired by the analyses of floristic centres and the intervals between them which Weimarck (1941) had carried out in the Cape Region and the mountains of Southern Africa. The subcentres of endemism proposed by White (1965) were all in the Zambezian Domain, viz. the Katanga Centre in South East Zaïre and the adjacent parts of Zambia, the Barotse Centre, covering the areas in Angola, Zambia and Botswana around the Caprivi Strip, and the Kariba Subcentre around the upper part of the Zambezi River, while the Sudanian Domain was floristically impoverished, and the Oriental Domain little known. The Sahel was given status as "Sahelian extension of the Oriental Domain" of the Sudano-Zambezian region.

White's later chorological works on woodlands and savannahs are either directly associated with the 'memoir' (White, 1983b), or other aspects of plant geography, such as historical plant geography (White, 1990). A precursor to the 'memoir' is the analysis of the Guineo-Congolian/Zambezia Transition Zone (White & Werger, 1978). It was concluded that this zone was characterised by many typical rain forest trees which occurred as marginal intruders, especially along rivers, and a much smaller number of Guineo-Congolian linking species. In other works on woodlands the emphasis is on historical plant geography; they are mentioned in the historical section below.

The chorology of Afromontane habitats

White's first sketch map of African chorology (White, 1965) indicated a distinct Afromontane region, and he stated in the text: "Afro-montane vegetation should be recognised as forming a distinct floristic region comparable to the Afro-alpine Region ...". Other students of African chorology had reached conclusions which resembled White's ideas about an Afromontane phytochorion. Troupin (1966) advocated the establishment of a distinct Afromontane region, rising out of various other phytochoria, and found it in "tous les endroits de l'Afrique atteignant et dépassant une altitude de 2000 m. Suivant certaine conditions locales, cette limite inférieure peut être située plus bas." He considered the Afroalpine Region a subdivision of the Afromontane and stated that three zones could be found on a large number of mountains: 1) the mountain forest zone; 2) the ericaceous zone; and 3) the afro-alpine zone. With regard to priority of his map he stated: "Le texte de ce mémoir a été déposé le 4 avril 1965. Peu après, nous avons eu connaissance de deux travaux relatifs aux divisions phytogéographique de l'Afriques ...". One of these two works is White (1965); Troupin further pointed out that he had produced a map showing the Afromontane region in his unpublished thesis

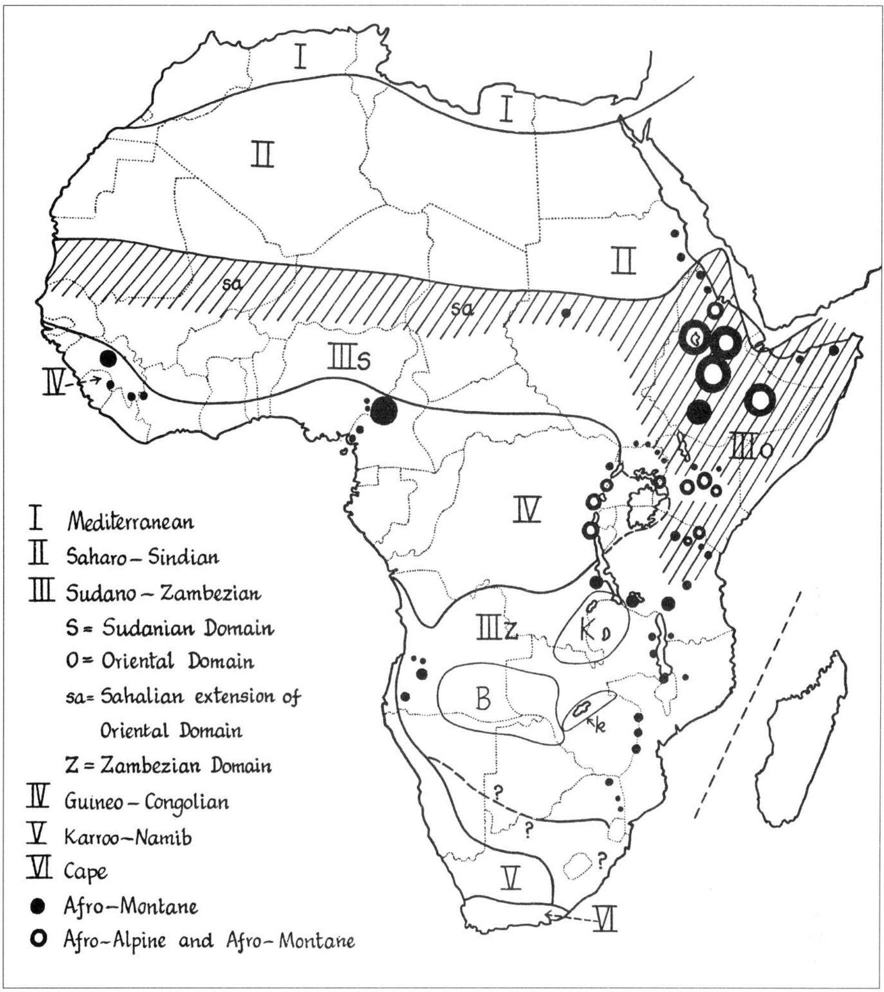

FIG. 5. The phytochoria of Africa, according to White (1965). Reproduced from White (1965). On the map three centres of endemism are indicated: **K** Katangan Centre; **B** Barotse Centre; **k** Kariban Subcentre. Reproduced from Webbia 19: 658 (1965), with permission.

from 1956 (this map is reproduced in Troupin 1966). Wickens (1976) analysed the flora of Jebel Marra, a mountain in Western Sudan surrounded by Sahel vegetation, and confirmed the existence of a distinct Afromontane Region there.

The major new feature of the next version of the chorological map of Africa (White in Chapman & White, 1970) was the southwards extension of the Afromontane region to sea-level in the Knysna Forest of the Cape province; this is a complete parallel with the classical chorology where montane phytochoria descend to near sea-level towards the poles. The Afromontane region, especially its woody flora, was further analysed by White (1978), when he compared the conditions and floras of seven blocks of mountains, from West Africa to Ethiopia, and south to the Drakensbergs. The distributions of 152 tree species were analysed, and it was shown that many Afromontane tree species were widespread, with only 37% restricted to a single mountain system. The percentage of Afromontane endemic species was high, that of endemic genera lower, and a strong element consisted of Afromontane endemic species in genera which had their main distribution in tropical lowlands, either in adjacent lowlands or in tropical lowlands on other continents (see historical section below).

The chorology of African lowland forests and Eastern Africa forest-woodland mosaics

White's first detailed study of African lowland forests was an analysis of the Indian Ocean Coastal Belt (Moll & White, 1978), previously termed the Usambara-Zululand Domain of the Guineo-Congolian Region (Chapman & White, 1970). The analysis concluded that the East African coastal belt was a mosaic in the sense of White (1976a), but had a high degree of endemism. Two phytochoria were recognised: a northern Zanzibar-Inhambane Regional Mosaic, and a southern Tongaland-Pondoland Regional Mosaic, separated by the Limpopo River. In the Zanzibar-Inhambane Regional Mosaic there were no endemic families, and only three endemic genera of forest trees, but 42% of the species in the sample were endemic. In the Tongaland-Pondoland Regional mosaic there was one near-endemic family, 23 endemic genera, and 40% of the species in the sample were endemic.

Only after this study of the East African coastal forests did White (1979) turn to an analysis of the Guineo-Congolian region. The study established the region as a major centre of specific endemism with 62.1% strictly endemic species, and a further 28.5% which only transgress into the surrounding transition zones. Only seven species of the sample of 288 species of trees were common to the Guineo-Congolian region and the Indian Ocean Coastal Belt, which confirmed the chorological distinction of the latter. The possibility of establishing subcentres was also studied from a sample of 277 species, and it was found that Lower Guinea (from South Nigeria through Cameroon to Gabon and western Zaïre) had the highest percentage of endemic species. Intervals in the region were also analysed.

In final comments on the chorological position of the Indian Ocean Coastal Belt, White again stressed that one should not over-emphasise the chorological connection between the Guineo-Congolian region and the Zanzibar-Inhambane Regional Mosaic. Diagnoses of phytochoria should be based on plant distributions alone, although some authors have used phylogenetic evidence to provide additional criteria.

The AETFAT chorological classification of Africa

The AETFAT chorological classification of Africa served as the framework for the description of the vegetation types shown on the UNESCO/AETFAT/UNSO vegetation map of Africa. The 'memoir' (White, 1983b) represented a synthesis of previous work, and had little theoretical discussion. It described African phytochoria named and partly discussed in previous papers: seven regional centres of endemism — Guineo-Congolian, Sudanian, Zambezian, Somalia-Masai, Karoo-Namib, Cape, and Mediterranean; (White (1976a) had tentatively recognised a Saharo-Sindian regional centre, but in the memoir that was termed a transitional zone); two archipelago-like centres of endemism (Afromontane and Afroalpine; the latter in the memoir referred to as a region of extreme floristic impoverishment); three regional mosaics (Lake Victoria Basin, Zanzibar-Inhambane, and Tongaland-Pondoland), and five regional transition zones (Guinea-Congo/Sudania, Guineo-Congolia/Zambezia, Sahel, Kalahari-Highveld, Sahara, and Mediterranean/Sahara). The most notable change between White (1976a) and White (1983a) is the status of Sahara (in White (1979) anything north of the Sahel is not named) which in the memoir is classified as a transition zone, so that three adjacent transition zones run parallel. This apparent anomaly is defended in the introduction (White, 1983b: 43) where it is explained that the Sahel and the

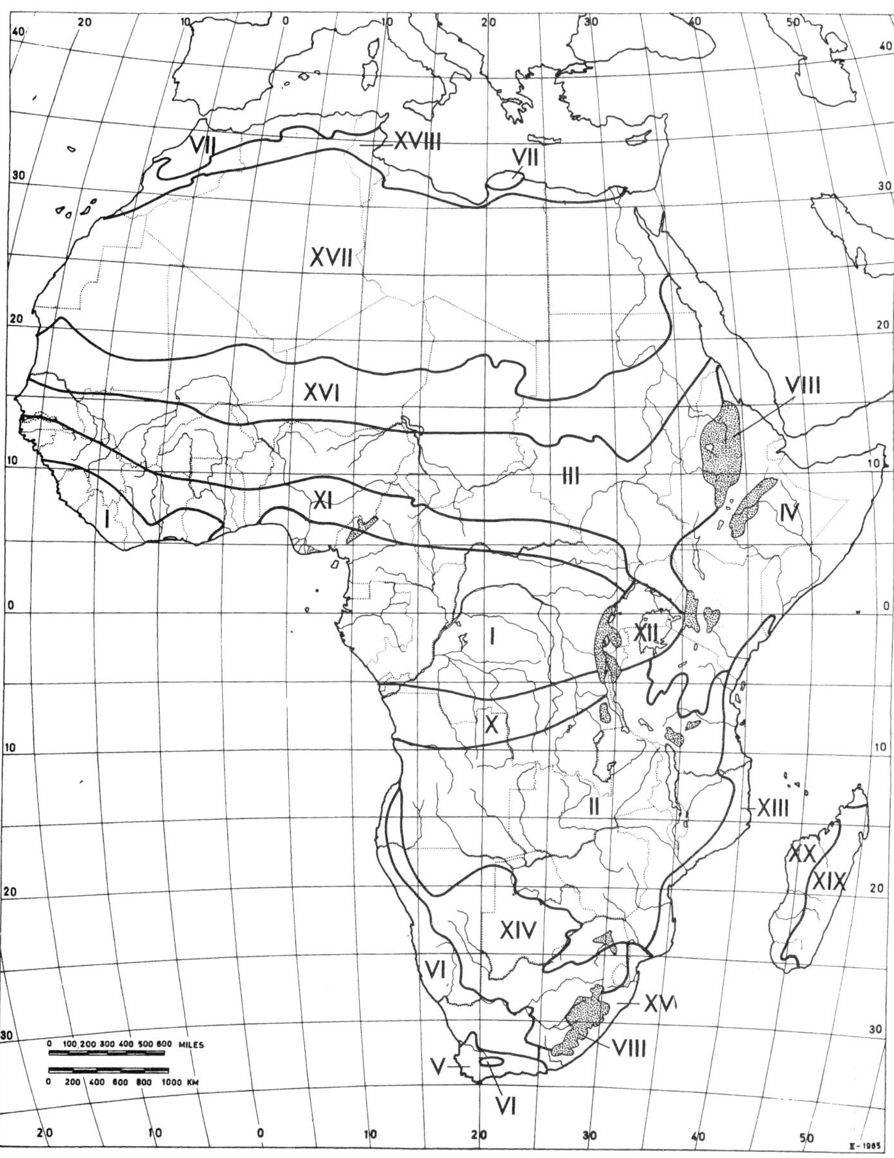

FIG. 6. The phytochoria of Africa and Madagascar, according to White (1993a). **I** Guineo-Congolian regional centre of endemism. **II** Zambezian regional centre of endemism. **III** Sudanian regional centre of endemism. **IV** Somalia-Masai regional centre of endemism. **V** Cape regional centre of endemism. **VI** Karoo-Namib regional centre of endemism. **VII** Mediterranean regional centre of endemism. **VIII** Afromontane Archipelago-like regional centre of endemism. **IX** Afroalpine archipelago-like region of extreme floristic impoverishment. **X** Guinea-Congolia/Zambezia regional transition zone. **XI** Guinea-Congolia/Sudania regional transition zone. **XII** Lake Victoria regional mosaic. **XIII** Zanzibar-Inhambane regional mosaic. **XIV** Kalahari-Highveld regional transition zone. **XV** Tongaland-Pondoland regional mosaic. **XVI** Sahel regional transition zone. **XVII** Sahara regional transition zone. **XVIII** Mediterranean/Sahara regional transition zone. **XIX** East Malagasy regional centre of endemism. **XX** West Malagasy regional centre of endemism. Reproduced from Bull Jard. Bot. Nat. Belg. 62: 237 (1993), with permission.

Mediterranean/Sahara zones have impoverished floras and few endemic species, while Sahara has a higher degree of endemism, but not enough for it to qualify as a Regional centre of endemism. The stability since the works of Engler of some of the phytochoria, notably the Guineo-Congolian region, the Karoo-Namib, and the Cape, is indeed striking, and illustrates the uniqueness of the flora in these areas. Fig. 6 shows the last version of the AETFAT chorological map of Africa published by White.

Historical plant geography

White's work on African plant geography was far from being restricted to chorology in the narrow sense of analytical plant geography. Already in White (1965) there were discussions of the various terms for phytogeographical elements, a subject of historical plant geography. In Chapman & White (1970) there is an analysis of the genetical elements of the Afromontane forest of Malawi, used in a historical analysis of the origin of the flora. White used the following categories of endemic elements in this analysis: a) the eu-Afromontane genetic element (without close relatives anywhere outside the Afromontane Region), b) the "nephew" element (with related species in the Guineo-Congolian Region), and c) the "orphan" element (with related species outside the African continent). Other analyses were concerned with marginally transgressing elements, and the last part considered species reaching their northern or southern limits in Malawi, or having a wide interval there. Arguments for a southern migration route between East and West Africa for species restricted to the Afromontane region were first presented briefly (White, 1978) and later developed (White, 1981).

White also showed strong interest in chorologically transgressing species in relation to historical plant geography. Possible explanations for widespread and partly disjunct species have been discussed in a number of White's papers, often in collaboration with others interested in the dispersing organisms, from birds to gorillas, for example White (1983b & c), Pannell & White (1988), and taxonomic and biological approaches combined, as in works on *Parkia* (Hopkins & White, 1984) or *Myrica* (White, 1993b). A monographic approach was applied to the Chrysobalanaceae, Ebenaceae and Meliaceae, e.g. White (1976b, 1988). Intervals in the distribution of species were discussed in a paper (White, 1990) dealing with disjunct species (*Ptaeroxylon obliquum, Ternstroemia polypetala* and others). This study compares transgressing taxa with continuous and discontinuous distributions, and reviews theories about Quarternary changes in African vegetation in the light of well documented discontinuous distributions. It is concluded that theories about past climate and vegetation must be compatible with what we know about present day plant distribution. In such papers, White often discussed theories about the palaeoenvironment of Africa, for example by Moreau (1966), in comparison with chorological findings. In one of his last papers, White (1993c) again reviewed current ideas about the history of the African flora and its implications for chorology.

A final status report

I have now followed the development of chorology from its founding fathers, the development of African chorology, and the work of White from his first studies to the synthesis in the AETFAT chorological classification of Africa. In 1993, White himself made a status report (White, 1993a), reviewing his work with African chorology. In that paper, which became his final synthesis, he mentioned that chorology had developed differently in different parts of the world, in some parts becoming a widely accepted tool, as among botanists working on the flora of Africa, in others developing into a range of conflicting classifications, as among botanists working on the European flora. To illustrate the situation in Europe White (1993a: 232–233) quoted a paper by Webb (1965) that in many ways echoed the critical words of A. De Candolle (1855) about chorology. Webb asked: 1) How are historical elements to be interpreted? 2) How do we avoid drawing arbitrary border lines between floras, and how is consensus on a classification reached? 3) How is consensus reached on methodology? 4) How to overcome the problem of weighting: "Is a *Euphrasia* as important as a *Quercus*? 5) Can a hierarchical classification be achieved at all, or is chorology "the incompetent in pursuit of the incomprehensible"?

White's replies to these questions were firmly in favour of the scientific nature of chorology: "In the author's opinion lack of agreement hitherto has partly stemmed from ignorance, partly from lack of clearly stated objectives, and partly from the insufficiently refined methods of analysis" (White, 1979). White pointed out that much had been achieved in African chorology through collaboration within the framework of AETFAT, and that this collaboration had lead to widespread consensus: "... recent chorological proposals for Africa did not differ radically, but that a regional framework was evolving, step by step, and that successive proposals involved emendation of previous ones and were based on increasing knowledge and perception ... that agreed principles were emerging, particularly concerning (a) the need for chorology to be based on sound taxonomy; (b) weighting and the use of non-chorological evidence in chorological classifications; and (c) the extent to which chorological classification should or should not be hierarchical ... [and] that international collaboration is essential if effective synthesis is to be achieved on a pan-African scale ..."

In a further section, White again reviewed the criteria for his phytochoria used in the memoir and concluded that they had stood the test of time. For some regional centres of endemism, for example for the Somalia-Masai Regional centre, recent figures for endemism were higher than previously indicated. Once again the three parallel transition zones north of 12°N were discussed. Since the memoir, the situation of the Sahara had been investigated by Léonard (1988, 1989) who placed Sahara in the Saharo-Sindian regional zone. The relationship between the vegetation map of Africa and the chorological testing of the phytochoria based on the map was carefully documented in the paper, and the importance of the more than 2000 distribution maps published before its conclusion was stressed. Since the publication of the AETFAT map more than 2000 additional distribution maps had been published; these maps were used by White (1993a) to test his chorological classification.

The confirmation of White's vegetation map by remote sensing was also mentioned (White, 1993a: 263). Since 1993, more confirmations have appeared, including Ehrlich & Lambin (1996) and Lambin & Ehrlich (1996), and Malingreau *et al.* (1996). That the methodology of the AETFAT chorological classification of Africa can be extended outside Africa has been demonstrated by White & Léonard (1991, 1994).

Criticism

Some criticism has occasionally been voiced against the AETFAT chorological division of Africa; the critical voices have chiefly dealt with the Afromontane region and its relation to other phytochoria. Hamilton (1975, 1989) pointed out that when comparing the woody forest flora of Uganda at various altitudinal levels he found constant coefficients of similarity between the flora on either side of the supposed lower border of the Afromontane region, and that there was therefore no sharp delimitation, for which reason he suggested that the region was arbitrarily defined. A poorly defined lower limit of the Afromontane region may be true in some areas with Afromontane vegetation, but certainly not in all cases as is demonstrated by Friis (1992), Friis & Lawesson (1993) and Friis (1994). Moreover, as we have seen in the historical review section it has been accepted since the days of Schouw that transition zones exist between all phytochoria, including between montane phytochoria and the surrounding lowlands. Very few will, for example, doubt the Arctic-Alpine discontinuous phytochorion in Europe because the lower boundary may be indistinct in some cases. White (1993a: 258) dealt briefly with aspects of the transition zone between the Afromontane region and the surrounding lowlands.

Lovett (1993a, b) also criticised the concept of an Afromontane Region because he believed it to be heterogeneous, and thus should, at least in Eastern Tanzania, be split between 1) an upper Afroalpine Region extended to include, for example, the Ericaceous belt and various montane grasslands and 2) a lower Eastern forest domain of the Guineo-Congolian region (Lovett, 1993b: 225). However, as shown previously by White (1978, 1983b: 161–162) the flora of the Afromontane region is so similar throughout its range from Mt. Cameroon to Ethiopia and south to the mountains of southern Africa that it appears as one clearly defined phytochorion on the continental scale, meeting White's criteria for a Centre of endemism. What Lovett (1993b) has demonstrated is that the forest vegetation in the mountains of East Tanzania has a strong element of endemics with related species (White's "nephews") in the Guineo-Congolian lowland forest, whereas the Ericaceous belt, the montane grassland and other open habitats include many species with relatives in the temperate zones or in the winter rainfall region of South Africa. The temperate affinity of the montane grasslands is in fact mentioned by White (1983b: 168) himself, and the distinction between the affinity of the African lowland (forest) grasses and the montane grasses was already pointed out by Hornemann (1819). Lovett's analysis appears correct as a partial study of the historical plant geography of an element in the Afromontane flora, but is not an argument for splitting the Afromontane Region, unless he wants to mix historical plant geography with analytical plant geography. Hedberg (1965) also demonstrated that the Afroalpine flora is composed of a number of genetic elements of very diverse affinities, but he did not propose a subdivision of the Afroalpine Region or a transfer in part to another phytochorion for that reason.

It should be noted, however, that Lovett's views are based on his long experience with the forests of the old basement complex mountains in Tanzania, and that he and others have now gathered much new information on these forests. The East African lowland forest mosaics are the ones which White managed to treat least satisfactorily. In this area, species of the woodland component are largely Sudano-Zambezian in distribution, while the forest species are either endemic or common with the Guineo-Congolian region. This is different from the Afromontane situation, where, apart from marginal transgressors from the phytochoria below, the flora is well marked. White unfortunately never returned to analyse the forests of the Indian Ocean Coastal Belt after the initial studies in Moll & White (1978).

Tasks for the future

White (1993a) outlined a number of tasks which still remain in relation to African chorology. There are still many areas of Africa which are poorly collected in relation to the richness of their flora. This is regularly demonstrated by the discovery and description of striking new taxa. Taxonomic revisions of the African flora are still far from adequate, although good progress has been made with some of the major floras in recent years. Surveys of transition zones, especially between Afromontane enclaves and their surrounding lowland phytochoria are much needed, and require careful analysis in order to be properly understood.

Remote sensing technology also opens up new possibilities for correlation between physical variables and biogeography. This is currently being tested by a group at the Centre for Tropical Biodiversity in Copenhagen examining the current ecoclimatic stability of 'hotspots' representing areas with rapid evolution in various biological groups (Fjeldså et al., 1997). 'Hotspots' are often defined from peak concentrations of neoendemics as well as geographical relict forms, but an attempt is being made to distinguish between the two categories, a problem already addressed theoretically by Engler (1882), but which can now be studied by the use of molecular systematics. In their pilot project, Fjeldså et al.(1997) found that areas with a predominance of 'old' species tended to be characterised by spatial uniformity in ecoclimatic variability, while areas where rapidly radiating groups dominate were spatially complex. However, such studies are currently best made on birds or other well studied vertebrates, as knowledge of the phylogeny of African plants is still very inadequate.

There are other opportunities for mapping species distributions by computer which open up new fields for plant geography. It is possible to test phytochoria by the detection of areas rich in endemic species (areas rich in species of restricted range), and to correlate these areas with particularly species-rich areas (areas with high species diversity). The computer program WORLDMAP is now being applied to various groups of African animals by J. Fjeldså, N. Burgess and others in Copenhagen, and for African trees by J. Lovett in York, in collaboration with the Copenhagen Centre for Tropical Biodiversity. However, this method should be applied with considerable caution, as discussed by Beentje (1996), and in the interpretation of the results it will be extremely useful to take into consideration many of White's comments, for example his careful consideration of the importance of sample size. Beentje, who had followed the compilation of the African data for the IUCN Centres of Plant Diversity project, also strongly warned against the use of collections which are not representative on a continental scale and stressed the need for much more field work.

The methods of Nelson & Platnick (1981) for testing theories about historical biogeography have so far been applied only to a few cases in, for example, Linder's analysis of the Disinae (Linder, 1983). A much better understanding of the biology of African plants and their interaction with animals will probably also provide useful results in the future. We must conclude that the application of the AETFAT chorological classification to historical plant geography is only just beginning.

Conclusions

Chorology should have a future among all these modern methods, both because it is a useful foundation for other research, and because some of the fundamental chorological questions are still with us, as we have seen from the historical review. White's chorology of Africa has been used by, for example, the IUCN review of protected areas in the tropics (MacKinnon & MacKinnon, 1986), the International Legume Database and Information Service (ILDIS) check-list of African legumes (Lock, 1989), and by the IUCN Centres of Plant Diversity project (Beentje, 1996).

I find it useful to compare chorology with the classification of plants into families, orders or taxa of even higher ranks, another field where there is also a lack of absolute consensus. Yet we can not be without this classification as a practical basis for other research. As in chorology, the traditional taxonomy of taxa of higher ranks is a scientific field where "perception" in White's sense has been widely applied, and where the achievements of less careful workers have occasionally brought their science into disrepute. Modern methods can be applied to the classification of flowering plants at the level of family or higher, but in cases where we can find neither a fully resolved cladogram, nor DNA-sequences that show us an unambiguous phylogeny, we still have to classify flowering plants by perception, and we need a classification where a reasonable consensus among serious botanists can be reached.

As we need a constantly improved taxonomic classification of the African flora, so we also need a gradually improving chorological classification. White has brought us towards a chorological classification that many botanists can accept, at least in its broad outlines. I have tried to show that White produced this classification in harmony with the ideas of classical chorology, and that he developed his new ideas with a deep insight into the theory of the discipline.

White stated that African chorology progressed so well because of the collaborative spirit of the botanists working on African flora and vegetation. African chorology has also made such progress because of White's leadership. He had the uncommon facility of both grasping the trends in large amounts of data, and expressing these in succinct language illustrated by informative maps and diagrammatic representations. He had the will and ability to generalise his observations in clear-headed and broad theoretical statements as in the classical aphorisms of the fathers of chorology. He will be sadly missed in African chorology, as in many other fields!

References

Aubreville, A. (1949a). Climats, forêts et désertification de l'Afrique tropicale. Société d'éditions géographiques, maritimes et coloniales, Paris.

Aubreville, A. (1949b). Contribution à la paléohistoire des forêts de l'Afrique tropicale. Société d'éditions géographiques, maritimes et coloniales, Paris.

Aubreville, A. (1949c). Flore forestière soudano-guinéenne du Sénégal à l'Oubangui-Chari. Société d'éditions géographiques, maritimes et coloniales, Paris.

Beentje, H. J. (1996). Centers of plant diversity in Africa. In: L.J.G. van der Maesen, X.M. van der Burgt & J.M. van Medenbach de Rooy (editors). The biodiversity of African plants. Pp. 101–109. Kluwer Academic Publishers, Dordrecht.

Böcher, T.W. (1977). Convergence as an evolutionary process. *Bot. J. Linn. Soc.* **75**: 1–19.

Brown, R. (1814). General remarks, geographical and systematical on the botany of Terra Australis. In: M. Flinders. A voyage to Terra Australis. Vol. 2. G. & W. Nicol, London.

Brown, R. (1818). Observations, systematical and geographical, on the herbarium collected by Professor Christian Smith, in the vicinity of the Congo, during the expedition to explore that river, under the command of Captain Tucky, in the year 1816. In: J.H. Tucky. Narrative of an expedition to explore the River Zaire, usually called the Congo, ... John Murray, London.

Buffon, G.-L. Leclercq Comte de (1761). Histoire naturelle générale et particulière, avec la description du Cabinet du Roi. Vol. 9. Paris.

Chapman, J.D. and White, F. (1970). The evergreen forests of Malawi. Commonwealth Forestry Institute, Oxford.

Chevalier, A. (1933). Le territoire géo-botanique de l'Afrique tropicale nord-occidentale et ses subdivisions. *Bull. Soc. Bot. France* **80**: 4–26.

De Candolle, A. (1855). Géographie botanique raisonnée ou exposition des faites principaux et des lois concernant la distribution géographique des plantes de l'époque actuelle. Vol. 1 & 2. Victor Masson, Paris; J. Kessmann, Genève.

De Candolle, A.-P. (1813). Théorie élementaire de la botanique, ou exposition des principes de la classification naturelle et de l'art de décrire et d'étudier les végétaux. Déterville, Paris.

De Candolle, A.-P. (1821). Géographie botanique. In: G.-F. Cuvier (editor). Dictionaire des Sciences Naturelles. Vol. 18: 359–422. Le Normant, Strasbourg and Paris.

Drude, O. (1890). Handbuch der Pflanzengeographie. Engelhorn, Stuttgart.

Du Rietz, G.E. (1931). Life-forms of terrestrial flowering plants. *Acta Phytogeogr. Suec.* **3**: 1–95.

Ehrlich, D. and Lambin, E.F. (1996). Broad scale land cover classification and interannual climatic variability. *Int. J. Remote Sensing* **17**: 845–862.

Eig, A. (1931a). Les éléments et les groupes phytogéographiques auxillaires dans la flore palestinienne. *Repert. Spec. Nov. Regni. Veg. Beih.* **63(1)**: 1–201 & **63(2)**: 1–120.

Eig, A. (1931b). Quelques faits de la phytogéographie palestinienne précédés par des remarques sur les notions phytogeographiques. *Bull. Soc. Bot. France* **78**: 297–305.

Eig, A. (1933). Ecological and phytogeographical observations on Palestine plants. *Beih. Bot. Centralbl.* **50**: 470–496.

Engler, A. (1882). Versuch einer Entwichlungsgeschichte der Pflanzenwelt, inbesondere der Florengebiete seit der Tertiärperiode. 2. Theil. Die extratropischen Gebiete der südlichen Hemisphäre und die tropischen Geiete. W. Engelmann, Leipzig.

Engler, A. (1899). Die Entwicklung der Pflanzengeographie in den letzten hundert Jahren und weitere Aufgaben derselben. Humboldt-Zentenarschrift der Gesellschaft für Erdkunde zu Berlin, Berlin.

Engler, A. (1908). Pflanzengeographische Gliederung von Afrika. *Sitzungsber. Königl. Preuss. Akad. Wiss.* 1908: 781–837.

Engler, A. (1910). Die Pflanzenwelt Afrikas, inbesondere seiner tropischen Gebiete. I. Allgemeiner Ueberblick über die Pflanzenwelt Afrikas und ihrer Existenzbedingungen: Allgemeiner Ueberblick über die Vegetationsverhältnisse von Afrika. In: A. Engler and O. Drude (editors). Die Vegetation der Erde. Vol. IX, 1, 1: 1870. W. Engelmann, Leipzig.

Engler, A. and Gilg, E. (1912). Syllabus der Pflanzenfamilien. Ed. 7. Gebr. Borntraeger, Berlin.

Fjeldså, J., Ehrlich, D., Lambin, E.F. and Prins, E. (1997). Are biodiversity "hotspots" correlated with current ecoclimatic stability? — A pilot study using NOAA-AVHRR remote sensing data. *Biodiversity and conservation* 6: 401–422.

Fries, R.E. and Fries, T.C.E. (1948). Phytogeographical researches on Mt. Kenya and Mt. Aberdare, British East Africa. *Kongl. Svenska Vetenskapsakad. Handl.*, ser. 3, **25(5)**: 1–83.

Friis, I. (1986). Ethiopia in regional phytogeography. *Symb. Bot. Upsal.* **26(2)**: 68–85.

Friis, I. (1992). Forests and Forest Trees of Northeast Tropical Africa. Their Natural Habitats and Distribution Patterns in Ethiopia, Djibouti and Somalia. *Kew Bull. Addit. Ser.* **15:** i–iv & 1–396.

Friis, I. (1994). Some general features of the Afromontane and Afrotemperate floras of the Sudan, Ethiopia and Somalia. In: J.H. Seyani and A.C. Chikuni (editors). Proceedings of the XIIIth plenary meeting of AETFAT, Zomba, Malawi, 2–11 April 1991. Vol. 2: 953–968. National Herbarium and Botanic Gardens of Malawi, Zomba.

Friis, I. and Lawesson, J.E. (1993). Altitudinal zonation of the forest tree flora of Northeast Tropical Africa. *Opera Bot.* **121**: 125–127.

Friis, I. and Vollesen, K. (1998). Flora of the Sudan-Uganda border area east of the Nile. I. Catalogue of vascular plants, 1st part. *Kongel. Danske Vidensk. Selsk. Biol. Skr.* **51(1)**: 1–389.

Friis, I. and Vollesen, K. (in prep.). Flora of the Sudan-Uganda border area east of the Nile. II. Catalogue of vascular plants, 2nd part; vegetation and phytogeography. *Kongel. Danske Vidensk. Selsk. Biol. Skr.* **51(2)**.

Gmelin, J.G. (1747–1769). Flora sibirica sive historia plantarum Sibiriae. Vol. 1–4. Academia Scientarum, St. Petersburg.

Good, R. (1947). The Geography of Flowering Plants. Longman, London.

Gräbner, P. (1910). Lehrbuch der allgemeine Pflanzengeographie nach entwichlungs- geschitlichen und physiologish-ökologischen Gesichspunkten. Quelle und Mayer, Leipzig.

Grisebach, A.H.R. (1871). Die Vegetation der Erde. Vol. 1–2. W. Engelmann, Leipzig.

Häckel, E. (1866). Generelle Morphologie der Organismen. Allgemeine Grundzüge der organischen Formen-Wissenschaft, mekanischer begründet durch die von Charles Darwin reformierte Descendenz-Theorie. Vol. 1–2. G. Reimer, Berlin.

Hamilton, A.C. (1975). A quantitative analysis of altitudinal zonation in Uganda forests. *Vegetatio* **30**: 99–106.

Hamilton, A. C. (1989). African forests. In: H. Lieth and M.J.A. Werger (editors). Tropical Rain Forest Ecosystems, biogeographical and ecological studies. Ecosystems of the world 14B: 155–182. Elsevier, Amsterdam.

Hauman, L. (1955). La "Region Afroalpine" en phytogeographe centro-africaine. *Webbia* **11**: 467–469.

Hedberg, O. (1951). Vegetation belts of the East African mountains. *Svensk Bot. Tidskr.* **48**: 140–202.

Hedberg, O. (1955). Altitudinal zonation of the vegetation on the East African mountains. *Proc. Linn. Soc. Lond.* **165**: 134–136.

Hedberg, O. (1957). Afroalpine vascular plants. A taxonomic revision. *Symb. Bot. Upsal.* **15(1)**: 1–411.

Hedberg, O. (1961). The phytogeographical position of the Afro-alpine flora. *Recent Advances Botany* **1**: 914–919.

Hedberg, O. (1964). Features of Afro-alpine plant ecology. *Acta Phytogeogr. Suec.* **49**: 1–144.

Hedberg, O. (1965). Afro-alpine flora elements. *Webbia* **19**: 519–529.

Hopkins, H. C. and White, F. (1984). The ecology and chorology of *Parkia* in Africa. *Bull. Jard. Bot. Belg.* **54**: 235–366.

Hornemann, J.W. (1819). De indole plantarum Guineensium observationes. Aniversaria ...Universitate Hauniense. University of Copenhagen, Copenhagen.

Humboldt, A. von (1806). Essai sur la geography des plantes; accompagné d'un tableau physique des régions équinoxales, accompagné d'un tableau physique des régions équinoctiales, ... Schoel & Co., Paris.

Humboldt, A. von (1807). Ideen zu einer Geographie der Pflanzen, nebst einem Naturgemälde der Tropenländer, ... F.G. Cotta, Tübingen.

Humboldt, A. von (1817). De distributione geographica plantarum secundum coeli temperiem et altitudinem montium, prolegomena. Libraria graeco-latino-germanica, Paris.

Humboldt, A. von, Bonpland, A. and Kunth, C.S. (1816). Nova genera et species plantarum quas in peregrinatione orbis novi collegerunt, ... Amat. Bonpland et Alex. de Humboldt. Vol. 1. Libraria graeco-latino-germanica, Paris.

Jussieu, A.L. de (1789). Genera plantarum secundum ordines naturales disposita, juxta methodum in horto regio parisiensi exaratum. V. Herissant, Paris.

Knapp, R. (1973). Die Vegetation von Afrika. Gustav Fischer, Jena.

Lambin, E.F. and Ehrlich, D. (1996). The surface temperature-vegetation index space for land cover and land-cover change analysis. *Int. J. Remote Sensing* **17**: 463–487.

Lebrun, J. (1947). La végétation de la plaine alluviale au su du lac Edouard. Exploration du Parc Nationale d'Albert; Miss. J. Lebrun (1937–38). Fascicule 1: 1–800. Inst. Parc Nat. Congo Belge, Bruxelles.

Léonard, J. (1988). Contribution à l'étude de la flore et de la vegetation des déserts d'Iran. Pt. 8 (Etude des aires de distribution. Les phytochories. Les chorotypes). Jardin Botaniqu National Belgique, Bruxelles.

Léonard, J. (1989). Contribution à l'étude de la flore et de la vegetation des déserts d'Iran. Pt. 9. (Considérations phytogéographique sur les phytochories irano-touranienne, saharo-sindienne et de la Somalie-pays Masai). Contribution à l'étude de la flore et de la vegetation des déserts d'Iran,

Linder, H.P. (1983). The historical phytogeography of the Disinae (Orchidaceae). *Bothalia* **14**: 565–575.

Linnaeus, C. (1736). Fundamenta botanica quae majorum operum prodromi instar theoriam scientiae botanices per breves aphorismos tradunt ... Salomon Schouten, Amsterdam.

Linnaeus, C. (1744). Oratio de telluris habitabilis incremento. Et Andreae Celsii Oratio de mutationibus generalioribus ... Habita cum Medicinae licentiatum Johannem Westermannum medicinae doctorem ... renunciaret. Cornelius Haak, Leiden.

Lock, J.M. (1989). Legumes of Africa — a check-list. Royal Botanic Garden, Kew.

Lovett, J.C. (1993a). Eastern arc moist forest flora. In: J.C. Lovett and S.K. Wasser (editors). Biogeography and ecology of the rain forests of eastern Africa. Pp. 33–56. Cambridge University Press, Cambridge.

Lovett, J.C. (1993b). Temperate and tropical floras in the mountains of eastern Tanzania. *Opera Bot.* **121**: 217–227.

Mackinnon, J. and Mackinnon, K. (1986). Managing protected areas in the tropics. IUCN, Gland, Switzerland.

Malingreau, J.-P., Ehrlich, D. and Lambin, E.F. (1996). Are there detectable discrete boundaries between African Biomes? *Int. J. Remote Sensing* **17**: 841–844.

Mattick, F. (1964). Uebersicht über die Florenreiche und Florengebiete der Erde. In: H. Melchior (editor). A. Engler's Syllabus der Pflanzenfamilien. 12. Auflage. Vol. 2. Angiospermen. Gerbrüder Borntraeger, Berlin.

Meyen, F.J.F. (1836). Grundriss der Pflanzengeografie. Haude und Spener, Berlin.

Moll, E.J. and White, F. (1978). The Indian-Ocean coastal belt. In: M.J.A. Werger (editor). Biogeography and Ecology of Southern Africa. Vol. 1: 561–598. W. Junk, The Hague.

Monod, T. (1957). Les grandes division chorologique de l'Afrique. C.S.A./C.C.T.A. Publication no. 24: 1–150. Conseil Scientifique pour l'Afrique au sud du Sahara/ Scientific Council for Africa South of the Sahara, London.

Moreau, R. E. (1966). The bird faunas of Africa and its islands. Academic Press, New York, London.

Nelson, G. and Platnick, N. (1981). Systematics and Biogeography. Cladistics and vicariance. Columbia University Press, New York.

Pannell, C.M. and White, F. (1988). Patterns of speciation in Africa, Madagascar, and the tropical Far East: Regional faunas and cryptic evolution in vertebrate-dispersed plants. *Monogr. Syst. Bot. Missouri Bot. Gard.* **25**: 639–659.

Raunkiær, C. (1904). Om biologiske Typer, med Hensyn til Planternes Tilpasning til at overleve ugunstige Aarstider. *Bot. Tidsskr.* **26**: XIV.

Raunkiær, C. (1905). Types biologiques pour la géographie botanique. *Overs. Kongel. Danske Vidensk. Selsk. Forh. Medlemmers Arbeider* 1905: 347–438.

Robyns, W. (1948). Flore des spermatophytes du Parc National d'Albert. I. Gymnospermes et Choripétales. Inst. Parc Nat. Congo Belge, Bruxelles.

Schimper, A.F.W. (1898). Pflanzengeographie auf physiologischer Grundlage. Gustav Fischer, Jena.

Schnell, R. (1970–71). Introduction à la phytogéographie des pays tropicaux. Vol. 1–2. Les problèmes généraux. Gauthier-Villars, Paris.

Schnell, R. (1976–77). Introduction à la phytogéographie des pays tropicaux. Vol. 3–4. La flore et végétation de l'Afrique tropicale. Gauthier-Villars, Paris.

Schouw, J.F. (1816). Dissertatio de sedibus plantarum originalis. Sectio prima. De pluribus cujusquis speciei individuis originariis statuendis. H.F. Popp, Copenhagen.

Schouw, J.F. (1822). Grundtraek til en almindelig Plantegeographie. Gyldendal, Copenhagen.

Schouw, J.F. (1823a). Grundzüge einer allgemeinen Pflanzangeographie. G. Reimer, Berlin.

Schouw, J.F. (1823b). Pflanzengeographischer Atlas zur Erlüterung von Schouws Grundzügen einer allgemeinen Pflanzengeographie. G. Reimer, Berlin.

Schouw, J.F. (1824). Plantegeografisk Atlas, Henhörende til sammes Grundtræk til en almindelig Plantegeographie. J.H. Schultz, København.

Schouw, J.F. (1833). Erindrings-Ord til en Forelæsning over de plantegeografiske Riger. Brünniche, Copenhagen.

Schouw, J.F. (1834). Momente zur einer Vorlesung über die pflanzengeographischen Reiche. *Linnaea* **8**: 625–652.

Takhtajan, A. (1969). Flowering plants. Origin and Dispersal. Oliver and Boyd, Edinburgh.

Takhtajan, A. (1986). Floristic regions of the world. University of California Press, Berkley.

Thonner, F. (1908). Die Blütenpflanzen Afrikas. Eine Anleitung Zum Bestimmen der Gattung der Afrikanischen Siphonogamen. R. Friedlander & Sohn, Berlin.

Tournefort, J.P. (1717). Relation d'un voyage du Levant, fait par ordre du Roi, contenant l'histoire ancienne & moderne de plusieur isles de l'Archipel, de Constantinople, des Côtes de la Mer Noire, de l'Arménie, de la Georgie, des fontières de Perse & de l'Asie mineure... Vol. 1–2. Paris.

Troupin, G. (1966). Étude phytocénologique du Parc National de l'Akagera et du Rwanda Oriental. Recherches d'une method d'analyse appropriée a la vegetation d'Afrique Intertropicale. Institut National de la Recherche Scientifique Butare, Republique Rwandaise. Publication No. 2: 1–293. Musée Royale de l'Afrique Centrale, Tervuren.

Wahlenberg, G. (1812). Flora lapponica exhibens plantas geographice et botanice consideratas... Taberna Libraria Scholae realis, Berlin.

Wahlenberg, G. (1813). De vegetatione et climate in Helvetiae septentrionali ... Orell, Fuessli et Soc., Zürich.

Wahlenberg, G. (1814). Flora Carpathorum principalium ... Vanderhöck et Ruprecht, Göttingen.

Warming, E. (1892). Lagoa Santa. Et bidrag til den biologiske Plantegeografi. *Kongel. Danske Vidensk. Selsk. Naturvidensk. Math. Afh. Ser. 6*, **6,3**: 153–454.

Warming, E. (1908a). The structure and biology of Arctic Flowering Plants. *Meddel. Grønland* **36**: 1–71, 169–236, 229–342.

Warming, E. (1908b). Om Planterigets Livsformer. Københavns Universitet (Festskrift I Anledning af Kongens Fødselsdag 1908), Copenhagen.

Warming, E. (1923). Økologiens Grundformer. Udkast til en systematisk Ordning. *Kongel. Danske Vidensk. Selsk. Naturvidensk. Math. Afh., Ser. 8*, **4,2**: 119–187.

Webb, D.A. (1965). Some difficulties in the establishment of phytogeographical divisions. *Rev. Roumaine Biol., Sér. Bot.* **10**: 33–41.

Weimarck, H. (1941). Phytogeographical groups, centres and Intervals within the Cape flora. *Acta Univ. Lund.*, n.s. **37(5)**.

Werger, M.J.A. (1978). Biogeographical division of southern Africa. In: M.J.A. Werger (editor). Biogeography and Ecology of Southern Africa. Vol. 1: 145–170. W. Junk, The Hague.

White, F. (1965). The savanna-woodlands of the Zambesian and Sudanian Domains — an ecological and phytogeographical comparison. *Webbia* **19**: 651–681.

White, F. (1971). The taxonomic and ecological basis of chorology. *Mitt. Bot. Staatssamml. München* **10**: 91–112.

White, F. (1976a). The vegetation map of Africa: the history of a completed project. *Bossiera* **24**: 659–666.

White, F. (1976b). The taxonomy, ecology and chorology of African Chrysobalanaceae (excluding *Acioa*). *Bull. Jard. Bot. Belg.* **46**: 265–350.

White, F. (1978). The Afromontane Region. In: M.J.A. Werger (editor). Biogeography and Ecology of Southern Africa. Vol. 1: 463–513. W. Junk, The Hague.

White, F. (1979). The Guineo-Congolian Region and its relationships to the other phytochoria. *Bull. Jard. Bot. Belg.* **49:** 11–55.

White, F. (1981). The history of the Afromontane archipelago and the scientific need for its conservation. *Afr. J. Ecol.* **19**: 33–54.

White, F. (1983a). Long-distance dispersal and the origins of the Afromontane flora. *Sonderb. Naturwiss. Vereins Hamburg* **7**: 87–116.

White, F. (1983b). The Vegetation of Africa: a descriptive memoir to accompany the UNESCO/AETFAT/UNSO vegetation map of Africa. *Natural Resources Research* **20**. UNESCO, Paris.

White, F. (1983c). Long distance dispersal, overland migration and extinction in the shaping of the African floras. *Bothalia* **14(3 & 4):** 395–403.

White, F. (1988). The taxonomy, ecology and chorology of African Ebenaceae II. The non Guinean-Congolian species of *Diospyros* (excluding sect. Royena). *Bull. Jard. Bot. Belg.* **58**: 325–448.

White, F. (1990). *Ptaeroxylon obliquum* (Ptaeroxylaceae), some other disjuncts, and the Quarternary history of African vegetation. *Bull. Mus. Natl. Hist. Nat. Paris. Sér. 4*, 12, Sect. B, *Adansonia*: 139–185.

White, F. (1993a). The AETFAT chorological classification of Africa: history, methods and application. *Bull. Jard. Bot. Belg.* **62**: 225–281.

White, F. (1993b). African Myricaceae and the history of the Afromontane flora. *Opera Bot.* **121**: 173–188.

White, F. (1993c). Refuge theory, ice-age aridity and the history of tropical biotas: an essay in plant geography. *Fragm. Florist. Geobot.* Suppl. 2 [Festschrift Jan Kornas]: 385–409.

White, F. and Léonard J. (1991). Phytogeographical links between Africa and southeast Asia. *Fl. Veg. Mundi* **9**: 229–246.

White, F. and Léonard, J. (1994). Rapports phytogeographiques entre l'Afrique et le Sud-ouest Asiatique. In: J.H. Seyani and A.C. Chikuni (editors). Proceedings of the XIIIth plenary meeting of AETFAT, Zomba, Malawi, 2–11 April 1991. Vol. 2: 1085–1103. National Herbarium and Botanic Gardens of Malawi, Zomba.

White, F. and Werger, M.J.A. (1978). The Guineo-Congolian transition zone to Southern Africa. In: M.J.A. Werger (editor). Biogeography and Ecology of Southern Africa. Vol. 1: 599–620. W. Junk, The Hague.

Wickens, G.E. (1976). The Flora of Jebel Marra (Sudan Republic) and its geographical affinities. *Kew Bull. Addit. Ser.* **5**: 1–368.

Willdenow, C.L. (1792). Grundriss der Kräuterkunde. Haude und Spener, Berlin.

Willdenow, C.L. (1798). Grundriss der Kräuterkunde. 2. Ausg. Haude und Spener, Berlin.

Clarke, G.P. (1998). A new regional centre of endemism in Africa. In: C.R. Huxley, J.M. Lock and D.F. Cutler (editors). Chorology, Taxonomy and Ecology of the Floras of Africa and Madagascar. Pp. 53–65. Royal Botanic Gardens, Kew.

4. A NEW REGIONAL CENTRE OF ENDEMISM IN AFRICA

G. P. CLARKE

The Hermitage, Hermitage Street, Crewkerne, Somerset, TA18 8ET, U.K.

Abstract

Recent botanical surveys and the ongoing publication of further volumes of the *Flora of Tropical East Africa* and *Flora Zambesiaca* have provided much new floristic data on White's Zanzibar-Inhambane regional mosaic, which were not available at the time of publication of *The Vegetation of Africa*. The data from these floras, supplemented by additional data from checklists, taxonomic monographs and herbarium records, are now sufficient to present a representative and near-complete list of the endemic vascular plant species in the phytochorion, and indicate a far higher level of floristic endemism along the tropical eastern African coastal strip than had previously been realised. An examination of the patterns in this endemism confirm an earlier observation by White that most of the endemic vascular plant species of this phytochorion are concentrated in the region to the north of Moçambique town. These results challenge the existing classification of the area as a single regional mosaic, and suggest that the name 'Zanzibar-Inhambane' is also inappropriate. White's Zanzibar-Inhambane regional mosaic is therefore reclassified as two regions, the Swahilian regional centre of endemism in the north, and the Swahilian/Maputaland regional transition zone in the south.

Introduction

The phytogeographical classification of the eastern African coastal zone has seen a plethora of different opinions as to whether this area should be delineated as a separate region/domain in its own right, or whether it should be considered as an outlier of another region (compare Brenan, 1978; Lebrun, 1960; Leroy, 1978; Lovett, 1993; Milne-Redhead, 1955; Monod, 1957). White (1976) eventually recognized the Zanzibar-Inhambane region as a regional mosaic, owing to its perceived depauperate flora containing 'a few hundred endemic species' and just four endemic genera (White, 1983a), two of which are now recognised as belonging to Lovett's Eastern Arc flora (Lovett, 1993). Estimates of the number of endemic vascular plant species and genera in the Zanzibar-Inhambane regional mosaic have subsequently been revised upwards to 450 species and 25 genera by Davis *et al.* (1994: 132), although a slightly earlier estimate reached 50 endemic genera (Vollesen, 1992). White however maintained the figure of "a few hundred endemic species and a few genera" in his final treatment on African phytogeography (White, 1993). This paper tests the integrity of White's classification of the Zanzibar-Inhambane region, based on real data, rather than on estimates.

Method

The status of the Zanzibar-Inhambane domain was analysed by cataloguing as far as possible the entire endemic vascular plant flora of the region, using published floras (Thulin *et al.*, 1993; Turrill & Milne-Redhead *et al.*, 1952; Exell & Wild *et al.*, 1960), supplemented by reviews of recent taxonomic literature (all volumes post 1950 of *Kew Bulletin*, all volumes of *Distributiones Plantarum Africanarum* (1969) and *Kirkia* (1975); selected issues of *Botanisk Notiser* and *Botanisk Tidsskrift*), as well as published check-lists and monographs (Adams & Holland, 1978; Beentje, 1988, 1990 & 1994; Brenan & Greenway, 1949; Clarke, 1995; Dowsett-Lemaire, 1990; Drummond, in prep.; Faden, 1991; Friis, 1991; Friis & Vollesen, 1989; Goldsmith, 1976; Gomes e Sousa, 1966 & 1967; Greenway with Rodgers *et al.*, 1988; Haines & Lye, 1983; Hawthorne, 1984 & 1993; Iversen, 1991; Johansson, 1978; Kapuya, 1994 & 1995; Manktelow, 1996; Medley, 1992; Mwasumbi *et al.*, 1994; Mziray, 1992; Palgrave, 1977; Pennington, 1991; Robertson & Luke, 1993; Rodgers *et al.*, 1983; Ruffo, 1991; Seyani, 1991; Temu, 1990; Verdcourt, 1996; Vollesen, 1980; Vollesen & Bidgood, 1992; White, 1988), electronic databases (East African Herbarium (Nairobi) LEAPMASTER database) and the examination of a few herbarium specimens and notes (at the Royal Botanic Gardens, Kew, and the University of Dar es Salaam) where no other data sources were available.

Species endemic to the Zanzibar-Inhambane regional mosaic were identified through an iterative process based on their geographical and altitudinal distributions (as listed in the aforementioned floras and other taxonomic literature). An initial search was conducted for all species whose distribution is limited to those geographical divisions of the *Flora of Tropical East Africa* and *Flora Zambesiaca* which approximate to the extent of the Zanzibar-Inhambane *sensu lato* (to allow for the outlying satellites mapped in White 1983b), i.e. Southern Somalia; K7, marginally intruding into the K1 and/or K4 divisions of Kenya; T3, T6 and/or T8, marginally intruding into the T2 divisions of Tanzania; Zanzibar Island; Pemba Island; coastal divisions of Mozambique, including Maputo (formerly Lourenço Marques) Province and Natal for species which also occur to the north of these divisions; Eastern Zimbabwe (Melsetter District) and southern Malawi. Within this geographical distribution a varying altitudinal limit was imposed (to 400 m in areas of high, regular rainfall, e.g. at the base of the East Usamabara mountains, Tanzania (contrary to White's classification of this area), going up to 1,000 m in areas with low, strongly seasonal rainfall, e.g. on the Rondo Plateau in south-eastern Tanzania), reflecting that the Zanzibar-Inhambane flora is essentially adapted to seasonal rainfall conditions and that a single altitudinal limit throughout its extent is unrealistic. Finally, collection localities were checked to determine whether these occurred within the limits of White's (1983b) Zanzibar-Inhambane regional mosaic, for which purpose a personal knowledge of the Tanzanian, Kenyan and southern Somali coastal districts (gained from spending five years in these areas) proved invaluable.

Results

A total of 1,356 vascular plant species were found to be endemic to the Zanzibar-Inhambane regional mosaic *sensu lato* (full analysis of data and annotated species list published in Burgess & Clarke, in press), far exceeding previous estimates (e.g. Davis *et al.*, 1994; White 1983a & 1993), and well in excess of the required minimum of 1,000 species for the region to qualify as a regional centre of endemism *sensu* White (1979, 1983a & 1993). Further botanical surveys and taxonomic syntheses may raise this figure by at least 150 species (Clarke *et al.*, in press).

The distribution of these endemics is not however evenly spread throughout the mosaic: 940 species (69%) are restricted to the northern part of the mosiac, along the 1,440 km coastline length of southern Somali, Kenya and Tanzania, compared to just 149 species (11%) confined to the 2,480 km of the Mozambique coastline. Reanalysing the data to factor out the effect of the different floral divisions being of different size shows a strikingly consistent high number of Zanzibar-Inhambane regional endemics per unit of coastline length (or per unit of latitude) along the

Kenyan and Tanzanian coasts, compared to a low number along the Somali, Mozambique and Natal coasts (Figs. 1–6). Although differences in the intensity of botanical survey undoubtedly influence this pattern, the Somali, Natal and southern Mozambique coasts are comparatively well surveyed relative to south-eastern Tanzania (T8 division of the FTEA), which has the highest number of Zanzibar-Inhambane endemic species confined to it (187 species). Further botanical surveys in the less well known areas of the Zanzibar-Inhambane regional mosaic (i.e. in Niassa Province of Mozambique, and in south-eastern Tanzania) would both reinforce the observed patterns of endemism by increasing the number of regional endemics recorded from T8, and also significantly increase the number of Zanzibar-Inhambane regional endemics recorded from Niassa Province, and would thereby extend the area of high endemism south into Cabo Delgado Region of Mozambique. This pattern of floristic endemism confirms White's earlier findings that (a) most of the Zanzibar-Inhambane endemics are confined to the region north of Moçambique town (Moll & White, 1978), and (b) a floristically impoverished area (the 'Malawi Interval') exists south of Lake Malawi (White, 1990).

Patterns in endemic species are matched by the endemic genera: 33 genera were found to be endemic to the Zanzibar-Inhambane region *sensu lato*, of which 25 (76%) are confined to the area north of Moçambique town (Table 1). Only the coastal divisions of Kenya (K7), Tanzania (T3, T6 and T8) and Niassa and Zambesia Provinces of Mozambique contain endemic Zanzibar-Inhambane genera which are confined to a single division, with the highest concentration in T8 (six genera).

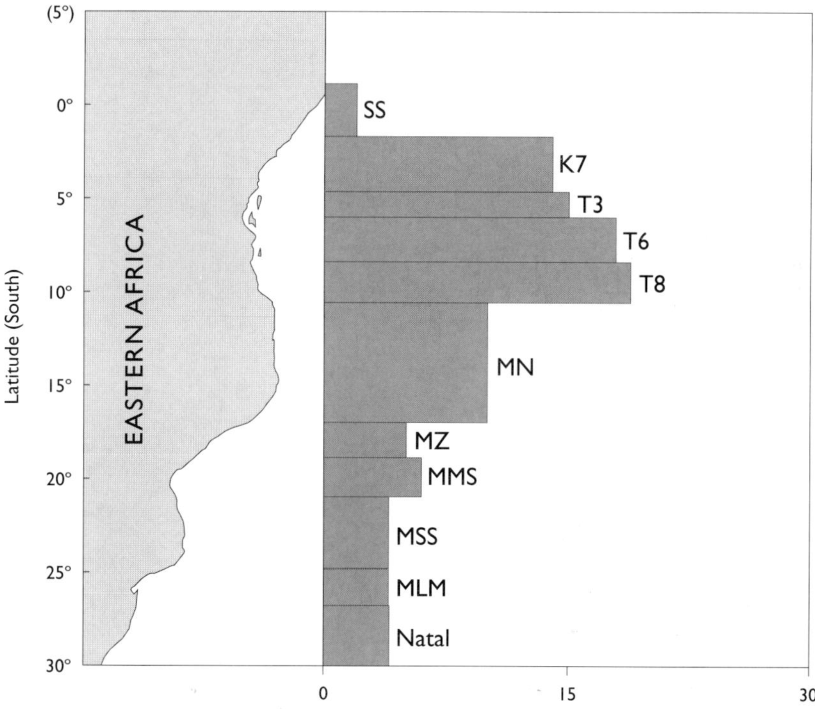

FIG. 1. Number of Zanzibar-Inhambane endemic genera present in each geographical division of the *Flora of Tropical East Africa* and *Flora Zambesiaca* plotted against the coastline of eastern Africa on the left.

TABLE 1. Vascular plant genera endemic to the Zanzibar-Inhambane regional mosaic *sensu lato*. Distributions as per the geographical divisions in the *Flora of Tropical East Africa* and *Flora Zambesiaca*.

Genus	Family	Distribution
Aerisilvaea[1, c, §]	Euphorbiaceae	T6 (Kimboza Forest)
Asteranthe[a, §]	Annonaceae	K7; T3,6,8; Z
Baptorhachis[b, §]	Poaceae	MN
Burrtdavya[a]	Rubiaceae	T3,6,8; Z; MN,MT,MMS; S. Malawi
Callopsis[a, ○]	Araceae	K7; T3,6 & possibly Cameroon
Cladoceras[a, §]	Rubiaceae	K7; T3,6; Mafia Island
Dielsothamnus[a]	Annonaceae	T8; MN, MZ; Central Malawi
Farrago[a, §]	Poaceae	T8 (Nachingwea)
Grandidiera[a]	Flacourtiaceae	South Somalia; K7; T3,6,8; Z; MMS
Hyalocalyx[2, d, §]	Turneraceae	T8; MN
Hystrichophora[e, §]	Asteraceae	T8 (Rondo)
Lamprothamnus[a, §]	Rubiaceae	Somalia; K1,7; T3,6,8; Mafia Island
Lettowianthus[a, §]	Annonaceae	K7; T3,6,8; Mafia Island
Mkilua[a, §]	Annonaceae	K7; T3,6,8; Z; P
Ophrypetalum[a, §]	Annonaceae	K7; T3,6,8
Paranecepsia[a, §]	Euphorbiaceae	T6; MN
Phellocalyx[a]	Rubiaceae	T8; MN; South Malawi
Pseudobersama[a]	Meliaceae	K7; T3,6,8; MMS,MSS,MLM; Natal
Sanrafaelia[f, §]	Annonaceae	T3 (E. Usambaras at Kwangumi)
Schlecterina[a]	Passifloraceae	K7; T3,6,8; Z; MN,MZ,MMS,MSS,MLM; Natal
Stephanostemma[g, §]	Apocynaceae	T6 (Gongolamboto)
Streptosiphon[h, §]	Acanthaceae	T8 (Rondo & Litipo)
Stuhlmannia[a, §]	Fabaceae	K7; T3,8
Thespesiopsis[b, §]	Malvaceae	MN
Trichaulax[i, §]	Acanthaceae	K7; T3,6
Vismianthus[a, i, §]	Connaraceae	T8 (Rondo & Makonde Plateau)
Zamioculcas[a]	Araceae	K1,7; T3,6; Z; P; Moz; Zimbabwe; Malawi; Natal
gen. nov.[j, §]	Acanthaceae	K7 (Gongoni)
gen. nov. of FTEA[a, §]	Rubiaceae	T8 (Rondo)
gen. unknown of FZ[b]	Rubiaceae	MZ (Milange)
gen. indet. of FTEA[a, §]	Annonaceae	T6 (Kimboza)
gen. indet.[k, §]	Annonaceae	T6 (Pugu)
gen. indet.[l, §]	Fabaceae	T8 (Selous)

Sources: [a]Turrill & Milne-Redhead *et al.*, 1952–; [b]Exell & Wild *et al.*, 1960–; [c]*Kew Bull.* 45: 147–156; [d]Leroy, 1978; [e]*Kew Bull.* 43: 249; [f]Verdcourt, 1996; [g]Hawthorne, 1984; [h]*Kew Bull.* 49: 401–407; [i]Vollesen, 1992; [j]Robertson & Luke, 1993; [k]University of Dar es Salaam Herbarium; [l]Vollesen, 1980.

This list only includes the endemic genera which could be identified from existing publications; a full list of all disjunct genera awaits the publication of complete floras for East and south-central Africa.

[1] Considered by Lovett (1993) to be an Eastern Arc endemic. This species is however limited to *Pandanus rabaiensis* swamp forest at 500 m altitude in Kimboza forest (Clarke & Dickinson, 1995: 98), which is similar to the swamp forest at Jozani on Zanzibar Island. It is therefore included as a Zanzibar-Inhambane endemic.

[2] Considered to be cultivated on Madagascar, where it is also recorded (see *Kew Bull.* 5: 335).

§ Genus endemic to the Swahilian regional centre of endemism (defined in this paper).

The genus *Cleistochlamys* (Annonaceae) has been cited as a Zanzibar-Inhambane endemic (e.g., in Vollesen, 1992; Davis *et al.*, 1994), but extends somewhat further inland (Distribution T6, 8; MN, MZ, MT, MMS, MSS; E. Zambia, N. Zambia; E. Zimbabwe, S. Zimbabwe; Malawi).

The genus *Primularia* (Melastomaceae) has been cited as a Zanzibar-Inhambane endemic (from the Rondo plateau in T8) but is now considered to be congeneric with *Cincinnobotrys*.

The Asclepiadaceae gen. indet. in Robertson & Luke (1993) is now recognised to be a new species of *Dregea*. Likewise the Rubiaceae gen. nov. aff. *Coffea* in Vollesen (1980) is now recognised to be *Psilanthus semseii*.

All genera listed are monotypic except *Asteranthe* (3 spp.) and *Lettowianthus* (2 spp.).

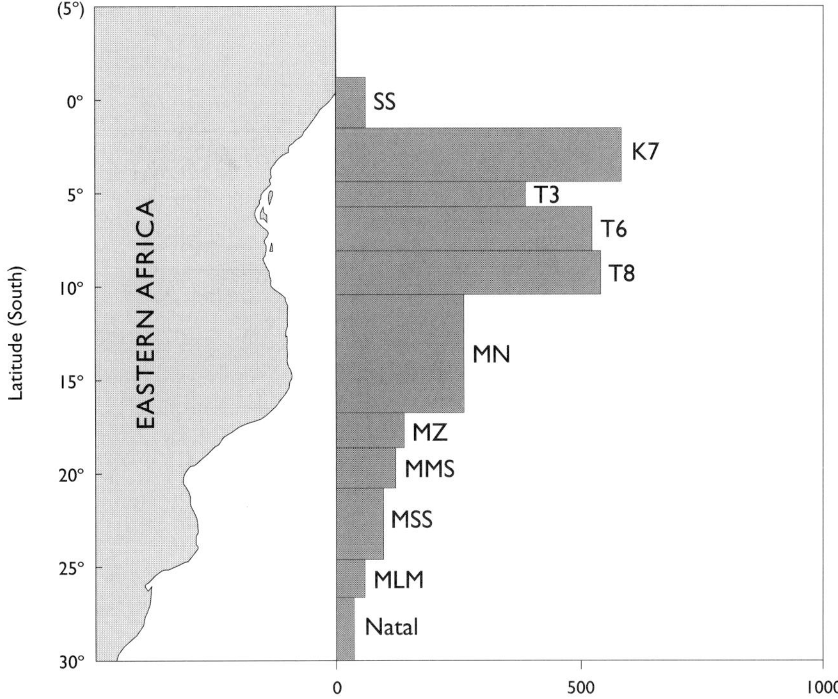

FIG. 2. Number of Zanzibar-Inhambane endemic species present in each geographical division of the *Flora of Tropical East Africa* and *Flora Zambesiaca* plotted against the coastline of eastern Africa on the left.

Classification

Under White's chorological classification, a regional centre of endemism has more than 50% of its phanerogamic species confined to it and a total of more than 1,000 endemic or near-endemic species of phanerogam (White, 1979, 1983a, 1993). These criteria were nonetheless relaxed in the case of the Sudanian regional centre of endemism, where White's most recent estimate was for about 1,250 endemic species which accounted for about 33% of the total flora, and only a few endemic genera (White, 1993, but see Davis *et al.*, 1994 and Table 2 for a lower endemic species total). In addition, the most recent estimate for the Somalia-Masai regional centre of endemism gives 1,240 endemic vascular plant species, accounting for 31% of the flora (Davis *et al.*, 1994).

A sufficiently high number of endemic plant species is present both in the northern part of the Zanzibar-Inhambane, as well as throughout the regional mosaic to warrant its reclassification as a regional centre of endemism. Given the uneven distribution of these endemic species across the region, it is proposed to split the Zanzibar-Inhambane regional mosaic into two smaller regions: (1) a regional centre of endemism along the Kenyan, Tanzanian and northern Mozambique coasts, marginally intruding into southern Somalia, and (2) a regional transition zone between this regional centre of endemism and the neighbouring Maputaland-Pondoland regional mosaic (*sensu* van Wyk, 1994), occurring along the Mozambique coast and extending into southern Malawi and eastern Zimbabwe. An estimated 4,000 species are present in the regional centre of

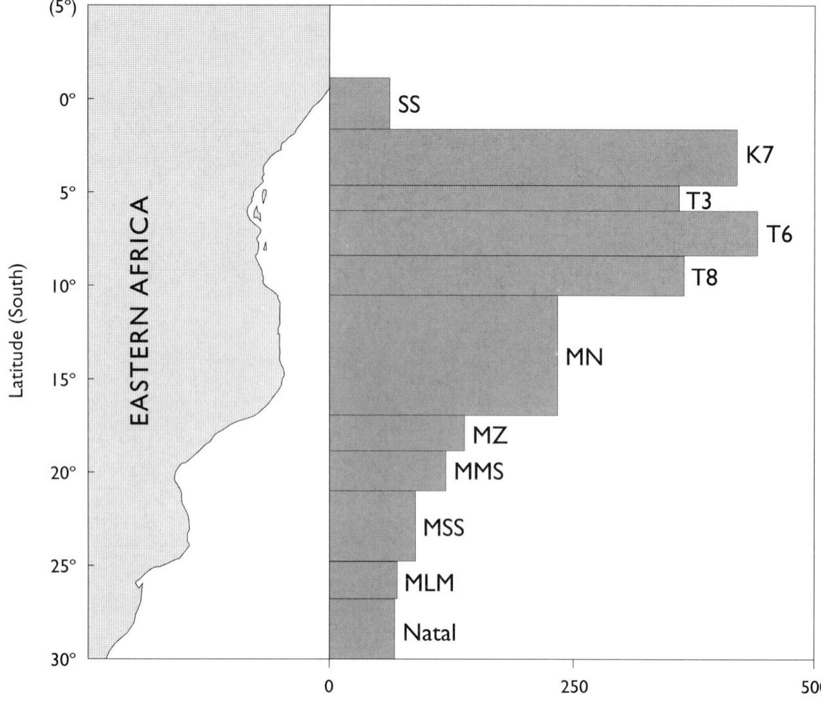

FIG. 3. Number of Zanzibar-Inhambane endemic species present in two or more geographical divisions of the *Flora of Tropical East Africa* and *Flora Zambesiaca*, i.e. excluding endemic species which are confined to a single geographical division; plotted against the coastline of eastern Africa on the left.

TABLE 2. Comparison of the regional centres of endemism in Africa, showing area, species richness, species endemism and percentage endemism of phanerogams.

Regional centre	Size (Km^2)[a]	Total no. of species[b]	No. of endemic species[b]	% endemism[b]
Guineo-Congolian	2,800,000	12,000	6,400	53[c]
Zambezian	7,770,000	8,500	4,590	54
Sudanian	3,731,000	2,750	960	35
Somalia-Masai				
excl. Arabia	1,873,000	4,000	1,250	31
Cape	71,000	8,600	5,870	68
Karroo-Namib	661,000	7,000	<3,500	35–50
Mediterranean				
excl. Eurasia	330,000	4,000	800	20
Afromontane	715,000	4,000	3,000	75
Swahilian	250,000 (est)	4,000	1,200	30

Sources: [a] White, 1983a; [b] Beentje with Adams *et al.*, 1994.
[c] On pages 106 and 115 of Beentje with Adams *et al.* (1994) a figure of 6,400 endemic species is given for the Guineo-Congolian regional centre of endemism, representing a 53% rate of endemism compared to the 80% species endemism rate given.

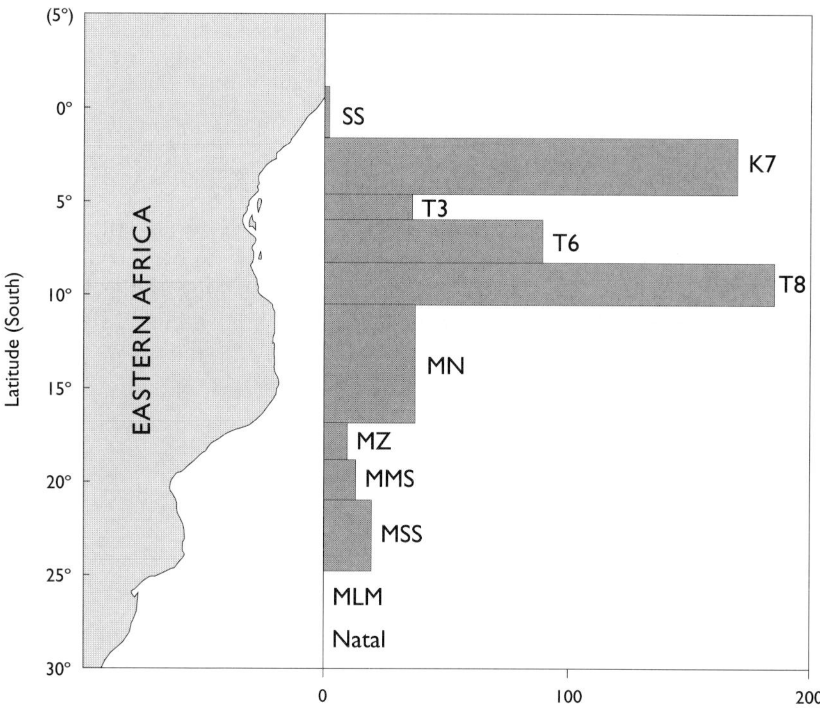

Fig. 4. Number of Zanzibar-Inhambane endemic species confined to a single geographical division of the *Flora of Tropical East Africa* and *Flora Zambesiaca* plotted against the coastline of eastern Africa on the left.

endemism, given Robertson & Luke's (1993) figure of 3,040 taxa present in non upland, coastal areas of the K7 geographical division (some of which belong to the neighbouring Somalia-Masai regional centre of endemism), together with the occurrence of a further 499 endemic species from the data which are neither found in the K7 geographical division, nor south of Moçambique town. Of these an estimated 1,200 species are endemic (given further botanical surveys and taxonomic syntheses over and above the 1,077 species identified in this study), giving an approximately 30% rate of endemism in the regional centre, or a 32% rate of endemism throughout the region *sensu lato*, assuming an additional 500 species are present in the southern part of the region.

The relatively low rate of endemism in the core area is probably a reflection of its comparatively small size (Table 2), such that edge effect interactions with the surrounding phytochoria will contribute large numbers of non-endemic species into this phytochorion. Furthermore, the long history of human settlement and cultivation in this area, a possibly even more ancient history of anthropic bush fires (Clarke & Karoma, in press), and its location on the eastern continental seaboard where trans-oceanic air and sea currents increase the opportunity for long-distance dispersals, mean that this area will contain a greater proportion of widespread and pan-tropical species than might be expected in the other regional centres of endemism. The requirement for 50% endemism can then be justifiably relaxed; it is even possible that this region would have recorded a 50% level of regional endemism prior to the large-scale extinctions of endemic species and colonisations of exotic species that are expected to have resulted from the extensive human modifications to the environment.

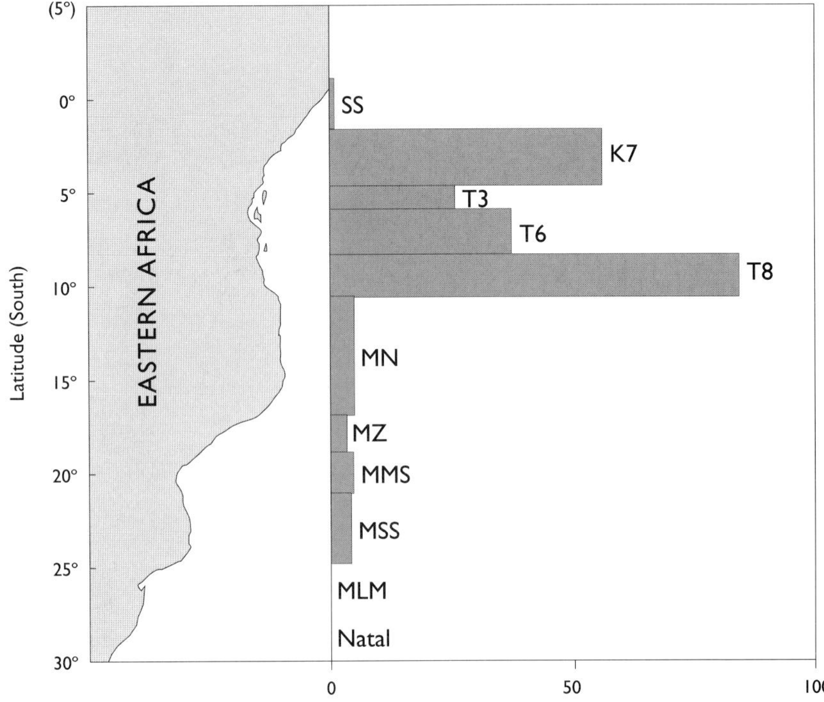

Fig. 5. Number of Zanzibar-Inhambane endemic species confined to a single geographical division of the *Flora of Tropical East Africa* and *Flora Zambesiaca*, divided by the number of degrees of latitude crossed by that division, plotted against the coastline of eastern Africa on the left.

Nomenclature

No reason was given by White (in Moll & White, 1978) for his choice of 'Zanzibar-Inhambane' although this may have been influenced by the geographical location of Zanzibar Island and Inhambane town towards the respective northern and southern limits of the region. However, the data demonstrate that Zanzibar Island and Inhambane Province (Mozambique) together contain a mere 10% of the endemic flora of the Zanzibar-Inhambane regional mosaic, and 1.5% of this endemic flora is strictly confined to these areas. 'Zanzibar-Inhambane' is therefore a misleading and unrepresentative name for this region, and is additionally impossible to maintain given the proposal to split the region in recognition of the far higher levels of floristic endemism that are present in the northern part.

The Zanzibar-Inhambane domain will therefore be renamed as the 'Swahilian region' in recognition of the fact that the proposed regional centre of endemism in the northern part of this area (Fig. 7) is similar in extent to the area which flourished under the Swahili civilisation from 800 to 1500 AD (which stretched from the Querimba Islands in northern Mozambique to Mogadishu; maps 2 & 4 in de Vere Allen, 1993), and almost coincides with those areas where Swahili was spoken as the mother-tongue (as opposed to a second language) during the last century (map 7 in Nurse & Spear, 1985). Phytochorion XIII of White (1983a) is hereby split as follows :

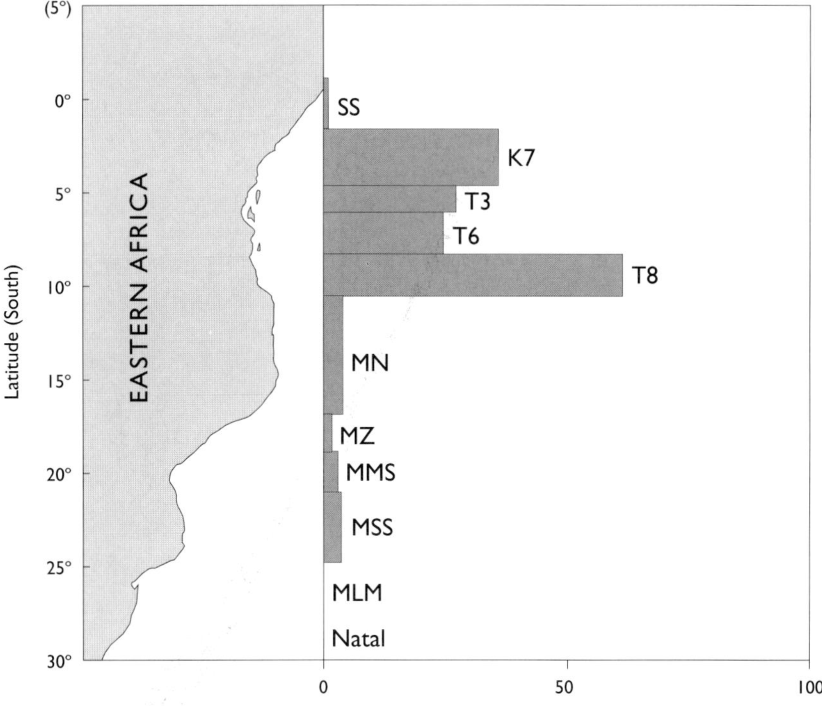

FIG. 6. Number of Zanzibar-Inhambane endemic species confined to a single geographical division of the *Flora of Tropical East Africa* and *Flora Zambesiaca*, divided by the coastline length of that division (per 100 km), plotted against the coastline of eastern Africa on the left.

XIIIa. Swahilian regional centre of endemism

Northern part of White's Zanzibar-Inhambane regional mosaic (White, 1976, 1983a & 1993) to Moçambique Island (cf. Moll & White, 1978). This regional centre includes 4,000 vascular plant species, of which an estimated 1,200 are endemic and 25 endemic genera. A further 287 species and eight genera can be considered as near-endemics, extending into the neighbouring Swahilian/Maputaland regional transition zone, a few of which marginally intrude into northern Natal.

XIIIb. Swahilian/Maputaland regional transition zone

Southern part of White's Zanzibar-Inhambane regional mosaic (White, 1976, 1983a & 1993) south of Moçambique Island (cf. Moll & White, 1978). This transition zone contains 3,300 species of which an estimated 100 are endemic (73 identified in the data). A further 40 species intrude into the Maputaland-Pondoland regional mosaic south to Natal. A single endemic genus (Rubiaceae, genus unknown of *Flora Zambesiaca* ined.) occurs in the extreme north of this regional transition zone.

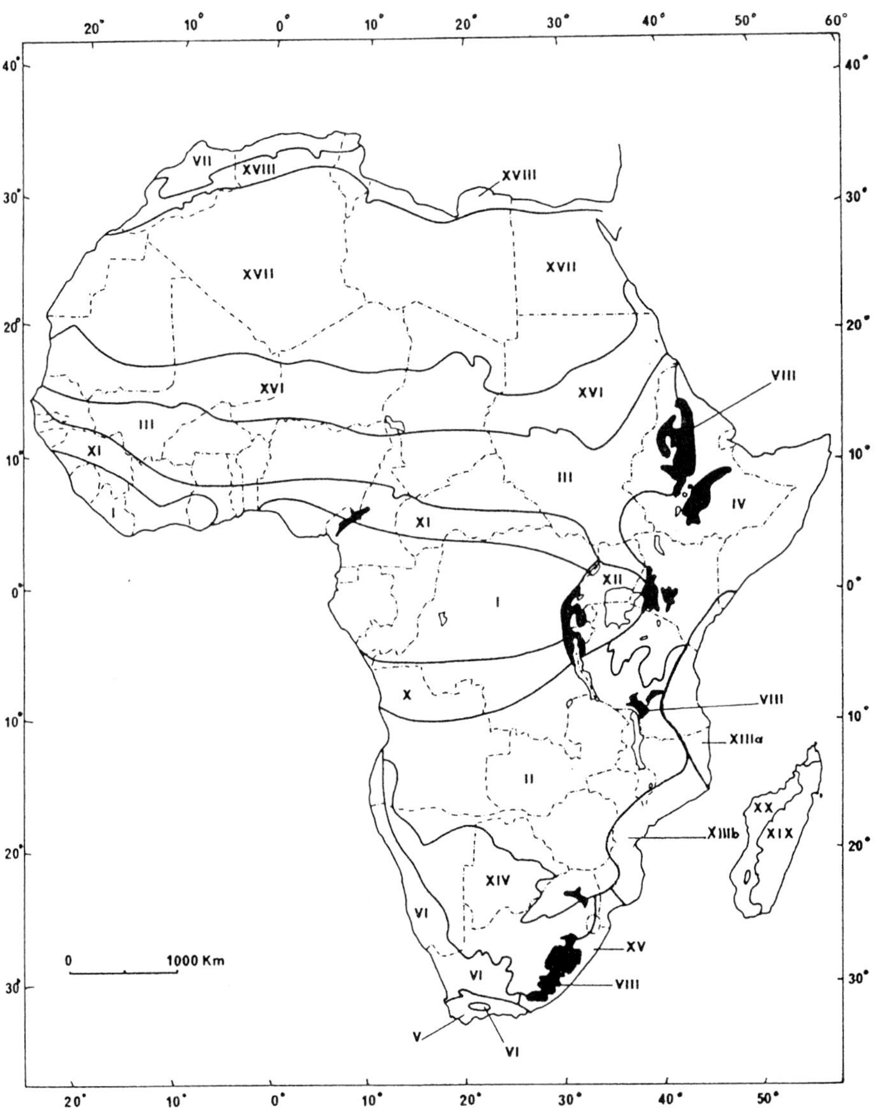

FIG. 7. The phytochoria of Africa (after White, 1983a), showing the extent of the proposed Swahilian regional centre of endemism (phytochorion XIIIa) and of the Swahilian/Maputaland regional transition zone (phytochorion XIIIb).

I. Guineo-Congolian regional centre of endemism. II. Zambezian regional centre of endemism. III. Sudanian regional centre of endemism. IV. Somalia-Masai regional centre of endemism. V. Cape regional centre of endemism. VI. Karoo-Namib regional centre of endemism. VII. Mediterranean regional centre of endemism. VIII. Afromontane archipelago-like regional centre of endemism, including IX, Afroalpine archipelago-like region of extreme floristic impoverishment (not shown separately). X. Guinea-Congolia/Zambezia regional transition zone. XI. Guinea-Congolia/Sudania regional transition zone. XII. Lake Victoria regional mosaic. XIIIa. Swahilian regional centre of endemism. XIIIb. Swahilian/Maputaland regional transition zone. XIV. Kalahari-Highveld regional transition zone. XV. Maputaland-Pondoland regional mosaic. XVI. Sahel regional transition zone. XVII. Sahara regional transition zone. XVIII. Mediterranean/Sahara regional transition zone. XIX. East Malagasy regional centre of endemism. XX. West Malagasy regional centre of endemism.

Acknowledgements

Many thanks to Alex Dickinson who first suggested that I carry out a taxonomic analysis of the Tanzanian Coastal Forest flora, to Neil Burgess who later encouraged me to generate a list of all the endemic Zanzibar-Inhambane plant species (at a time when we expected that list to be just 500 species long!), to Jon Lovett who provided the use of his extensive library and for many discussions on the phytogeography of Africa, to Kaj Vollesen and Sally Bidgood for assistance during visits to the Royal Botanic Gardens, Kew, and for identifying a number of the species cited, to Leonard Mwasumbi, Frank Mbago and Haji Suleiman for assistance during visits to the herbarium of the University of Dar es Salaam, to G. Freeman-Grenville for references on the Swahili civilisation, to Braam van Wyk for suggesting the name for the Swahilian /Maputaland regional transition zone, and to Henk Beentje, Diane Bridson, David Goyder, Roger Polhill, Gerald Pope, Sylvia Phillips and Bernard Verdcourt for help in filling some of the gaps in the data.

References

Adams, B.R. and Holland, R.W.K. (1978). The genus *Sarcostemma* in East Africa. *Cact. Succ. J. (Los Angeles)* **50**: 107–111.

Beentje, H.J. (1988). Atlas of rare trees of Kenya. *Utafiti* **1**: 71–120.

Beentje, H.J. (1990). A reconnaissance survey of Zanzibar forests and coastal thicket. Zanzibar Environmental Study Series Number 7 of 1990. Zanzibar: The Commision for Lands and Environment.

Beentje, H.J. (1994). Kenya trees, shrubs and lianas. National Museums of Kenya, Nairobi.

Beentje, H.J. with Adams, B., Davis, S.D. and Hamilton, A.C. (1994). Regional Overview: Africa. In: S.D. Davis, V.H. Heywood and A.C. Hamilton (editors). Centres of Plant Diversity. A Guide and Strategy for their Conservation.: WWF & IUCN, Godalming & Cambridge.

Brenan, J.P.M. (1978). Some aspects of the phytogeography of tropical Africa. *Ann. Missouri Bot. Gard.* **65**: 437–478.

Brenan, J.P.M. and Greenway, P.J. (1949). Check-list of the forest trees and shrubs of the British Empire. 5 Tanganyika Territory. Parts 1 and 2. Imperial Forestry Institute, Oxford.

Burgess, N.D. and Clarke, G.P. (in press). Coastal Forests of Eastern Africa. IUCN, Gland & Cambridge.

Clarke, G.P. (1995). Tanzanian Coastal Forest Research Programme, Checklist of the vascular plants from 13 Coastal Forests in Tanzania. Frontier-Tanzania Technical Report No. 15. London & Dar es Salaam: The Society for Environmental Exploration/the University of Dar es Salaam.

Clarke, G.P. and Dickinson, A. (1995). Status Reports for 11 Coastal Forests in Coast Region, Tanzania Frontier-Tanzania Technical Report No. 17. London & Dar es Salaam: The Society for Environmental Exploration/The University of Dar es Salaam.

Clarke, G.P. and Karoma, J.A. (in press). History of anthropic disturbance. In: N.D. Burgess and G.P. Clarke (editors). Coastal Forests of Eastern Africa. Chapter 5.1. IUCN Forest Conservation Series. IUCN Publications Unit, Cambridge.

Clarke, G.P., Vollesen, K. and Mwasumbi, L.B. (in press). Vascular plants. In: N.D. Burgess and G.P. Clarke (editors). Coastal Forests of Eastern Africa. Chapter 4.1 and Appendix 3. IUCN Forest Conservation Series. IUCN Publications Unit, Cambridge.

Davis, S.D., Heywood, V.H. and Hamilton, A.C. (1994). Centres of Plant Diversity. A Guide and Strategy for their Conservation. WWF & IUCN, Godalming & Cambridge.

de Vere Allen, J. (1993). Swahili Origins. Swahili Culture and the Shungwaya Phenomenon. James Currey, E.A.E.P. and Ohio University Press London, Nairobi and Athens.

Dowsett-Lemaire, F. (1990). The flora and phytogeography of the evergreen forests of Malawi II: Lowland forests. *Bull. Jard. Bot. Belg.* **60**: 9–71.

Drummond, R. (in prep.). Botanical checklist for the Haroni and Rusitu Forests, Melsetter District, Zimbabwe.

Exell, A.W. and Wild, H. *et al.* (1960). Flora Zambesiaca. Crown Agents for Overseas Governments and Administrations: M.O. Collins (Pvt.) Ltd. London & Harare.

Faden, R.B. (1991). The Morphology and Taxonomy of *Aneilema* R. Brown (Commelinaceae). *Smithsonian Contr. Bot.* **76**: 1–166.

Friis, I. (1991). Skovtraesfloraen i det nordostlige tropiske Afrika (Etiopien, Djibouti og Somalia) — en sammenfattende redegorelse for tidligere studier. Botanisk Museum, Copenhagen.

Friis, I. and Vollesen, K. (1989). Notes on the vegetation of southernmost Somalia, with some additions to the flora. *Willdenowia* **18**: 455–477.

Goldsmith, B. (1976). The trees of Chirinda Forest. *Rhodesia Sci. News* **10**: 40–50.

Gomes e Sousa, A. (1966 & 1967). Dendrologia de Moçambique. Volumes I & II. Lourenco Marques: Instituto de Investigação Agronomica de Moçambique.

Greenway, P.J. with Rodgers, W.A., Wingfield, R.J. and Mwasumbi, L.B. (1988). The vegetation of Mafia Island, Tanzania. *Kirkia* **13**: 197–238.

Haines, R.W. and Lye, K.A. (1983). The Sedges and Rushes of East Africa. East African Natural History Society, Nairobi.

Hawthorne, W.D. (1984). Ecological and biogeographical patterns in the Coastal Forests of East Africa. Oxford: Unpublished D. Phil. thesis, University of Oxford.

Hawthorne, W.D. (1993). East African coastal forest botany. In: J.C. Lovett and S.K. Wasser (editors). Biogeography and ecology of the rain forests of eastern Africa. Pp. 57–99. Cambridge University Press, Cambridge.

Iversen, S.J. (1991). The Usambara Mountains, NE Tanzania: phytogeography of the vascular plant flora. *Symb. Bot. Upsal.* **29(3)**: 1–234.

Johansson, D.R. (1978). *Saintpaulias* in their natural environment with notes on their present status in Tanzania and Kenya. *Biol. Conservation* **14**: 45–62.

Kapuya, J.A. (1994). Phenological and island vegetation studies of Pangani Falls. Report to NorConsult, Dar es Salaam.

Kapuya, J.A. (1995). Salvage of the African Violet (*Saintpaulia tongwensis*) community at Pangani Falls. Report to NorConsult, Dar es Salaam.

Lebrun, J. (1960). Sur la richesse de la flore de divers territoires africains. *Bull. Séances Acad. Roy. Sci. Outre Mer* **6**: 669–690.

Leroy, J.F. (1978). Composition, origin and affinities of the Madagascan vascular flora. *Ann. Missouri Bot. Gard.* **65**: 535–589.

Lovett, J.C. (1993). Eastern Arc Moist Forest Flora. In: J.C. Lovett and S.K. Wasser. (editors). Biogeography and ecology of the rain forests of eastern Africa. pp 33–55. Cambridge University Press, Cambridge.

Manktelow, M. (1996). *Phaulopsis* (Acanthaceae) — a monograph. *Symb. Bot. Upsal.* **31(2)**: 1–183.

Medley, K.E. (1992). Patterns of forest diversity along the Tana River, Kenya. *J. Trop. Ecol.* **8**: 353–371.

Milne-Redhead, E. (1955). Distributional ranges of flowering plants in tropical Africa. *Proc. Linn. Soc. London* **165**: 25–35.

Moll, E.J. and White, F. (1978). The Indian Ocean Coastal Belt. In: M.J.A. Werger (editor) with assistance of A.C. van Bruggen. Biogeography and Ecology of Southern Africa. Pp. 563–598. W. Junk, The Hague.

Monod, T.L. (1957). Les grandes divisions chorologiques de l'Afrique. London: CCTA/CSA. 24: 1–147.

Mwasumbi, L.B., Burgess, N.D. and Clarke, G.P. (1994). The vegetation of Kiono and Pande forests, Tanzania. *Vegetatio* **113**: 71–81.

Mziray, W. (1992). Taxonomic studies in Toddalieae Hook.f. (Rutaceae) in Africa. *Symb. Bot. Upsal.* **30**: 1–95.

Nurse, D. and Spear, T. (1985). The Swahili. Reconstructing the History and Language of an African Society, 800–1500. Philadelphia: University of Pennsylvania Press.

Palgrave, K.C. (1977). Trees of Southern Africa. C. Struik Publishers, Cape Town and Johannesburg.

Pennington, T.D. (1991). The genera of Sapotaceae. Royal Botanic Gardens, Kew & New York Botanical Garden.

Robertson, S.A. and Luke, W.R.Q. (1993). Kenya Coastal Forests. Report of the NMK/WWF Coast Forest Survey. Worldwide Fund for Nature, Nairobi.

Rodgers, W.A., Mwasumbi, L.B., Griffiths, C.J. and Vollesen, K. (1983). The Conservation Values and Status of Kimboza Forest Reserve, Tanzania. Forest Conservation Working Group, University of Dar es Salaam.

Ruffo, C.K. (1991). A report on species identification for Jozani and Ngezi forest inventory. Lushoto: Unpublished report by the Tanzania Forestry Research Institute.

Seyani, J.H. (1991). The Genus *Dombeya* (Sterculiaceae) in Continental Africa. National Botanic Garden Belgium, Meise.

Temu, R.P.C (1990). Taxonomy and biogeography of woody plants in the Eastern Arc Mountains, Tanzania: case studies in *Zenkerella, Scorodophloeus,* and *Peddiea. Compreh. Summ. Uppsala Diss. Fac. Sci.* **286**: 1–68.

Thulin, M. *et al.* (1993). Flora of Somalia.Volume 1. Royal Botanic Gardens, Kew.

Turrill, W.D. and Milne-Redhead, E. *et al.* (1952). Flora of Tropical East Africa.: A.A. Balkema, Rotterdam.

van Wyk, A.E. (1994). Maputaland-Pondoland Region. In: S.D. Davis, V.H. Heywood and A.C. Hamilton (editors). Centres of Plant Diversity. A Guide and Strategy for their Conservation. Pp. 227–235. WWF & IUCN, Godalming & Cambridge.

Verdcourt, B. (1996). *Sanrafaelia* a new genus of *Annonaceae* from Tanzania. *Garcia de Orta, Sér. Bot.* **13**: 43–44.

Vollesen, K. (1980) Annotated check-list of the vascular plants of the Selous Game Reserve, Tanzania. *Opera Bot.* **59**: 1–117.

Vollesen, K. (1992). *Trichaulax* (Acanthaceae: *Justicieae*), a new genus from East Africa. *Kew Bull.* **47**: 613–616.

Vollesen, K. and Bidgood, S. (1992). Kew Expedition to Tanzania & Malawi Jan–April 1991. Unpublished report to the Royal Botanic Garden, Kew.

White, F. (1976). The vegetation map of Africa: the history of a completed project. *Boissiera* **24**: 659–666.

White, F. (1979). The Guineo-Congolian Region and its relationships to other phytochoria. *Bull. Jard. Bot. Belg.* **49**: 11–55.

White, F. (1983a). The Vegetation of Africa: a descriptive memoir to accompany the UNESCO/AETFAT/UNSO vegetation map of Africa. *Natural Resources Research* **20**. UNESCO, Paris.

White, F. (1983b). UNESCO/AETFAT/UNSO vegetation map of Africa. Scale 1:5,000,000 (in colour). UNESCO, Paris.

White, F. (1988). The taxonomy, ecology and chorology of African *Ebenaceae*. II. The non-Guineo Congolian species of *Diospyros* (excluding sect. Royenna*). Bull. Jard. Bot. Belg.* **58**: 325–448.

White, F. (1990). *Ptaeroxylon obliquum* (Ptaeroxylaceae), some other disjuncts, and the Quaternary history of African vegetation. *Bull. Mus. Natl. Hist. Nat., B, Adansonia* **12(2)**: 139–185.

White, F. (1993). The AETFAT chorological classification of Africa: history, methods and applications. *Bull. Jard. Bot. Belg.* **62**: 225–281.

Linder, H.P. (1998). Numerical analyses of African plant distribution patterns. In: C.R. Huxley, J.M. Lock and D.F. Cutler (editors). Chorology, Taxonomy and Ecology of the Floras of Africa and Madagascar. Pp. 67–86. Royal Botanic Gardens, Kew.

5. NUMERICAL ANALYSES OF AFRICAN PLANT DISTRIBUTION PATTERNS

H.P. Linder

Bolus Herbarium, Botany Department, University of Cape Town, Rondebosch 7700, South Africa.

Abstract

Distributional data for 794 species from 98 genera were used to test Frank White's African phytochoria. The data are presence records for 248 2.5 × 2.5 degree grid squares. Summing the number of species recorded for each grid provides an estimate of species richness patterns across sub-Saharan Africa, and this reveals that the most speciose areas are the rain forest belts around the Gulf of Guinea, and in eastern Congo. The data also indicate those areas as yet too poorly sampled to give a robust indication of their phytochorological affinities: especially the Horn of Africa, and the arid areas of Namibia and Botswana. The patterns of similarities in species composition among the grids was determined with the Jaccard coefficient, and this retrieved the phytochoria of White fairly accurately. This shows that White's phytochoria were indeed based on the patterns of species distribution, and not on vegetation structure. The patterns in Eastern Africa were not retrieved: possibly because the grids are too coarse to differentiate the closely interdigitating islands of Afromontane vegetation, and the shrublands of the Somali-Masai centre. Analysis of the frequency occurrence of range-restricted species shows very distinct 'centres of endemism', the most prominent of which are centered on Mt. Cameroun, on Kivu and on the Tanzanian coastline. The search for a historical explanation of these patterns is still in its infancy, but several lines of investigation would repay further work: the patterns of disjunction, and a search for a cladistic biogeographical analysis of the continental flora. The latter is hampered by a great shortage of phylogenetic hypotheses on the African flora.

Nevertheless there are indications that historical plant geography in Africa is about to enter a vigorous and fruitful phase.

White, 1971.

Introduction

AETFAT, during the years from its founding until c. 1980, was an organisation with a mission. It set itself many practical tasks, but one issue that dominated recurrent meetings was the relationship between the floras of the different parts of Africa. Reading back through the proceedings of these meetings, one senses the enthusiasm and excitement with which new insights were received. Although there is no record of the drafting of the vegetation map, it is not difficult to imagine what discussion this must have generated. Part and parcel of this process was Frank White. He appeared to dominate the phytogeographical thinking, separating vegetation and floristic mapping, looking for broad phytogeographical regions, for a broad vision that could encompass the whole of the African flora. This approach culminated in the monumental *The Vegetation of Africa* (White, 1983), with its 'theoretical justification' appearing in 1993 (White, 1993).

This great productivity was only possible because White had a clear idea of what he wanted, and was vigorous in his criticism of other approaches. He did not like hierarchical phytogeographical classifications, and frequently stated so. He convinced the Africanists at AETFAT, most of whom had started by producing such classifications. He was opposed to a crude number-crunching approach, although to my knowledge there have been very few attempts at a quantitative approach to African phytogeography. He generally criticised attempts to generate explanations for the phytogeographical patterns, especially if they were based on palaeopalynology. Maybe he could be caricatured as an empiricist, who was looking for patterns in the distribution of African plants, and who did not like to incorporate other sources of information.

The central planks of White's phytogeography are the phytochoria. These, and most other concepts in descriptive phytogeography, are based on the notion of endemism. Areas of endemism are difficult to define, and White changed his definitions over the years, summarising his various definitions in 1993. There are two concepts current in his work. The first is the 'Regional centre of endemism', which he later (1983, 1993) regarded as the 'fundamental' unit, and which is largely equivalent to a phytochorological unit. He used the levels of endemism in the regions as a form of 'ranking criterion', in that chorological units which have more than a defined level of endemism should be recognized as regional centres of endemism.

The second level at which centres of endemism were used, was the 'local centre of endemism', often (e.g. White, 1965; 1979) simply called a 'centre of endemism.' These are areas rich in narrow endemics, currently often termed 'hotspots', and may be equivalent to 'centres of plant diversity' (Beentje, 1996). Although White did not pay much attention to these local centres, they have recently received substantial attention.

Frank White's work laid the foundations upon which we need to build African phytogeography. Like any good research, it raises many questions. The first set of questions is about the corroboration of the phytochoria:

A. Are White's phytochoria based entirely on floristic distribution patterns, or are vegetation characteristics also included?

B. Can these phytochoria be retrieved by numerical analysis?

C. Are the same phytogeographical areas produced by an analysis of all species (wides as well as endemics) as are produced by an analysis of endemics only?

The second set of questions concerns issues beyond the phytochoria:

D. Are there local centres of endemism in the African flora?

E. Are there disjunctions in the African flora?

F. How are these centres of endemism and disjunctions analysed and interpreted?

Methods

Distributional Data

Distributional data were collected for 685 species or infraspecific taxa in 99 genera (see Appendix) published in the series *Distributiones Plantarum Africanarum.* The distributions were scored as presence-absence data on a grid of 2.5×2.5 degrees, thus forming a matrix of 685 species by 248 grid squares. The size of grid is of some importance. The advantage of smaller grid sizes is that they have a finer resolution, suitable for detecting small-scale pattern, like an 'island' of montane species in a savanna area. There are numerous disadvantages of making the grid size too small: they generate numerous false absences, increase the number of errors in data recording, and generate data bases which may be difficult to handle with conventional spreadsheets and commercially available analytical programs. The published distribution maps have a 5 degree grid on them, thus making 2.5 degree squares rather convenient for data recording.

Phytochoria

A similarity matrix for the grids was calculated using Jaccard's coefficient, which does not take shared absences into account. Shared absences of species are biogeographically meaningless. The grids were clustered using UPGMA. Both Jaccard and UPGMA were used as implemented in NTSys (Rohlf, 1994). Centres of richness were determined by summing the number of species recorded for each grid square, thus obtaining the grid diversity.

Local Endemism

Areas rich in local endemics were determined by extracting all species restricted to one or two grids from the data base. The number of such geographically restricted species for each grid was then obtained, by summing them for each grid. This does not delimit areas of endemism as such, but plots the geographical distribution of locally endemic species. Areas of endemism can be much larger, and can only be obtained by finding tension zones, characterised by a very high turnover in species composition. To establish that the patterns of endemic species are not simply a by-product of patterns of species richness, the percentage of the flora of each grid that are one- or two-grid endemics was calculated.

Disjunctions

As disjunctions are not amenable to statistical analysis, they were compiled from the literature. The two major sources were previous summaries of disjunctions, and the maps in *Distributiones Plantarum Africanarum*. No grid square definition of disjunctions was developed; they were compiled from visual inspection. However, I did not compile an exhaustive list of all disjunctions.

Historical Analysis

The historical relationships among the areas were investigated using component analysis (Humphries & Parenti, 1986) as implemented by Page (1993). The analysis was done by mapping widespread species (from more than one area) under Assumption 0, and by assuming that absences are uninformative. Assumption 0 groups together areas that share the same species. The trees located were combined using Adams Consensus, which does not allow 'aberrant' elements to collapse all the nodes, but collapses them to the base of the tree. Several criteria have to be met before a phylogeny can be biogeographically informative: all species should be included, sufficiently detailed distributional information should be available, the cladograms should have been obtained by an explicit and repeatable process, and at least one 3-area statement should be included. Only two studies meet these criteria: *Phaulopsis* (Manktelow, 1996) and *Justicia* sect. *Ansellia* (Ensermu, 1990). The study of the Toddalieae (Mziray, 1992) does not include all species in the group; *Justicia* sect. *Harnieria* (Hedrén, 1989) is not based on an explicit methodology, and the phylogeny of *Wellstedia* (Thulin & Johansson, 1996) covers only two areas.

The areas used for the cladistic analysis were based on the local centres of endemism, as located from the analysis above, and complemented by centres evident from the literature. There are four rain forest centres: Guinea, Cameroun, Kivu and the East African coast; two savanna centres: the Sudanian region (a rather wide, but homogeneous region) and the Congo-Zambezi watershed; two desert systems: Somalia and the Namibia-Kalahari region; and five Afromontane areas: Ethiopia, the East African highlands, Bamenda, eastern Zimbabwe and southern Malawi, and the Drakensberg.

Results

Centres of species richness

The grid richness is immensely variable, from a maximum of 165 species in the 2.5 degree square on Mt. Cameroun, to none (Fig. 1). Almost half of the grids have less than 20 species, and only 10 of the 248 grids have more than 100 species, while four have more than 140 species (Fig. 2).

Phytochoria

The area phenogram shows a remarkable degree of resolution. Eighteen groups are distinct at a similarity level of 0.118. Of these, eight groups are large, and when mapped form large coherent areas (Fig. 3). The remaining groups are small, and most contain scattered squares. With the exception of the two squares reflecting Namaqualand in South Africa, the rest are all coastal, containing largely water, suggesting that they are

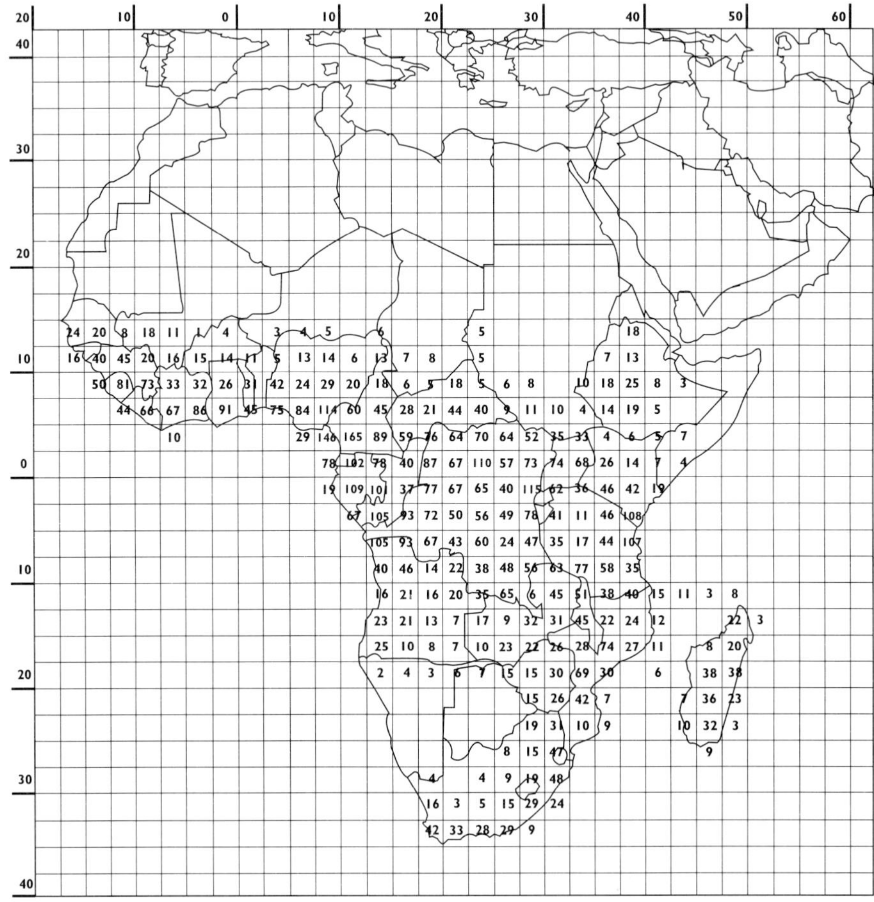

FIG. 1. Grid diversity in Subsaharan Africa, based on the sample of 794 species. Grid size is 2.5 by 2.5 degrees, and the numbers in each cell are the actual number of species recorded from that grid.

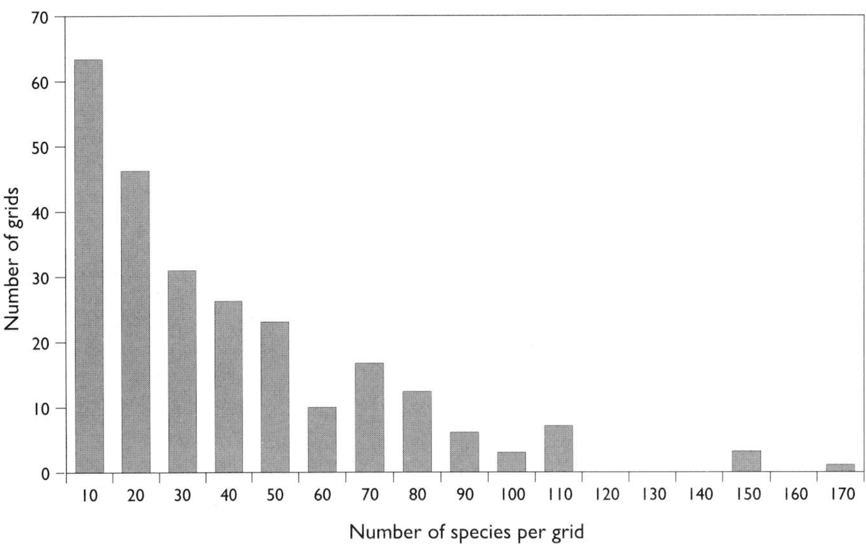

FIG. 2. Patterns of Subsaharan grid diversity, showing that most grids have less than 20 species.

in fact misplaced squares, due to inadequate information. They are therefore ignored in the interpretation of the phenogram. The differences among the groups are in many cases small, and the three clusters from North-east Africa are adjacent, and linked at a similarity of 0.093. It is likely that small changes in the species included in the study would result in these being combined into a single area under the criteria employed here.

There is substantial regional pattern especially within the larger groups representing the two large savanna phytochoria and the rain forest phytochorion. These have not been analysed in more detail.

Centres of local endemism

More than half the 685 species occur in fewer than six grids, so relatively few species have wide distribution ranges (Fig. 4). One hundred and four species are recorded from single grids; these can be defined as narrow endemics, and their distribution shows a scatter across equatorial Africa (Fig. 5). One hundred and nine species are known from two grids; combined with the narrow endemics this provides a data base of 213 species. This shows rather distinctive centres of endemism (Fig. 6), and essentially the same centres are retrieved if the percentage the flora made up of restricted species is used (Fig. 7). In sequence of species richness these are:

1. Cameroun, with outliers up the Sanaga river valley, and along the coast from the Congo to the Niger rivers.

2. The northeastern Tanzanian area, including the Pare, Usambara and Uluguru mountains.

3. Western Rift mountains from Lake Albert to Lake Tanganyika: this includes the Ruwenzori and the highlands around Lake Kivu.

4. Cape Floristic Region, centred around Cape Town at the southern tip of the continent.

5. Madagascar: especially the northern and southern extremes, but virtually every grid on the island contains 'narrow' endemics.

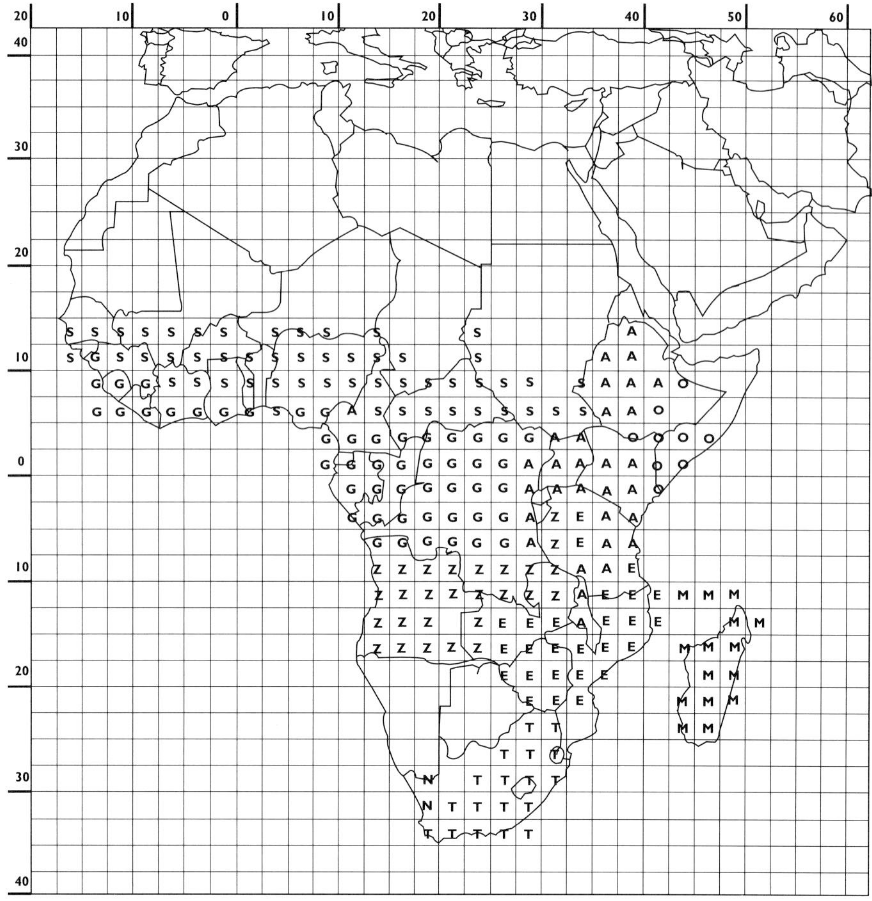

FIG. 3. Mapped phytogeographical areas, based on the Jaccard similarity among the grids. S = Sudanian, G = Guineo Congolean, A = broadly Afromontane, Z = Zambezian, E = eastern Zambezian, O = Somalian, M = Malagasy, T = South African and N = Namaqualand.

6. Mbala region of Northern Zambia, reaching eastwards to the Kipengeres.

7. Eastern Highlands of Zimbabwe and southern Malawi.

8. The headwaters of the Niger, from Fouta Djalon south-east to the Man plateau. The first four centres are substantially richer than the others, as is apparent from the figures.

Cladistic biogeography

A heuristic search of the two cladograms of the areas shown in Fig. 8 found 525 trees, with 12 leaves added; the Adams consensus tree found substantial structure (Fig. 9), while the strict consensus tree found none. This is presumably due to one area (possibly Bamenda) being placed into many different areas. These results are at best dubious, due to the lack of information. However, they may well be the best that can be done with the data, and may serve to illustrate the sort of questions that can be addressed by this method.

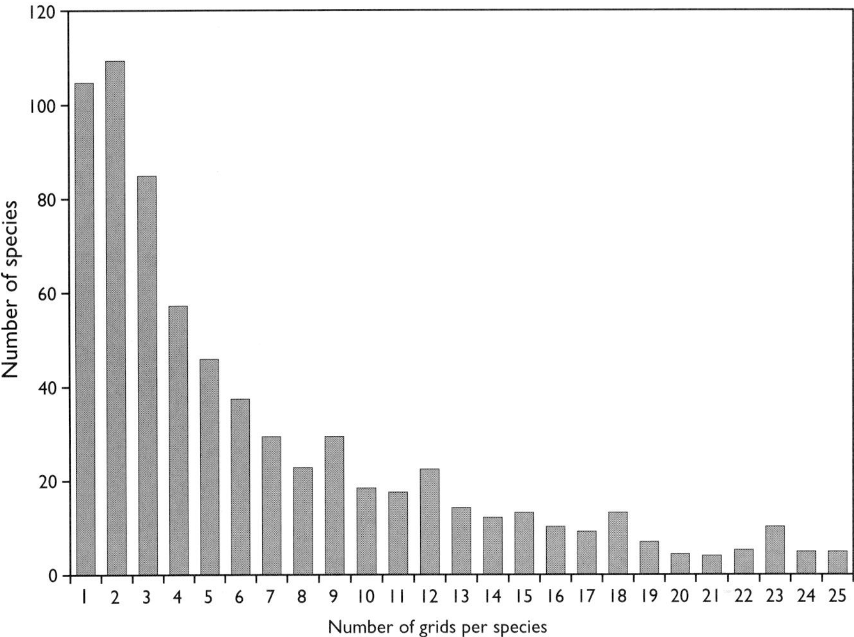

Fig. 4. Distribution ranges of Subsaharan species, as determined by the number of grid squares occupied. This indicates that most species are recorded from less than four grids.

Discussion

Patterns of species richness

The greatest grid diversity is in the area between the mouths of the Niger and Congo rivers, and generally the rain forests have a higher grid diversity than the savannas. These patterns are consistent with those reported from the literature, and probably agree with the generalisation that grid diversity is related to total rainfall (Gentry, 1988). However, these sorts of data would be very sensitive to biases in the data base, and I would not like to extrapolate any further from these results, except to draw attention to them. Inspection of the data suggests that there may be a bias towards rain forest and Afromontane taxa; taxa centered in the Cape Flora and the Somalian and Zambezian regions are under-represented. There is clearly a need to expand the data set to include such taxa (see also Fig. 1 in Barthlott and Porembski, this volume).

Phytochoria

The larger groups located by the analysis are almost perfectly congruent with the phytochoria of White (1983). The Guineo-Congolian phytochorion is retrieved almost exactly, even including the Dahomey Gap. The two savanna phytochoria (Zambesian and Sudanian) are also almost exactly retrieved. In both transitional regions between the rain forests and savannas the boundaries retrieved by this analysis lie closer to the rain forest boundary. At the level of similarity used, two areas for the Zambezian Region are retrieved. These, however, lie next to each other on the phenogram and, when combined, map the Zambezian Region of White (1983). These floristic phytochoria closely follow the vegetational pattern, which been recognized by all

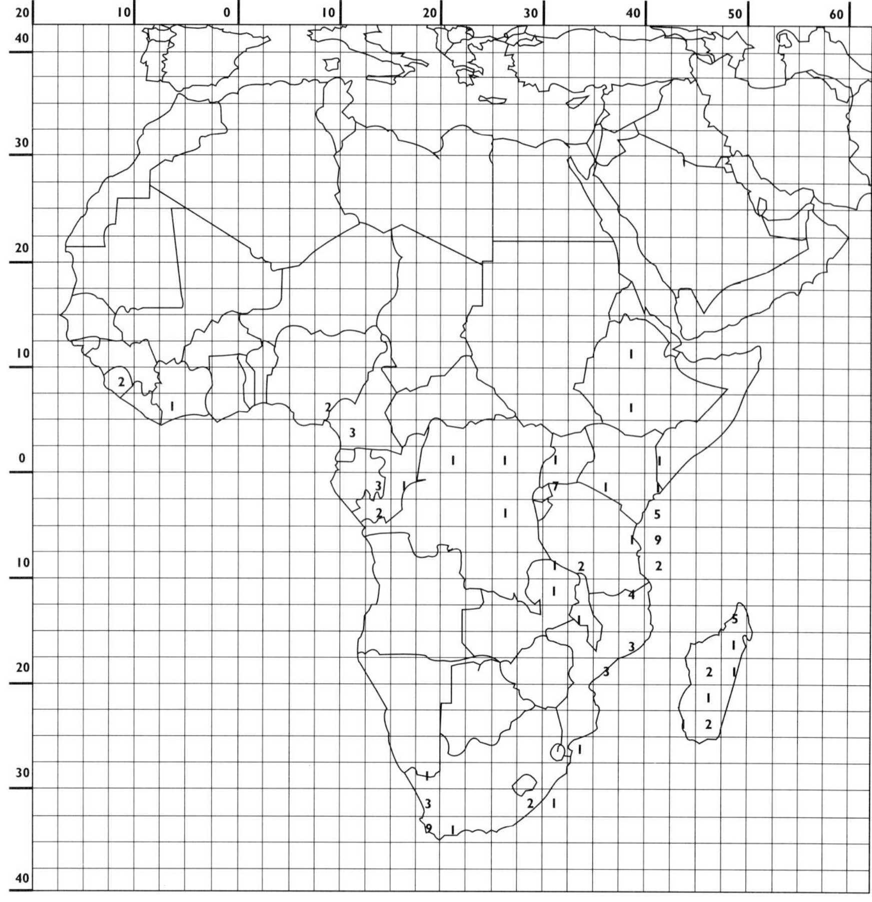

FIG. 5. Number of species recorded from a single 2.5 × 2.5 degree grid square for each square.
Blank squares have no one-grid endemics.

vegetation maps of Africa (Wickens, 1976; White, 1979; 1983, and further references cited in White, 1979).

The southern African phytochoria are poorly resolved. The Namib-Kalahari region and Cape Floristic region (Goldblatt, 1978) are not resolved, due to paucity of data; I did not attempt to incorporate data from these areas into the analysis. The grid squares from the Namib-Kalahari region were not included in the analysis, and only *Anthospermum* (Rubiaceae) of the very numerous Cape Floristic Region elements was included in the study. However, there is no doubt about the distinctness of the Cape Floristic Region, as it shows remarkably high levels of endemism: at specific level the endemism exceeds 90% (Goldblatt, 1978).

The Zanzibar-Inhambane regional mosaic was partially retrieved. This could be due to two reasons. The first is that a mosaic is by definition mixed with the neighbouring phytochorion, and is therefore likely to require smaller grid squares to detect by numeric means. In addition, the coastal mosaic is narrow, and the grids are therefore likely to contain the neighbouring Zambesian flora or the sea in many cases.

74

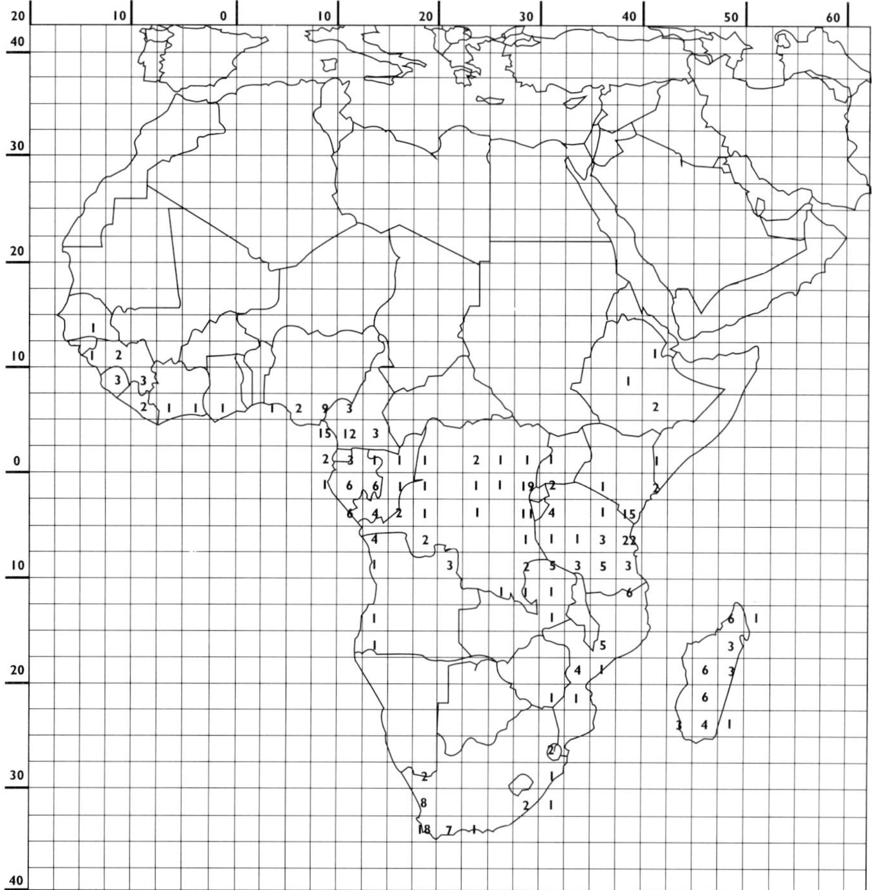

FIG. 6. Number of species recorded from a single or two 2.5 × 2.5 degree grid squares for each
square. Blank squares have no one- or two-grid endemics.

The distribution patterns in East Africa do not fit the phytochoria proposed by
White. He mapped a relatively large Somali-Masai phytochorion, which abuts onto
the Zambezian phytochorion to the south and the Sudanian to the north. My results
suggest a smaller Somali-Masai phytochorion, ranging from the mouth of the Tana
River to Marsabit, and north into the Horn. Most of Ethiopia, Uganda, the Tanzania
uplands and Malawi is combined into one region, which has an outlier in the
Bamenda highlands of Cameroun, thus suggesting that this is an Afromontane forest
phytochorion, and that those grids that contain this phytochorion partially or
entirely have been combined on this basis. It is substantially larger than usually
mapped, but does not include the Zimbabwean, Angolan or southern African
uplands. As such it is quite different from any previous mapping of this region, and
could therefore be interesting to explore further. It appears possible that any grids
that contain montane elements have been grouped together, thus obscuring the
distribution patterns suggested by the surrounding lowlands. In some of the analyses
with smaller data sets these grid squares were grouped on the basis of their lowland

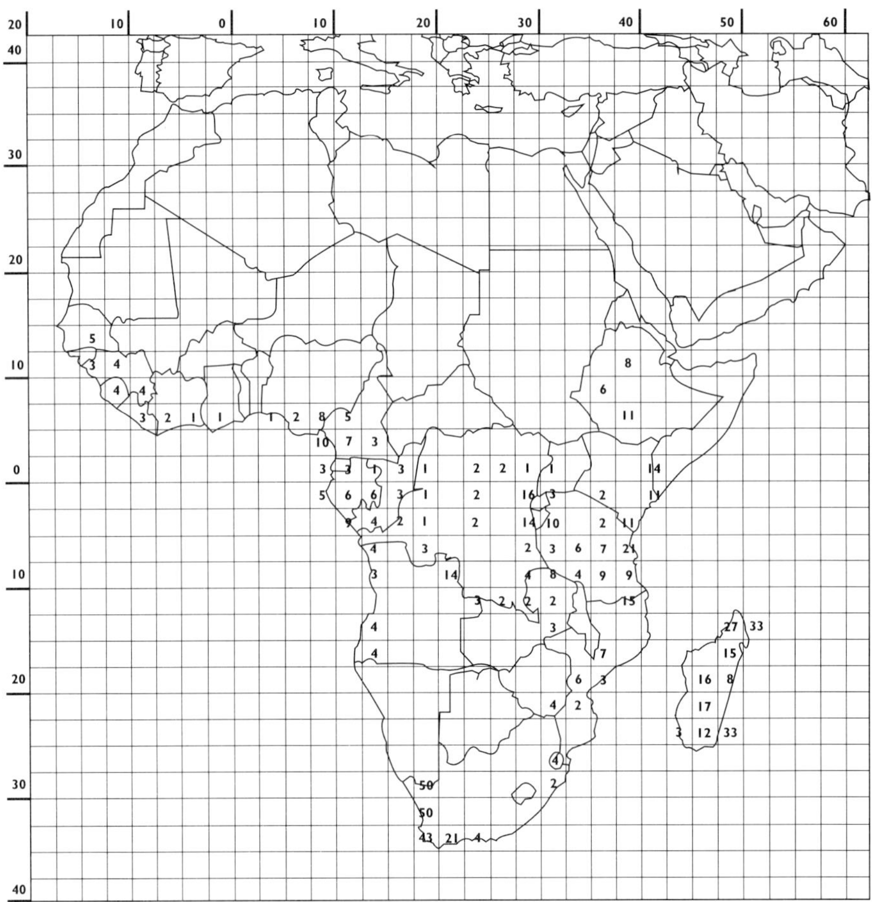

FIG. 7. Percentage of the species in grids with one- or two-grid endemics which are one- or two-grid endemics. Blank squares have no one- or two-grid endemics.

taxa, leaving no trace of the Afromontane region! Thus again, the size of the grid squares is causing a problem.

This quantitative analysis, with the riders added above, therefore largely corroborates the chorology established by White and others. This is rather impressive, as the sources cited by White (1993) are more limited than the ones used in this study. It appears likely that the exact boundaries of the phytochoria were based heavily on field experience and unquantified opinions of the numerous botanists who contributed to the map. White (1993) termed this 'perception', and noted that it played an important role in the evolution of the AETFAT chorological classification (see also White, 1979). However, White also noted that 'subsequent rigorous analysis' was also necessary. I am of the opinion that the quantified analysis will return a map more similar to White's phytochoria if more species are included and the grid size can be made smaller, at least for some areas. Equally important will be to examine congruence between different groups of plants: angiosperms, ferns, bryophytes, etc., and use a wide taxonomic base within the angiosperms, as this will give us an indication of the generality of these patterns.

76

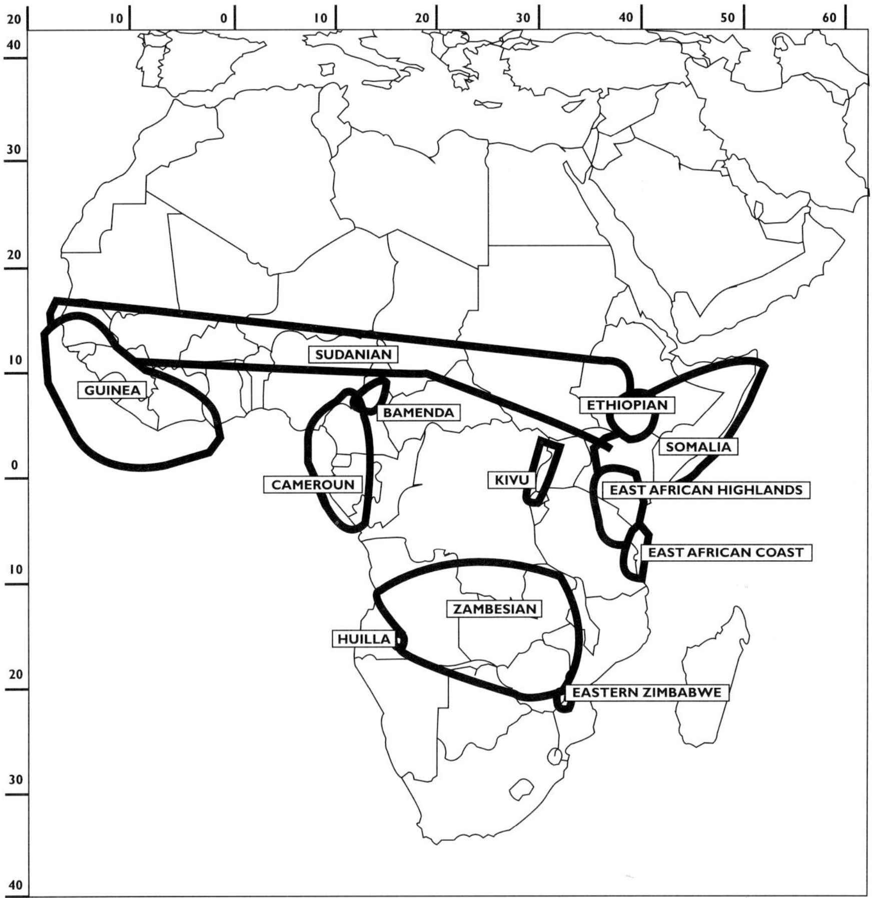

FIG. 8. Areas used for the cladistic biogeographical analysis, shown in Figure 9.

It is also evident from the results that the phytochoria reflect the patterns of species distribution. Although they are also congruent with vegetation types (savanna, rain forest, semi-desert areas, frost-influenced grasslands), this is incidental. This is illustrated by the uniqueness of Madagascar, and the two groups for the northern and southern savannas. It can therefore be inferred that although environment may be determining the extent of the phytochoria, historical factors are playing a role in their differentiation. If this were not the case, then the northern and southern savannas would form one region, and Madagascar would be divided into its constituent vegetation types, each linking more closely to its African counterpart. White (1993) stressed that the separation of chorological and ecological units was well understood by the AETFAT community, and the evidence before us is that despite the fact that opinion and experience, which could be easily swayed by vegetation appearance, has had so much to do with the delimitation of the phytochoria, this did not lead to an over-emphasis on ecology.

White (1965) argued against the simultaneous analysis of all the species in the flora: "If the distributions of all species are used as a basis for the subdivision of a Region, no

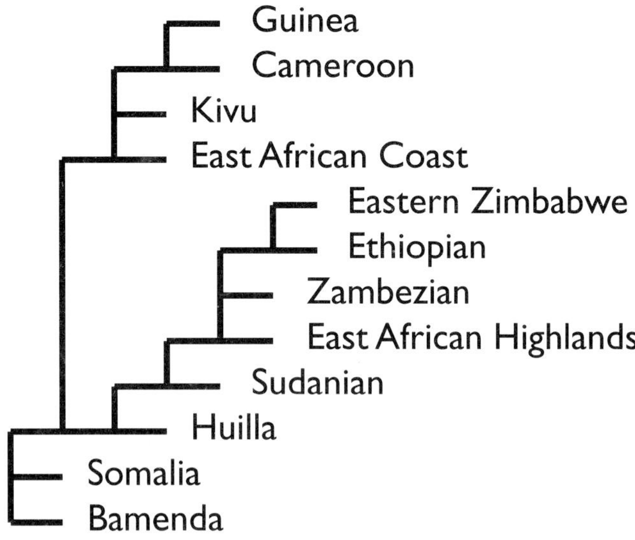

FIG. 9. The Adams consensus tree of the cladistic biogeographical analyses, showing the relationships among the areas.

clear picture will emerge or it will be considerably obscured, since the distinctive patterns of any particular ecological elements will be masked by those of other ecological elements or swamped in the mass of statistics produced by such a crude approach." (page 652). The results obtained here demonstrate some of the problems envisaged by White, in particular in the separation of the Afromontane and Somali-Masai elements of the flora. However, for large, extensive areas of relatively uniform vegetation, like the savannas and the rain forests, this problem does not emerge. I do not think that this is a problem intrinsic to a simultaneous analysis of all the species; it is rather related to the geographical units that are used. White recognized this to some extent when he referred to regional centres of endemism as having no boundaries (1993). In addition, it should be borne in mind that the separation of ecological elements implies some pre-judgment, some deviation from the strict pattern of floristic similarity, which White (1993) recognized to be important.

It is remarkable that such a small data base can produce the pattern intuitively derived for the continent. There is some lack of independence, in that the taxa studied during the period when the boundaries of the phytochoria were finalised are also the taxa analysed in this study; nonetheless, the results suggest that these phytochoria are 'real' as far as these species are concerned.

Centres of local endemism

Despite the small data set, clear-cut centres of narrow endemism were retrieved. The single grid method could lead to some underestimation of centres that are divided between two grids, and so the two-grid sample is preferred. Brenan (1978), in a careful country-by-country compilation, found very similar results to mine, with the high endemism reaching from eastern Nigeria, through Cameroun to Gabon, then becoming less southwards past the estuary of the Congo, and with largely the same areas of endemism scattered across Africa.

The absence of centres of endemism from the savanna and arid regions is striking. White (1965) noted the absence of any areas of endemism in the Sudanian region, with at best a gradual replacement of taxa from the west to the east. However, he recognized three centres in the Zambezian domain, which this analysis has failed to detect. However, this may be consistent with the rather extensive differentiation within the Zambezian zone detected by the Jaccard analysis. It is possible that this could be a sampling artifact, and this needs to be tested by studying a larger data set. Similarly, the southern African region was undersampled, as the emphasis of this study is on the tropical areas. Otherwise it could be expected that the Drakensberg region should emerge as a centre, as suggested by Phillips (1917) and Hilliard & Burtt (1987).

The three lowland rain forest centres suggested by Hamilton (1976) and White (1979), and also documented by Robbrecht (1996) for Rubiaceae, were also retrieved. The richest, Cameroun and surrounding areas, has been well documented and has a richness comparable to other equatorial regions (Gentry, 1988). Recently there have been attempts to define these areas of high richness and endemism in the Cameroun-Congo area more precisely, but with little apparent agreement among the various studies (Rietkerk *et al.*, 1996; Sosef, 1996).

The Western Rift mountains (Kivu centre) have not received much attention in the recent literature. The endemism is not restricted to the Ruwenzori Mountains, as is evidenced by the high richness of the next grid to the south. However, single-grid endemics are largely restricted to the grid with the Ruwenzori, indicating that the southern extension may be due to more widespread species. The endemism does not extend into Uganda, which according to Brenan (1978) almost lacks endemic taxa. Brenan did not locate a centre of endemism in the Western Rift, but instead recorded many species restricted to the central Congo basin. It is not immediately clear why this difference exists; it may have been caused by biases in the data sets analysed.

The region in West Africa from the coast to the Nimba mountains also emerges as an area rich in endemics, and this was also recognized by White (1979). This area has recently attracted more attention, and Van Rompaey (1996) showed that this is a complex centre, with most wet forest species restricted to the coastal plain, while drought tolerant species are found higher up on the Nimba mountains.

The high endemism of the Eastern Arc mountains was discussed in detail by Lovett (1988) and Iversen (1991), and is largely centred in the Usambara and other coastal mountains (Brenan, 1978; Lovett, 1988). The level of endemism to these mountains, for the montane forests, appears to be approximately 30%, out of a flora of 2173 species in 818 genera. Lovett (1988) suggests that the volcanic mountains (Mt. Kilimanjaro, Mt. Kenya) do not contribute much to this richness. There are several suggested explanations for this pattern: they might be too young, the substrates may be too unstable or the soils might be unsuitable. It is intriguing that the endemics are virtually restricted to the coastal grids. These are also well represented in narrow endemics, indicating that the endemism cannot be encapsulated by a single area in the region, but extends all along the coast. Brenan (1978) indicates that this centre of endemism ranges along the coast from southern Kenya to northern Mozambique (see also Clarke, this volume).

The Chimanimani mountains and their endemics attracted the attention of Phipps & Goodier (1962) and Wild (1964), while southern Malawi has attracted less attention. It is possible that the bulk of the Malawian endemism is associated with Mt. Mulanje, a large and botanically spectacular massif. It is striking that the endemism for this area is rather narrowly circumscribed. Brenan (1978) does not pay particular attention to this centre, noting only that the Chimanimani mountains contain perhaps the only concentration of endemics in Zimbabwe. Wild (1964) listed 41 species as endemic to the Chimanimani mountains (to which many more have been added), i.e. 4.6%, and 30 species endemic to Mt. Mulanje.

The watershed between the Congo and Zambesi has been identified as a rich area by Exell & Wild (1961) and White (1965), it has numerous endemic orchids

(Williamson, 1977; Linder, 1981) and Iridaceae (Goldblatt, 1977). Although the Mbala region shows up well, the rest of the watershed towards Mwinilunga is comparatively low in endemics. It is not clear whether this is a sampling artifact, or whether the area is genuinely low in endemism. The same argument applies to the Nyika Plateau, Haut Katanga (Brenan, 1978) and the eastern highlands of Angola. It appears likely that the sample analysed here is too small and preliminary.

North-eastern Africa, especially Somalia, is known to be a species-rich area (Thulin, 1993), and high in endemics. Brenan (1978) notes that 'the flora of the Somali Republic is a remarkable one with very many outstandingly distinct species found nowhere else.' This centre of endemism spills out into the Harar Province of Ethiopia, Northern Kenya and Socotra. Again my analysis failed to find this centre, and again I would suggest that this is due to the poor sampling.

Madagascar as a whole is a centre of endemism, but it is interesting the extent to which narrow endemics occur on the island. This is consistent with ecological information, which indicates that the island contains many diverse vegetation types so that the constituent species virtually have to be local endemics (see Du Puy and Moat, this volume).

Refugia

Hamilton (1976) suggested that these species-rich centres of local endemism may be refugia, where rain forests were able to survive the drier glacial periods. This hypothesis has recently received much more attention (e.g. Maley, 1996; Sosef, 1996), and it has also been extended to the East African coast, particularly as a refugium for montane rain forest taxa (Iversen, 1991). It is possible that these areas are indeed refugia. This might account for the lack of local centres of endemism in the Sudanian region, where the topography is relatively uniform, so that climatic change would result in a simple northwards or southwards displacement of the flora. In areas of more complex topography the possibilities of wetter (or drier) sites forming refugia where species survive are better. Indeed, there is a relationship between mountains and endemism, but it is weak. The areas with the highest levels of endemism are all centred on or at the base of mountains: Mt. Cameroun, Ruwenzori, Usambara, Chimanimani and Mt. Mulanje. But other mountains lack high levels of endemism: Mt. Elgon, Mt. Kenya, Mt. Kilimanjaro (these are all recent volcanoes). There is need for a correlative study between local endemism and the particular conditions pertaining to those sites. Mountains may play several roles, such as increasing the local rainfall and blocking the flow of hot, dry wind from the interior.

Disjunctions

There are numerous disjunctions in the African flora. The 'arid corridor' between Namibia and the Ethiopian region has received some attention (De Winter, 1966; 1971; Verdcourt, 1969; Thulin & Johansson, 1996). De Winter (1971) listed three families, eight genera and 41 species which are disjunct between these areas. However, many of them are also more widespread in other, more northerly, arid areas. In addition, there are nine species-pairs, with one species in Namibia, and a closely related congener in Northeast Africa. De Winter (1971) interprets these patterns as indicating a former arid corridor linking the two arid areas, rather than the process of long-distance dispersal.

African savannas can be divided into northern and southern regions. This disjunction was explored by White (1965), who showed in a detailed analysis that although 51% of the tree genera analysed were common to the Sudanian and Zambezian domains, only 14 Sudanian genera are restricted to the northern region, while 62 genera are endemic to the southern region, reflecting the different patterns of species richness. In addition there is a close link between the Sudanian savannas and the Guineo-Congolian rain forests. These links also largely apply at the specific level.

The Afromontane region was recognized during the 1950s (e.g. Gillett, 1955) following the early work of Weimarck (1933; 1936) and has been described in detail by White (1978). Wild (1964) suggested that the Afromontane and Cape floras are derived from a common original flora, an opinion also supported by Linder (1990). There are numerous taxa that illustrate this Africa-wide set of disjunct distributions, and the relationships among the various 'islands' of this flora have received some, but not enough, attention. Both vicariance explanations (e.g. Wild, 1964) and dispersalist explanations (e.g. Linder, 1994) have been advanced for these patterns. The vicariance explanation suggests that the temperate flora was once widespread throughout Africa, and that the current disjunctions are relictual. The extension of the Afromontane region into West Africa has few endemics, but with remarkable disjunctions, and with many of the species widespread in temperate areas in the rest of Africa (Morton, 1972). Morton carefully documented the patterns of disjunctions in the flora, and argued that these distributions are relics of a more widely distributed temperate flora, and that long-distance dispersal could not have led to this pattern. Exell & Wild (1961) and Wild (1964) documented several endemics in the mountains along the eastern escarpment of south-central Africa, and suggested that the flora of these mountains is related to the Cape flora and the Afromontane flora. Hamilton (1976) suggested that the Afromontane forests radiated from the Usambara region. This interpretation is consistent with that of Meadows & Linder (1993), who suggested that montane forests were much more restricted during the glacials, and that the montane regions were essentially grassland during this period. This accounts for the lack of endemism in the forest flora of these regions, as compared to the high levels of endemism in the grasslands. Despite the volume of work on the Afromontane disjunctions, we still do not have an explanation of the history of the region.

The Congolian-Tanzanian gap is demonstrated by several genera of the Guttiferae, e.g. *Vismia* subg. *Afrovismia, Allanblackia* and *Aoranthe* (Rubiaceae). A more general pattern extends south along the base of mountains, right down to the Natal coast. This pattern is demonstrated by *Oncinotis*. This is a pattern of fragmentation of forests, either rain forests or groundwater forests. This may be a 'tropical' or lowland forest pattern, as compared to the montane forest patterns shown by the Afromontane disjunctions. Hamilton (1976) published an excellent and detailed review of the forest distribution patterns, and indicated the disjunctions between the Liberian, Cameroonian, Eastern Congo, and the East African coastal forests. This pattern has been corroborated by numerous subsequent studies. Lovett (1988; 1993) presented ample evidence for the relationships between the Tanzanian forests and the Guineo-Congolian forests. Of the forest species not endemic to the Eastern Arc mountains, 82% are also found in Congo. In addition, many of the endemic species have their closest relatives in Congo. However, an analysis of 288 rain forest species by White (1979) showed that only seven species are common to Congo and the Indian Ocean forest enclaves, and on this basis White suggested that these forests not be treated as part of the Guineo-Congolian Region. A further larger study revealed that only c. 20% of the tree species from the Indian Ocean coastal forests are common to Congo.

There have been many explanations for these disjunctions. Lovett (1993) suggested that the disjunction is a result of the mid-Tertiary uplift, but Faden (1974) suggested several connections to account for generic and specific level disjunctions. Hamilton (1976) presented a reconstruction indicating that the western forests, from Lake Victoria to Guinea-Bissau, were continuous directly after the last glacial, some 8,000 yr BP, but had been fragmented into small enclaves along the West African coast, from Mt. Cameroon to Gabon, and around Lake Kivu during the glacials, with the modern distribution reflecting this complex pattern of once continuous distribution, and previous more severe fragmentation.

Cladistic analysis of centres of endemism

Cladistic biogeographical analysis is a powerful tool for analysing the relationships among the centres of endemism, and for suggesting the temporal sequence in which the disjunctions within Africa arose. This has been done effectively within the Australian flora (Crisp *et al.*, 1995; Ladiges *et al.*, 1992), and in establishing the continental breakup sequence in the South Pacific (Linder & Crisp, 1995). Africa is rich in discrete centres of endemism, therefore such an analysis should be productive. To date, only the Afromontane patterns, of many disjunctions, have been explored, using spider distributions (Griswold, 1991) and disoid orchids (Linder, 1983; 1994).

Recently, several cladograms of tropical African taxa have been published: *Phaulopsis* (Manktelow, 1996), the Toddalieae (Mziray, 1992) and two sections of *Justicia* (Hedrén, 1989; Ensermu, 1990). Of these studies, that of Mziray only included some 30% of the species in the cladogram, and did not provide detailed distributional information, while Hedrén did not do a critical cladistic analysis. This leaves only two papers on which a cladistic analysis could be based, and the information available is not adequate to resolve many of the areas. There are several cladistic analyses of largely Cape Flora elements, but these generally have at most one or two widespread species to the north.

The results of this analysis group the rain forest areas together, and the savanna and montane areas together. There are a number of pertinent questions which this analysis cannot answer, but which would be interesting to address:

- What is the relationship between the four rain forest centres? The current hypothesis is that the East African centre is most isolated, but the pattern among the three others is not known.
- What is the relationship between the rain forest flora, the montane flora, and the savannas?
- To what extent do the savanna floras share a common origin and to what extent do they derive independently from the rain forests?
- How good is the evidence for a common origin of the two arid floras, or what are they derived from?
- What are the relationships among the Afromontane centres?

Further progress with the understanding of the historical biogeography of tropical African plants will be dependent on more extensive cladistic studies.

Conclusion

From 1950 to 1975 there was a substantial body of descriptive work on African phytogeography. During this period high standards for the mapping of distributions, the delimitation of species, and the synthesis of the phytochoria were set, and the taxonomic and distributional foundations for chorological work laid down.

The vegetation map and the phytochoria, were synthesized out of the combined knowledge and experience of almost all botanists working on the African flora. This work culminated in the *Vegetation of Africa*. In the past decade, further analyses of the East and North-east African forests (Lovett, 1988, 1993; Lovett & Norton, 1989; Friis, 1992; Lovett & Friis, 1996) have added detail to this work. The recent interest in local centres of endemism can also be seen as part of this descriptive exercise, as is the documentation of disjunctions in the flora.

Historical interpretation, by contrast, has lagged behind. Hamilton (1976) interpreted the local centres of endemism as refugia from glacial aridification periods, and this line of argument has recently been developed further (Maley, 1996; Sosef, 1996). Disjunctions in the Afromontane flora (Wild, 1964), the rain forests (Lovett,

1993) and the arid floras (De Winter, 1971; Thulin & Johansson, 1996) have generally been referred to as vicariance events, indicating changes in the distribution patterns of the floras.

However, these interpretations have been largely anecdotal, and attached onto descriptive studies. The descriptive base for excellent historical studies is available, as White was aware. The scene is set to start synthesizing explanations for the diversification of the African flora, taking the excellent descriptive work of White to its historical conclusion.

Acknowledgements

This study was funded by the Foundation for Research Development (Pretoria), and the Vacation Research and Training Programme of the University of Cape Town, which provided funding for several students to collect the data.

References

Beentje, H.J. (1996). Centers of plant diversity in Africa. In: L.J.G. van der Maesen, X.M. van der Burgt and J.M. van Medenbach de Rooy (editors). The biodiversity of African plants. Pp. 101–109. Kluwer Academic Publishers, Dordrecht.

Brenan, J.P.M. (1978). Some aspects of the phytogeography of tropical Africa. *Ann. Missouri Bot. Gard.* **65**: 437–478.

Crisp, M.D., Linder, H.P. and Weston, P.H. (1995). Cladistic biogeography of plants in Australia and New Guinea: congruent pattern reveals two endemic tropical tracks. *Syst. Biol.* **44**: 457–473.

De Winter, B. (1966). Remarks on the distribution of some desert plants in Africa. *Palaeoecol. of Africa* **1**: 188–189.

De Winter, B. (1971). Floristic relationships between the northern and southern arid areas in Africa. *Mitt. Bot. Staatssamml. München* **10**: 424–437.

Ensermu, K. (1990). *Justicia* sect. *Ansellia* (Acanthaceae). *Symb. Bot. Upsal.* **29**: 1–96.

Exell, A.W. and Wild, H. (1961). A statistical analysis of a sample of the Flora Zambesiaca. *Kirkia* **2**: 108–130.

Faden, R.B. (1974). East African coastal- West African rain forest disjunctions. In: E.M. Lind and M.E.S. Morrison (editors). East African Vegetation. Pp. 202–203. Longman, London.

Friis, I. (1992). Forests and forest trees of Northeast Tropical Africa. HMSO, London.

Gentry, A.H. (1988). Changes in plant community diversity and floristic composition on environmental and geographical gradients. *Ann. Missouri Bot. Gard.* **75**: 1–34.

Gillett, J.B. (1955). The relation between the highland floras of Ethiopia and British East Africa. *Webbia* **11**: 459–466.

Goldblatt, P. (1977). Systematics of *Moraea* (Iridaceae) in Tropical Africa. *Ann. Missouri Bot. Gard.* **64**: 243–295.

Goldblatt, P. (1978). An analysis of the flora of Southern Africa: its characteristics, relationships, and origins. *Ann. Missouri Bot. Gard.* **65**: 369–436.

Griswold, C.E. (1991). Cladistic biogeography of Afromontane spiders. *Austral. Syst. Bot.* **4**: 73–89.

Hamilton, A. (1976). The significance of patterns of distribution shown by forest plants and animals in Tropical Africa for the reconstruction of upper Pleistocene palaeoenvironments: a review. *Paleoecol. Africa* **9**: 63–97.

Hedrén, M. (1989). *Justicia* sect. *Harnieria* (Acanthaceae) in tropical Africa. *Symb. Bot. Upsal.* **29 (1)**: 1–141.

Hilliard, O.M. and Burtt, B.L. (1987). The botany of the southern Natal Drakensberg. National Botanic Gardens, Cape Town.

Humphries, C.J. and Parenti, L.E. (1986). Cladistic biogeography. Oxford monographs in biogeography, Oxford.

Iversen, S.T. (1991). The Usambara mountains, NE Tanzania: phytogeography of the vascular plant flora. *Symb. Bot. Upsal.* **24 (3)**: 1–234.

Ladiges, P.Y., Prober, S.M. and Nelson, G. (1992). Cladistic and biogeographic analysis of the 'Blue Ash' Eucalypts. *Cladistics* **8**: 103–124.

Linder, H.P. (1981). Taxonomic studies in the Disinae (Orchidaceae) IV. A revision of *Disa* Berg. sect. *Micranthae* Lindl. *Bull. Jard. Bot. Belg.* **51**: 255–346.

Linder, H.P. (1983). The historical phytogeography of the Disinae (Orchidaceae). *Bothalia* **14**: 565–570.

Linder, H.P. (1990). On the relationship between the vegetation and floras of the Afromontane and the Cape regions of Africa. *Mitt. Inst. Allg. Bot. Hamburg* **23b**: 777–790.

Linder, H.P. (1994). Afrotemperate phytogeography: implications of cladistic biogeographical analysis. In: J.H. Seyani and A.C. Chikuni (editors). Proceedings of the XIIIth Plenary Meeting AETFAT, Malawi. Pp. 913–930. National Herbarium and Botanic Gardens, Zomba, Malawi.

Linder, H.P. and Crisp, M.D. (1995). *Nothofagus* and Pacific biogeography. *Cladistics* **11**: 5–32.

Lovett, J.C. (1988). Endemism and affinities of the Tanzanian montane forest flora. In: P. Goldblatt and P.P. Lowry (editors). Systematic Studies in African Botany. Pp. 591–598. Missouri Botanical Garden, St Louis.

Lovett, J.C. (1993). Eastern Arc moist forest flora. In: J.C. Lovett and S.K. Wasser (editors). Biogeography and ecology of the rain forests of eastern Africa. Pp. 33–55. Cambridge University Press, Cambridge.

Lovett, J.C. and Friis, I. (1996). Patterns of endemism in the woody flora of north–east and east Africa. In: L.J.G. van der Maesen, X.M. van der Burgt and J.M. van Medenbach de Rooy (editors). The biodiversity of African plants. Pp. 582–601. Kluwer Academic Publisher, Dordrecht.

Lovett, J.C. and Norton, G.W. (1989). Afromontane rain forest on Malundwe Hill in Mikumi National Park, Tanzania. *Biol. Conservation* **48**: 13–19.

Maley, J. (1996). Le cadre paléoenvironnemental des refuges forestiers africain: quelques données et hypothèses. In: L.J.G. van der Maesen, X.M. van der Burgt and J.M. van Medenbach de Rooy (editors). The biodiversity of African plants. Pp. 519–535. Kluwer Academic Publisher, Dordrecht.

Manktelow, M. (1996). *Phaulopsis* (Acanthaceae) — a monograph. *Symb. Bot. Upsal.* **31 (2)**: 1–184.

Meadows, M.E. and Linder, H.P. (1993). A palaeoecological perspective on the origin of Afromontane grasslands. *J. Biogeogr.* **20**: 345–355.

Morton, J.K. (1972). Phytogeography of the West African mountains. In: D.H. Valentine (editor). Taxonomy, Phytogeography, and Evolution. Pp. 221–236. Academic Press, London.

Mziray, W. (1992). Taxonomic studies in Toddalieae Hook.f. (Rutaceae) in Africa. *Symb. Bot. Upsal.* **30 (1)**: 1–94.

Page, R.D.M. (1993). COMPONENT, ver. 2.0. The Natural History Museum, London.

Phillips, E.P. (1917). A contribution to the flora of the Leribe plateau and environs. *Ann. S. African Mus.* **16**: 1–379.

Phipps, J.B. and Goodier, R. (1962). A preliminary account of the plant ecology of the Chimanimani Mountains. *J. Ecol.* **50**: 291–391.

Rietkerk, M., Ketner, P. and de Wilde, J.J.F.E. (1996). Caesalpinioideae and the study of forest refuges in Central Africa. In: L.J.G. van der Maesen, X.M. van der Burgt and J.M. van Medenbach de Rooy (editors). The biodiversity of African plants. Pp. 618–623. Kluwer Academic Publishers, Dordrecht.

Robbrecht, E. (1996). Geography of African Rubiaceae with reference to glacial rain forest refugia. In: L.J.G. van der Maesen, X.M. van der Burgt and J.M. van Medenbach de Rooy (editors). The biodiversity of African plants. Pp. 564–581. Kluwer Academic publishers, Dordrecht.

Rohlf, F.J. (1994). NTSYS-pc. Numerical Taxonomy and Multivariate Analysis System for the IBM PC microcomputer (and compatible), ver. 7. New York.

Sosef, M.S.M. (1996). Begonias and African rain forest refuges: general aspects and recent progress. In: L.J.G. van der Maesen, X.M. van der Burgt and J.M. van Medenbach de Rooy (editors). The biodiversity of African plants. Pp. 602–611. Kluwer Academic Publishers, Dordrecht.

Thulin, M. (1993). Flora of Somalia, vol. 1. Royal Botanic Gardens, Kew.

Thulin, M. and Johansson, A.N.B. (1996). Taxonomy and biogeography of the anomalous genus *Wellstedia*. In: L.J.G. van der Maesen, X.M. van der Burgt and J.M. van Medenbach de Rooy (editors). The biodiversity of African plants. Pp. 73–86. Kluwer Academic Publishers, Dordrecht.

Van Rompaey, R.S.A.R. (1996). Rain forest refugia in Liberia. In: L.J.G. van der Maesen, X.M. van der Burgt and J.M. van Medenbach de Rooy (editors). The biodiversity of African plants. Pp. 624–628. Kluwer Academic Publishers, Dordrecht.

Verdcourt, B. (1969). The arid corridor between the northeast and southwest areas of Africa. *Palaeoecol. Africa* **4**: 140–144.

Weimarck, H. (1933). Die Verbreitung einiger Afrikanisch-montanen Pflanzen-gruppen, I–II. *Svensk Bot. Tidskr.* **27**: 400–419.

Weimarck, H. (1936). Die Verbreitung einiger Afrikanisch-montanen Pflanzen-gruppen, III–IV. *Svensk Bot. Tidskr.* **30**: 36–56.

White, F. (1965). The savanna woodlands of the Zambezian and Sudanian domains. *Webbia* **19**: 651–681.

White, F. (1971). The taxonomic and ecological basis of chorology. *Mitt. Bot. Staatssamml. München* **10**: 91–112.

White, F. (1978). The Afromontane Region. In: M.J.A. Werger (editor). Biogeography and Ecology of southern Africa. Pp. 463–513. W. Junk, The Hague.

White, F. (1979). The Guineo-Congolian Region and its relationship to other phytochoria. *Bull. Jard. Bot. Belg.* **49**: 11–55.

White, F. (1983). The Vegetation of Africa: a descriptive memoir to accompany the UNESCO/AETFAT/UNSO vegetation map of Africa. *Natural Resources Research* **20**. UNESCO, Paris.

White, F. (1993). The AETFAT chorological classification of Africa: history, methods and applications. *Bull. Jard. Bot. Belg.* **62**: 225–281.

Wickens, G.E. (1976). The flora of Jebel Marra (Sudan Republic) and its geographical affinities. HMSO, London.

Wild, H. (1964). The endemic species of the Chimanimani mountains and their significance. *Kirkia* **4**: 125–157.

Williamson, G. (1977). The orchids of south central Africa. J.M. Dent & Sons, Ltd., London.

APPENDIX: Genera used for the quantitative analysis, taken from *Distributiones Plantarum Africanorum*, and the number of species in each.

Family	Genus	Number of species	Family	Genus	Number of species
Acanthaceae	*Alstonia*	2	Flacourtiaceae	*Camptostylus*	3
	Carvalhoa	1		*Lindackeria*	6
	Chamaeclitandra	1		*Poggea*	4
	Clitandra	1	Frullaniaceae (Bryo)	*Frullania*	38
	Cyclocotyla	1	Gentianaceae	*Faroa*	17
	Dewevrella	1	Guttiferae	*Allanblackia*	9
	Dictyophleba	5		*Endodesmia*	1
	Farquaharia	1		*Garcinia*	20
	Funtumia	2		*Hypericum*	10
	Holarrhena	3		*Lebrunia*	1
	Isonema	3		*Mammea*	1
	Justicia	9		*Pentadesma*	5
	Malouetia	4		*Psorospermum*	2
	Motandra	3		*Vismia*	7
	Oncinotis	7	Humiriaceae	*Sacoglottis*	1
	Pleioceras	5	Lecythidaceae	*Napoleonaea*	8
	Pycnobotrya	1		*Petersianathus*	1
	Schizozygia	1	Leguminosae	*Cynometra*	1
	Stephanostema	1		*Eriosema*	11
	Strophanthus	32		*Gilbertiodendron*	1
	Voacanga	7		*Parkia*	4
Araliaceae	*Cussonia*	15		*Pterocarpus*	1
	Polyscias	8		*Scorodophloeus*	1
	Schefflera	12	Mayacaceae	*Mayaca*	1
	Seemannaralia	1	Meliaceae	*Entandrophragma*	1
Aristolochiaceae	*Pararistolochia*	10	Najadaceae	*Najas*	17
Asteraceae	*Aedesia*	1	Ochnaceae	*Lophira*	1
	Brachythrix	1	Oleaceae	*Linociera*	2
Bombacaceae	*Bombax*	3	Palmae	*Saba*	3
	Rhodognaphalon	6	Rosaceae	*Parinari*	6
Burmanniaceae	*Burmannia*	1	Rubiaceae	*Alberta*	6
Caryophyllaceae	*Uebelinia*	8		*Anthospermum*	49
Chrysobalanaceae	*Bafodeya*	1		*Aoranthe*	5
	Chrysobalanus	1		*Carpacoce*	10
	Hirtella	6		*Carphalea*	12
	Licania	1		*Crocyllis*	1
	Magnistipula	18		*Galium*	8
	Maranthes	12		*Galopina*	4
	Neocarya	1		*Hymenocoleus*	15
Combretaceae	*Combretum*	2		*Nematostylis*	1
	Terminalia	1		*Nenax*	11
Cornaceae	*Afrocrania*	1		*Petitiocodon*	1
Dichapetalaceae	*Dichapetalum*	62		*Phylohydrax*	2
	Tapura	6		*Psychotria*	88
Ebenaceae	*Diospyros*	33		*Rubia*	2
Euphorbiaceae	*Martretia*	1		*Tricalysia*	61
	Necepsia	9	Thuidiaceae (Bryo)	*Pelekium*	1
	Spondianthus	2		*Rauiella*	1
Flacourtiaceae	*Buchenerodendrum*	2		*Thuidium*	16

Dowsett-Lemaire, F. & Dowsett, R.J. (1998). Parallels between F. White's phytochoria and avian zoochoria in tropical Africa: an analysis of the forest elements. In: C.R. Huxley, J.M. Lock and D.F. Cutler (editors). Chorology, Taxonomy and Ecology of the Floras of Africa and Madagascar. Pp. 87–96. Royal Botanic Gardens, Kew.

6. PARALLELS BETWEEN F. WHITE'S PHYTOCHORIA AND AVIAN ZOOCHORIA IN TROPICAL AFRICA: AN ANALYSIS OF THE FOREST ELEMENTS

F. DOWSETT-LEMAIRE AND R.J. DOWSETT

Rue des Lavandes 12, F-34190 Ganges, France

Abstract

This review relates the Afrotropical phytochoria to the distribution of birds (in this instance, forest species). There is a very close correlation between the patterns of forest bird species distribution and the following chorological categories: Guineo-Congolian, Afromontane, Eastern and linking (Pluriregional). The high rate of endemism found by White among Guineo-Congolian and Afromontane trees is confirmed for birds. A few examples are given of different distribution patterns; a few Afromontane near-endemic birds are still present in small satellite populations on the 'Southern Migratory track' between the mountain ranges of West and East Africa (as are several tree species), as well as along a 'Northern track'. The Eastern region is equally poor in endemic birds and trees. Frank White's phytochoria provide zoologists with an excellent environmental framework, a prerequisite for studies of biodiversity, endemism and species conservation.

Introduction

Ornithologists in Africa tend to describe the distribution of birds only in geographical terms, and almost never in terms of the phytochorology. This is in spite of there being an excellent pioneering study of the zoological divisions on the continent by Chapin (1932, especially pp. 83–264), which took full account of what was known of plant distributions at that time (and included an account of earlier studies). Moreau (1966) continued along the same lines, basing his vegetational background on the work of Keay (1959), itself a forerunner of the seminal work of White (1983a). Unfortunately neither Chapin nor Moreau provided lists of the species they considered to belong to each of their avifaunal divisions, and so it has been impossible to verify, quantify or otherwise develop their reviews. It is undoubtedly a consequence of this lack that the few subsequent reviews of the Afrotropical avifaunas have tended towards mathematical analyses without direct reference to underlying plant distributions. The publication of two important atlases covering the whole of the resident African avifauna (Hall & Moreau, 1970; Snow, 1978) encouraged statistical analysis of distribution data (e.g. Crowe & Crowe, 1982; Diamond & Hamilton, 1980), albeit limited to the maps of museum specimens only, with the several potential biases implied. Above all, we remain without a chorology for the Afrotropical avifauna, as witnessed by the lack of any mention in the ongoing standard work, *The Birds of Africa* (Urban *et al.*, 1997 and earlier volumes).

The idea of re-analysing bird distribution data in the light of Frank White's regional framework had been discussed with him on several occasions, but remains an unfinished project. This paper is a first step in this direction but deals with forest elements only. Our ornithological distribution data are the most up-to-date and complete possible, being based not only on published museum collections, but including many published and unpublished field observations by ourselves and various correspondents. Bird nomenclature follows Dowsett and Forbes-Watson (1993).

Chorological categories

Almost all evergreen forest birds in tropical Africa fall into one of four chorological categories: they are either Guineo-Congolian, Afromontane, Eastern or linking elements (Pluriregional). Whereas the definitions of Guineo-Congolian and Afromontane regions follow those of White (1978, 1979, 1983a) unchanged, a modification has to be brought to the treatment of the lowland forests of eastern Africa, following Dowsett-Lemaire and White in Dowsett-Lemaire *et al.* (in press). These were previously divided into two regional mosaics: Zanzibar-Inhambane and Tongaland-Pondoland (Moll & White, 1978; White, 1983a). Recent analyses of the forest flora of Malawi prompted Dowsett-Lemaire *et al.* (in press) to reconsider the status of the two regional mosaics of the Indian Ocean coastal belt and extend them somewhat inland into an archipelago-like Eastern (forest) regional mosaic. White (1983a: 186) had already classified the Malawi Hills in southern Malawi as an 'exclave' of Zanzibar-Inhambane lowland rain forest in the Zambezian region, but there are many other such areas, in Malawi and elsewhere, although they are usually too small to be mapped. The concept of an Eastern region has already been used by Dowsett-Lemaire in her studies of Malawi forest vegetation (1988a, 1989a & 1990) and avifauna (1989b). Geographically it encompasses, in addition to the two regional mosaics already cited, the lowland evergreen forest patches of the Somalia-Masai region, the fringes of the Lake Victoria basin, the eastern half of the Zambezian region and eastern fringes of the Sudanian and Kalahari-Highveld regions. A few elements of essentially eastern distribution but which extend into western Zambia (approximately west of 25°E) and Angola or into the Cape are referred to as Eastern near-endemics. The main vegetation types of the Eastern region are evergreen rain forest and related types (including riparian and scrub forest) and there is thus virtually no overlap with Zambezian or other drier vegetation — except in transitional stages, particularly transition woodland. The proportion of phanerogamic endemic elements is less than 50% and decreases from east to west. The qualification of 'regional mosaic' is retained here (cf. White, 1979: 19) as these Eastern forests have complex chorological relationships — i.e. non-endemic species may be shared not only with adjacent phytochoria but also more distant ones such as the Guineo-Congolian region.

Birdwise the **Guineo-Congolian forests** are the most diverse of the African tropics, with 251 endemic and near-endemic species (the latter including species extending into forest enclaves of the neighbouring transition zones); these represent c. 80% of all the bird species present in Guineo-Congolian forest, a very high proportion, comparable to the 90% of White's sample of 277 trees (1979: 26). Within the region, two main subcentres of endemism can be recognized: Upper Guinea (29 endemics) and Lower Guinea plus the Congo basin (68 endemics). The number of species confined to either Lower Guinea or Congolia is rather low, contrary to the proportions observed in plants, which show the highest level of local endemism in Lower Guinea (White, 1979): in a sample of 277 tree species belonging to 11 families White found as many as 71 (25.6%) are confined to Lower Guinea. Bird species endemic to Lower Guinea are by contrast very few, with only five strict endemics, all passerines: the swallow *Hirundo fuliginosa*, flycatcher *Batis minima*, rockfowl *Picathartes oreas*, and weavers *Ploceus batesi* and *Malimbus racheliae*, while two more, confined to Lower Guinea in the region, reappear in Angola (the bulbul *Chlorocichla falkensteinii* and the flycatcher *Dyaphorophyia chalybea*).

Examples of Upper Guinea endemics replaced by close relatives in Lower Guinea-Congolia include guineafowls of the genus *Agelastes*, warblers of the genus *Macrosphenus* and weavers of the genus *Malimbus* (cf. distribution maps of *Agelastes meleagrides-niger* superspecies in Snow (1978), and of *Macrosphenus kempi-flavicans* and *Malimbus scutatus-cassini* superspecies in Hall and Moreau (1970)). The geographical limit between Lower and Upper Guinea species is either in the Dahomey Gap (e.g. *Agelastes* spp., *Picathartes* spp.) or near the Cameroon-Nigeria border (e.g. the passerines cited above).

Most Guineo-Congolian species are widely distributed within the region, often extending into the neighbouring transition zones. The forests of Uganda and western Kenya (Kakamega) are part of the L. Victoria regional mosaic as defined by White (1983a) but are dominated by Guineo-Congolian elements. Birdwise these forests do not contain any local endemic, one species is essentially Eastern (the flycatcher *Trochocercus cyanomelas*), while most of the others are Guineo-Congolian near-endemics, also occurring in eastern Zaïre if not further west.

Some of the most abundant bird species in Guineo-Congolian forests are not endemics but Guineo-Congolian linking elements (i.e. Pluriregional), extending widely into forest patches elsewhere on the continent. Two examples are the sunbird *Nectarinia olivacea* and the bulbul *Andropadus virens* which extend to the eastern side of Africa in coastal forest and as far south as southern Malawi/northern Mozambique (*A. virens*) and the eastern Cape (*N. olivacea*).

A special category of Guineo-Congolian linking elements are the Guineo-Congolian/Afromontane linking species: outside the Guineo-Congolian forests these are found essentially in montane forest, mainly on the eastern side of Africa. Examples can be found in trees and birds alike, e.g. *Maesa lanceolata*, *Parinari excelsa* and *Zanthoxylum gilletii*, and the ibis *Bostrychia olivacea*, the francolin *Francolinus squamatus* and the thrush *Alethe poliocephala*. Further details on distribution patterns of Guineo-Congolian forest birds are given elsewhere (Dowsett-Lemaire & Dowsett, in press), with full species lists.

The **Afromontane archipelago** is made up of 'islands' widely scattered on the continent and adjacent SW Arabia. White (1978: Fig. 1) grouped these islands into seven regional systems on the basis of inter-island distances. Carcasson (1964) and Moreau (1966) — the latter followed by Dowsett (1986) — used the distribution of montane butterflies and birds respectively to determine tentative boundaries between mountain systems: they recognize seven groups, broadly similar to those of White. But, unlike White, zoologists recognize the Angolan highlands as a separate, albeit impoverished, entity — White (1983a: 161) considered the inclusion of this area as problematic and the flora certainly needs more study before this point can be settled. The other main difference is that the 'Chimanimani' group of east Zimbabwe is so impoverished faunistically that it is lumped with the most similar system of the Drakensberg into an enlarged South-eastern group (Dowsett, 1986, excluding however the mountains of southern Malawi which clearly belong to the Tanzania-Malawi group: Dowsett-Lemaire, 1989b: 55).

The Afromontane region is characterized by 166 forest bird species that are endemic or near-endemic to it — excluding of course those species confined to montane shrubland and grassland, but including some 16 Sub-Afromontane species occurring at medium altitudes (cf. Dowsett-Lemaire & Dowsett, in press). These represent c. 50 % of the local forest avifauna and the proportion increases with altitude (Dowsett-Lemaire, 1989b: 59–60).

The relative species richness and level of local endemism in the montane birds of different groups have been discussed in several recent papers including Dowsett (1986), Dowsett-Lemaire (1989b, especially the Tanzania-Malawi group), Dowsett-Lemaire & Dowsett (1989, Cameroon group), Dowsett-Lemaire & Dowsett (1990: Albertine Rift or Kivu-Ruwenzori group). A similar analysis is not available for forest

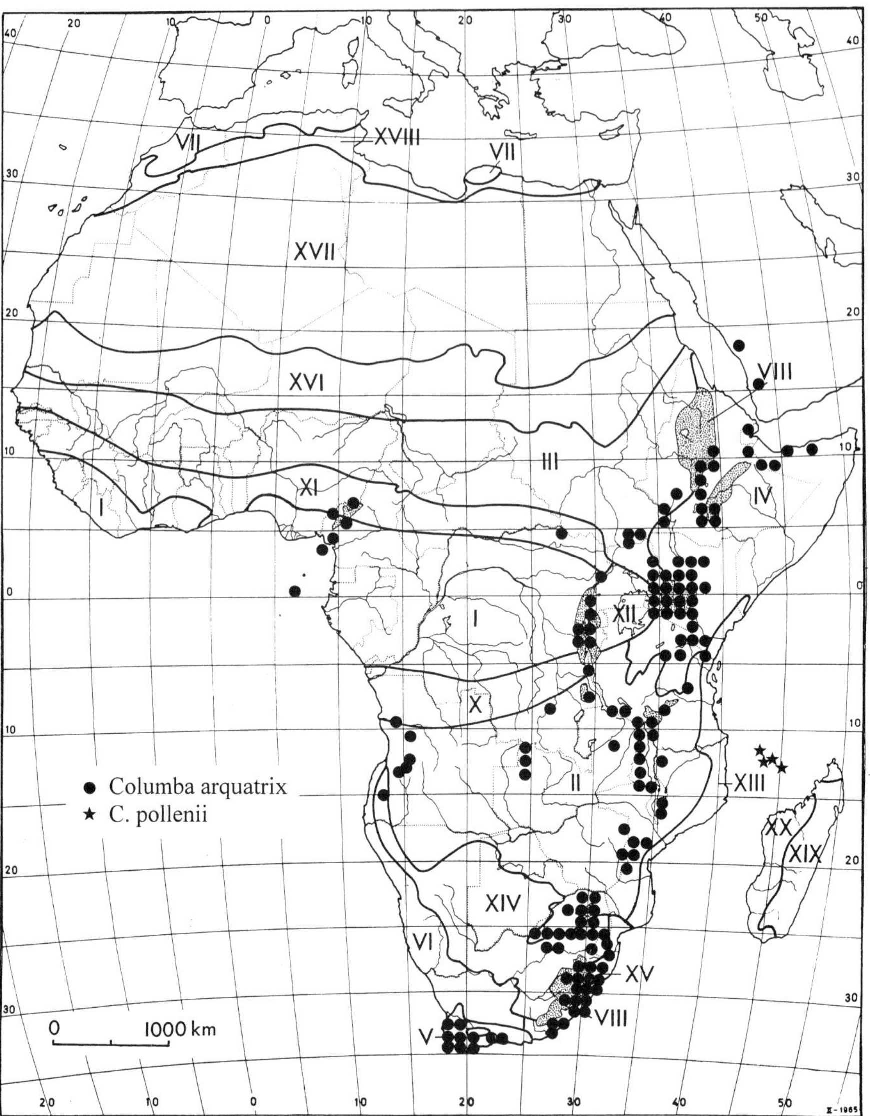

FIG. 1. Distribution map of the Rameron pigeon *Columba arquatrix* and its close relative the Comoros Rameron pigeon *C. pollenii*, based on regional atlases, Snow (1978 etc.).

trees, but some striking differences apparently exist, especially as regards to the Cameroon-Nigeria highlands where the level of local endemism in trees is very low (cf. Dowsett-Lemaire, 1989c: 9). While this is paralleled by low levels of endemism in the butterflies (i.e. only 2–3 of 25 montane wide species recorded in the forests of Obudu and Mambilla are endemic to the group (Dowsett *et al.*, 1989, see also Carcasson 1964)), the birds, however, show nearly 50% local endemism, the highest proportion recorded in any montane forest group (Dowsett-Lemaire & Dowsett, 1989).

Notwithstanding these varying patterns of local endemism, one of the most striking characteristics of Afromontane flora and fauna alike is the wide distribution of many species across the continent. One Afromontane endemic bird, the swift *Schoutedenapus myoptilus*, one of the few migrants in montane forest birds, is present in all seven regional groups. A small sedentary babbler, *Pseudoalcippe abyssinica,* is in six of the seven groups and the trogon *Apaloderma vittatum* in five of them — and both occur in the most isolated highlands of Cameroon-Nigeria. The trogon is also essentially a sedentary species, although some individuals are known to undertake inter-montane movements of up to 100 km (Dowsett-Lemaire, 1989b: 70).

The high proportion of species of plants, butterflies and birds shared between the West and East African mountains, some 2,000 km apart, has led biologists to postulate the existence of connecting migratory routes during colder or wetter climatic phases along the edges of the Zaïre basin. White (1981, 1983b) documented the occurrence of several Afromontane near-endemic trees in small satellite populations along the Zaïre-Zambezi watershed and other mid-altitude forest patches in western Africa and coined the term 'Southern Migratory track'. Much the same route has been used by several Afromontane near-endemic birds, including doves and warblers. An example of particular interest is the Rameron pigeon *Columba arquatrix,* a bird present in all seven regional systems and (with the possibly conspecific Comoros Rameron pigeon *C. pollenii*) some offshore islands whose overall distribution (Fig. 1) is extremely close to that of *Olea capensis,* one of its main fruit plants (Fig. 2). Both tree and bird occur along the Zaïre-Zambezi watershed, the latter turning up on passage for example in north-western Zambia. The Rameron pigeon is one of the most specialized frugivores among montane forest birds in Africa (Dowsett-Lemaire, 1988b) and there is little doubt that the wide distribution of several Afromontane near-endemic trees (including *Myrica humilis, Podocarpus latifolius* and *Polyscias fulva*) is in part the result of close bird-tree interdependence and of the migratory habits of the pigeon. Other Afromontane birds with satellite populations along the Southern Migratory Track include two warblers (*Apalis cinerea, Bradypterus lopezi*) which are apparently sedentary in habit. It is not, however, difficult to imagine that in climatic conditions not very different from those of today, these satellite populations could have been more numerous and closer to each other than they are now. In the forest-savanna mosaic of the Teke Plateau of middle Congo (600–700 m) there is palynological evidence that a montane *Ilex-Olea-Podocarpus-Rapanea* forest existed during the last glaciation (Elenga & Vincens, 1990), and similar *Olea* deposits exist elsewhere in the Guineo-Congolian region.

Although a 'Northern Migratory route' has apparently not been used by montane trees in recent times (White, 1981), some Afromontane or Sub-Afromontane near-endemic birds have left relict populations along the northern edge of the Guineo-Congolian rain forest, notably the two warblers *Apalis jacksoni* and *Phylloscopus budongoensis* recently discovered on some hills and plateaux of northeast Gabon, northern Congo and southern Cameroon (Dowsett-Lemaire & Dowsett, 1996).

The **Eastern forest region** has 40 endemic or near-endemic bird species. As little has been published on this subject we include the list of species in an appendix. The impoverishment of this forest avifauna is no doubt due to the extreme fragmentation of the habitat and its isolation from the Guineo-Congolian forests. Several species are highly localized within the existing archipelago and are very vulnerable to extinction (examples appear in Fig. 3).

The overall percentage of Eastern elements in lowland forest avifauna is well below 50%: in Malawi it varies from 11 to 20% in different areas of lowland and mid-altitude forest. For forest trees and shrubs this proportion is between 20 and 34% in lowland forest below 1,100 m alt. (Dowsett-Lemaire in Dowsett-Lemaire *et al.*, in press), the highest figure being found in the Malawi Hills at 600–940 m . Here the equivalent figure for birds is 20%, based on Table 4 in Dowsett-Lemaire 1989b with three species

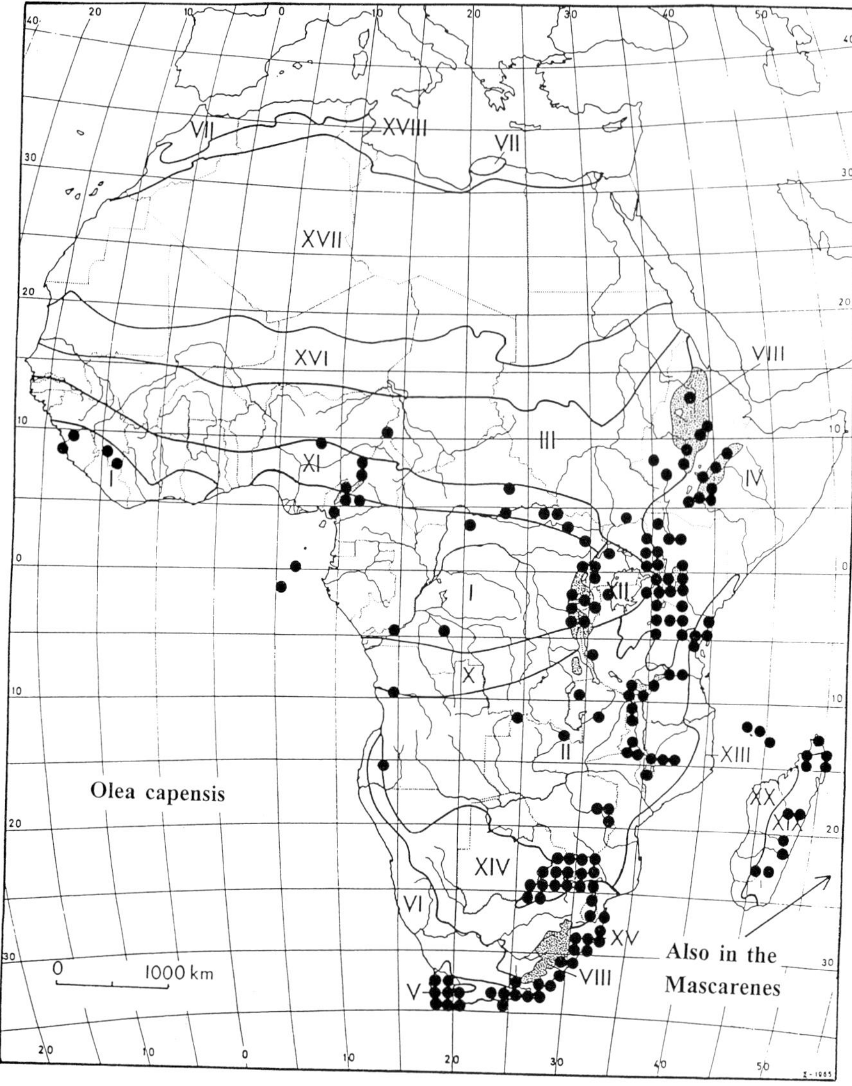

FIG. 2. Distribution map of *Olea capensis*, prepared by F. White and F. Dowsett-Lemaire.

(*Phyllastrephus cerviniventris, Trochocercus cyanomelas* and *Nicator gularis*) now being treated as Eastern rather than Pluriregional. A similar picture is presented by butterflies: in the lake-shore forests some 20% of species in the best-studied families (i.e. excluding Lycaenidae and Hesperiidae) are Eastern elements (Dowsett in MS).

The proportion of Eastern elements is expected to be higher nearer the coast: thus at least 16 Eastern bird species are present in the lowland forest of the Uzungwa Mountains in Tanzania, which is about 25% of the whole avifauna below the altitude of 1,000 m (figures from Stuart *et al.* (1987) excluding some montane species recorded at low altitudes during the winter months and unlikely to breed that low).

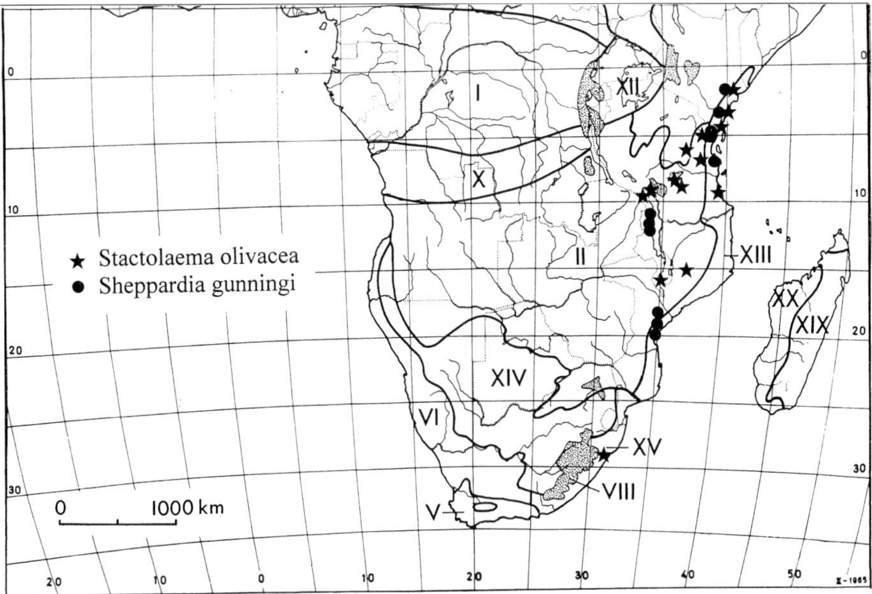

Fig. 3. Distribution map of two birds endemic to the Eastern region.

Eastern endemic birds and trees are by definition characteristic of lowland forest but many ascend the mountains of the Eastern Arc to medium altitudes: e.g. the barbets *Stactolaema leucotis*, *S. olivacea* and *Pogoniulus simplex*, warbler *Apalis chariessa*, robin *Sheppardia gunningi* and flycatcher *Batis mixta* all occur, at least locally, in the forests of the coastal belt but ascend the mountain slopes often to c. 1400–1500 m, in places higher (Stuart *et al.*, 1987; Dowsett-Lemaire, 1989b). With increasing altitude the proportion of Eastern elements decreases while that of Afromontane elements increases. In addition there are two examples of wide-ranging species represented by different populations (the cuckoo *Cercococcyx montanus*) or even races (*Phyllastrephus flavostriatus*) in lowland and montane forests and these may be more appropriately called Afromontane/Eastern linking species.

Many Eastern species have their closest relatives in lowland vegetation elsewhere in Africa, as in the Zambezian and especially Guineo-Congolian regions — the genera *Tauraco*, *Otus*, *Campethera*, *Phyllastrephus*, *Sheppardia*, *Macrosphenus*, *Erythrocercus*, *Trochocercus*, *Batis* and *Oriolus*. Thus the two Eastern *Tauraco* species (*fischeri* and *corythaix*) have their closest relatives in the Guineo-Congolian forests (*T. persa* and *schuetti*) or the dense woodlands of the Zambezian region and neighbouring areas (*T. schalowi*), while *T. livingstonii* is an Eastern/Zambezian/Afromontane linking species — cf. Dowsett-Lemaire and Dowsett, 1988. The closest relative of the rare and local Sokoke Scops Owl *Otus ireneae* seems to be *O. icterorhynchus*, which is widely distributed in the Guineo-Congolian rain forests. The crested flycatcher *Trochocercus cyanomelas* is a member of a superspecies with *T. nitens*, the latter replacing it in the whole of the Guineo-Congolian region. In some cases, the geographical separation between western and eastern forms has not yet led to speciation, as seems to be the case for the forest owl *Bubo poensis* (represented in the Tanzanian forests by the local race *vosseleri*); the ant-thrush *Neocossyphus rufus* and babbler *Illadopsis rufipennis* are also widely distributed in the Guineo-Congolian region, with isolated populations confined to some forests of

93

the coastal belt of East Africa. These three species can be conveniently designated as Guineo-Congolian/Eastern linking species. Finally the finch *Spermophaga ruficapilla*, represented by a small isolated population in the Usambaras is otherwise widespread on the periphery of the Guineo-Congolian region in the transition zones from Angola to the Lake Victoria basin, a distribution that defies a simple term.

The close taxonomic relationships between the birds (and also butterflies: Carcasson, 1964: 136–7) of Eastern and Guineo-Congolian forests are somewhat in contrast with the situation in trees: White (1979: 35) cited six species of Eastern endemic trees whose closest relatives are in the Guineo-Congolian forests, but these are apparently outnumbered by species of less pronounced Guineo-Congolian affinity — though this point was not elaborated any further.

The Afromontane tree and bird communities have complex relationships with other phytochoria: for many species the closest relatives are in the lowlands of Africa, but for some they have to be looked for further afield, in temperate latitudes (e.g. *Juniperus* and *Afrocrania* now considered congeneric with *Cornus*, and among the birds, species of *Buteo* and *Phylloscopus*) or elsewhere in the tropics. Examples of montane forest birds whose closest relatives are to be found in tropical Asia include the owl *Phodilus prigoginei* of the Albertine Rift (related to *P. badius* of south-east Asia), the broadbill *Pseudocalyptomena graueri* of the same subregion (whose nearest relatives are also in south-east Asia) and the recently-discovered francolin of the Uzungwa Mountains of Tanzania, *Xenoperdix udzungwensis*, whose affinities are with the hill-partridges of the Himalayas *Arborophila* spp. (Dinesen *et al.*, 1994). A number of Guineo-Congolian bird species also have their closest relatives in Asia.

Conclusion

This preliminary analysis confirms that White's phytochoria, based on plant distribution data that fit into dominant vegetation types, provide zoologists with an excellent environmental framework. Our own field experience convinces us that the relationship between phytochoria and avian zoogeography is real and important, and that it should no longer be neglected by ornithologists. The forest zoochoria examined here can be equated to forest phytochoria with very few, if any, alterations of geographical limits. The exercise should be continued with non-forest elements too. We plan the present essay as a start towards the project Frank White envisaged, and will present aspects of it in future publications, as our knowledge expands and time allows.

References

Carcasson, R.H. (1964). A preliminary survey of the zoogeography of African butterflies. *E. African Wildlife J.* **2**: 122–157.

Chapin, J.P. (1932). The birds of the Belgian Congo. Part I. *Bull. Amer. Mus. Nat. Hist.* **65**: 1–756.

Crowe, T.M. and Crowe, A.A. (1982). Patterns of distribution, diversity and endemism in Afrotropical birds. *J. Zool., London* **198**: 417–442.

Diamond, A.W. and Hamilton, A.C. (1980). The distribution of forest passerine birds and Quaternary climatic change in tropical Africa. *J. Zool., London* **191**: 379–402.

Dinesen, L., Lehmberg, T., Svendsen, J.O., Hansen, L.A. and Fjeldsa J. (1994). A new genus and species of perdicine bird (Phasianidae, Perdicini) from Tanzania; a relict form with Indo-Malayan affinities. *Ibis* **136**: 3–11.

Dowsett, R.J. (1986). Origins of the high-altitude avifauna of tropical Africa. In: F. Vuilleumier and M. Monasterio (editors). High altitude tropical biogeography. Pp. 557–585. Oxford University Press, New York & Oxford.

Dowsett, R.J. and Forbes-Watson, A. (1993). Checklist of birds of the Afrotropical and Malagasy Regions. Tauraco Press, Liège.

Dowsett, R.J., Heck, J. and Knoop, D.P. (1989). Ecological notes on two collections of butterflies (Lepidoptera) from eastern Nigeria. *Tauraco Res. Rep.* **1**: 31–37.

Dowsett-Lemaire, F. (1988a). The forest vegetation of Mt Mulanje (Malawi): a floristic and chorological study along an altitudinal gradient (650–1950 m). *Bull. Jard. Bot. Belg.* **58**: 77–107.

Dowsett-Lemaire, F. (1988b). Fruit choice and seed dissemination by birds and mammals in the evergreen forests of upland Malawi. *Rev. Écol.* **43**: 251–285.

Dowsett-Lemaire, F. (1989a). The flora and phytogeography of the evergreen forests of Malawi I: Afromontane and mid-altitude forests. *Bull. Jard. Bot. Belg.* **59**: 3–131.

Dowsett-Lemaire, F. (1989b). Ecological and biogeographical aspects of forest bird communities in Malawi. *Scopus* **13**: 1–80.

Dowsett-Lemaire, F. (1989c). Physiography and vegetation of the highland forests of eastern Nigeria. *Tauraco Res. Rep.* **1**: 6–12.

Dowsett-Lemaire, F. (1990). The flora and phytogeography of the evergreen forests of Malawi II: Lowland forests. *Bull. Jard. Bot. Belg.* **60**: 9–71.

Dowsett-Lemaire, F. and Dowsett, R.J. (1988). Vocalisations of the green turacos (*Tauraco* species) and their systematic status. *Tauraco* **1**: 64–71.

Dowsett-Lemaire, F. and Dowsett, R.J. (1989). Zoogeography and taxonomic relationships of the forest birds of the Cameroon Afromontane region. *Tauraco Res. Rep.* **1**: 48–56.

Dowsett-Lemaire, F. and Dowsett, R.J. (1990). Zoogeography and taxonomic relationships of the forest birds of the Albertine Rift Afromontane region. *Tauraco Res. Rep.* **3**: 87–109.

Dowsett-Lemaire, F. and Dowsett, R.J. (1996). Découverte de *Phylloscopus budongoensis* et autres espèces à caractère montagnard dans les forêts d'Odzala (Cuvette congolaise). *Alauda* **64**: 364–367.

Dowsett-Lemaire, F. and Dowsett, R.J. (in press). African forest birds: patterns of endemism and species richness. In: W. Weber, A. Vedder, H. Simmons-Morland and L. White (editors). African rain forest ecology and conservation. Yale University Press.

Dowsett-Lemaire, F., White, F. and Chapman, J.D. (in press). Evergreen Forest Flora of Malawi. Royal Botanic Gardens, Kew.

Elenga, H. and Vincens, A. (1990). Paléoenvironnements quaternaires récents des Plateaux Bateke (Congo): étude palynologique des dépôts de la dépression du bois de Bilanko. In: R. Lanfranchi and D. Schwartz (editors). Paysages quaternaires de l'Afrique centrale atlantique. Pp. 271–282. ORSTOM, Paris.

Hall, P. and Moreau, R.E. (1970). An atlas of speciation in African passerine birds. British Museum (Nat. Hist.), London.

Keay, R.W.J. (1959). Vegetation map of Africa. Oxford University Press, London.

Moll, E.J. and White, F. (1978). The Indian Ocean Coastal Belt. In: M.J.A. Werger (editor). Biogeography and ecology of Southern Africa. Pp. 561–598. W. Junk, The Hague.

Moreau, R.E. (1966). The bird faunas of Africa and its islands. Academic Press, London & New York.

Snow, D.W. (editor) (1978). An atlas of speciation in African non-passerine birds. British Museum (Nat. Hist.), London.

Stuart, S.N., Jensen, F.P. and Brogger-Jensen, S. (1987). Altitudinal zonation of the avifauna in Mwanihana and Magombera Forests, eastern Tanzania. *Gerfaut* **77**: 165–186.

Urban, E.K., Fry, C.H. and Keith, S. (editors) (1997). The birds of Africa. Vol. 5. Academic Press, London.

White, F. (1978). The Afromontane Region. In: M.J.A. Werger (editor). Biogeography and ecology of Southern Africa. Pp. 463–513. W. Junk, The Hague.

White, F. (1979). The Guineo-Congolian region and its relationships to other phytochoria. *Bull. Jard. Bot. Belg.* **49**: 11–55.
White, F. (1981). The history of the Afromontane archipelago and the scientific need for its conservation. *African J. Ecol.* **19**: 33–54.
White, F. (1983a). The Vegetation of Africa: a descriptive memoir to accompany the UNESCO/AETFAT/UNSO vegetation map of Africa. *Natural Resources Research* **20**. UNESCO, Paris.
White, F. (1983b). Long-distance dispersal and the origins of the Afromontane flora. In: K. Kubitzki (editor). Dispersal and Distribution. *Sonderb. Naturwiss. Vereins Hamburg* **7**: 87–116.

APPENDIX: List of species (near-) endemic to the Eastern forest region

Southern banded snake eagle	*Circaetus fasciolatus*
Knysna lourie	*Tauraco corythaix* (near-endemic)
Fischer's turaco	*Tauraco fischeri*
Sokoke scops owl	*Otus ireneae*
Mangrove kingfisher	*Halcyon senegaloides*
Silvery-cheeked hornbill	*Bycanistes brevis*
White-eared barbet	*Stactolaema leucotis*
Green barbet	*Stactolaema olivacea*
Eastern green tinkerbird	*Pogoniulus simplex*
Brown-breasted barbet	*Lybius melanopterus*
Knysna woodpecker	*Campethera notata* (near-endemic)
Mombasa woodpecker	*Campethera mombassica*
Sokoke pipit	*Anthus sokokensis*
Sombre bulbul	*Andropadus importunus* (near-endemic)
Grey-olive gulbul	*Phyllastrephus cerviniventris* (near-endemic)
Fischer's greenbul	*Phyllastrephus fischeri*
Tiny greenbul	*Phyllastrephus debilis*
Gunning's akalat	*Sheppardia gunningi*
Brown scrub robin	*Erythropygia signata*
Eastern bearded scrub robin	*Erythropygia quadrivirgata* (near-endemic)
Kretschmer's longbill	*Macrosphenus kretschmeri*
Rudd's apalis	*Apalis ruddi*
White-winged apalis	*Apalis chariessa*
Black-headed apalis	*Apalis melanocephala*
Forest batis	*Batis mixta*
Woodwards' batis	*Batis fratrum*
Livingstone's flycatcher	*Erythrocercus livingstonei* (near-endemic)
Little yellow flycatcher	*Erythrocercus holochlorus*
Blue-mantled flycatcher	*Trochocercus cyanomelas* (near-endemic)
Plain-backed sunbird	*Anthreptes reichenowi*
Uluguru violet-backed sunbird	*Anthreptes neglectus*
Amani sunbird	*Anthreptes pallidigaster*
Grey sunbird	*Nectarinia veroxii*
Neergaard's sunbird	*Nectarinia neergardi*
Green-headed oriole	*Oriolus chlorocephalus*
White-throated nicator	*Nicator gularis*
Chestnut-fronted helmet shrike	*Prionops scopifrons*
Black-bellied glossy starling	*Lamprotornis corruscus* (near-endemic)
Clarke's weaver	*Ploceus golandi*
Pink-throated twinspot	*Hypargos margaritatus*

Near-endemic = extending west of the Eastern region into western Zambia or the Cape.

Du Puy, D.J. & Moat, J. (1998). Vegetation mapping and classification in Madagascar (using GIS): implications and recommendations for the conservation of biodiversity. In: C.R. Huxley, J.M. Lock and D.F. Cutler (editors). Chorology, Taxonomy and Ecology of the Floras of Africa and Madagascar. Pp. 97–117. Royal Botanic Gardens, Kew.

7. VEGETATION MAPPING AND CLASSIFICATION IN MADAGASCAR (USING GIS): IMPLICATIONS AND RECOMMENDATIONS FOR THE CONSERVATION OF BIODIVERSITY

DAVID J. DU PUY AND JUSTIN MOAT

Royal Botanic Gardens, Kew, Richmond, Surrey, TW9 3AB, UK

Abstract

A map of the 'Remaining Primary Vegetation' in Madagascar has been derived (Map 1) and divided into broad vegetation zones. It is based on the vegetation cover map of Faramalala (1988, 1995), produced from satellite imagery, and the classification of Humbert (1955). A map of the 'Simplified Geology', derived from Besairie (1964), has also been produced, with the geological categories grouped into broad rock types which are thought to have a strong influence on the vegetation they support and its species composition. These two base maps have been superimposed to show the extent and distribution of the remaining primary vegetation in Madagascar, classified firstly into broad vegetation 'zones' and secondly by the underlying geology into vegetation 'types'(Maps 2 to 5; see also Du Puy & Moat, 1996). These maps have then been compared to the map of 'Protected Areas' (COEFOR/CI, 1993), and analysed using Geographical Information Systems (GIS). Histograms produced show the extent of the remaining primary vegetation (in km²) classified according to the underlying geology, and the areas of each which fall within the current system of protected areas (Figs 1 to 5). These maps and data are discussed, and the major omissions in the current system of protected areas are highlighted. Conservation priorities are identified for vegetation types which are not currently protected or have only minimal protection, and the remaining areas of these vegetation types are mapped to show their current extent and distribution (Maps 2 to 5).

Introduction

Madagascar is singled out by the international scientific and conservation community as one of the richest countries in the world in terms of biodiversity, endemism and range of habitats. Its flora is diverse and unique. Of approximately 10,000 native higher plant species, about 8,000 species are thought to be endemic to the island. As a comparison, Madagascar is about 2.5 times as large as Britain, which has about 1,200 species of which only 10 to 20 are endemic. The value of the flora of Madagascar, both to the local peoples and in a global sense, is potentially immense. Despite its importance, this flora is under serious threat. Over 80% of the island has already been stripped of its native vegetation cover (Fig. 1); the majority of this area is now very species-poor secondary grassland which is burnt annually and is subject to intense erosion. The heritage of biological diversity in Madagascar is probably under greater threat than in any other country. This unique diversity, combined with the threats to the remaining native vegetation, puts Madagascar amongst the highest conservation priority areas in the world.

In response to the Convention on Biological Diversity (resulting from the 'Rio Summit'), a conservation strategy is being implemented as part of Madagascar's Environmental Action Plan: part of this Action Plan is to increase the number of protected areas. Data on phytodiversity are rarely used in conservation planning, because of the paucity and incompleteness of information concerning the distribution and rarity of the vast majority of plant species. These data are time-consuming to acquire and, given the rapid rate at which primary vegetation is being destroyed, the only means of ensuring that informed decisions are made concerning the conservation of as many plant species as possible, is through methods of rapid biodiversity assessment. This paper presents one such method, which maps the remaining primary vegetation, and classifies it in a way which mirrors patterns of phytodiversity distribution. The extent of each vegetation type is quantified (in km^2), as is the amount of each type which is included within the current system of protected areas.

Recommendations concerning the conservation of phytodiversity in Madagascar are made on the basis of these analyses. For the first time, such recommendations are supported by statistical data. Priority areas are identified for further reserves which would optimise the range of vegetation types (and by implication the range of plant species) included in the system of protected areas. It should be recognised, however, that although a system of protected areas will offer theoretical protection to a portion of the remaining diversity, many species will inevitably be excluded. Areas with herbaceous or succulent vegetation, which may also be rich in endemic species (such as on the inselbergs and rocky outcrops of central Madagascar), are also missing from these analyses as the vegetation is too sparse to be recognised as primary vegetation in satellite images, and these rock outcrops are generally too small in area to be mapped on a national scale.

Derivation of the base maps 'Remaining Primary Vegetation' and 'Simplified Geology', and the composite map of 'Remaining Primary Vegetation Classified by the Underlying Geology'

Field work has indicated that the species composition of vegetation alters radically with major changes in substrate. Different vegetation types, with distinct species compositions, occur on different rock types. It can be demonstrated that individual species are frequently confined to a particular rock type, or that they avoid one or more rock types (see below). A more informative vegetation map can therefore be produced by subdividing the broad primary vegetation zones into individual vegetation types on the basis of the rock type on which they occur, echoing the patterns demonstrated for individual species distributions.

The map of 'Remaining Primary Vegetation' (Map 1), initially derived from satellite imagery by Faramalala (1988, 1995), broadly maintains Humbert's (1955) main phytogeographic zones. These zones were slightly modified following Faramalala. A further modification is that the 'Sambirano' in the north-west was not recognised as a distinct zone but rather as a continuation of the 'evergreen humid forest: low altitude': it constitutes an area of high local endemism, perhaps due to the change in the underlying geology from the widespread 'basement rocks (metamorphic and igneous)' to the much more restricted 'sandstones' in this area (see Du Puy & Moat, 1996). The geology map of Besairie (1964) was digitised and then simplified into broad rock types which are thought to strongly influence the vegetation they support, resulting in the 'Simplified Geology' map (see Du Puy & Moat, 1996). These two maps were superimposed, resulting in the composite map of the 'Remaining Primary Vegetation, classified by the Underlying Geology'. This map provides new insights into the patterns of variation within the vegetation zones and the distributions of individual plant species, especially in western and southern Madagascar where the geology varies substantially (see Maps 3–5).

These maps have already been published (Du Puy & Moat, 1996*), and the preparation of the 'Simplified Geology' map, the 'Remaining Primary Vegetation' map, the 'Remaining Primary Vegetation, classified by the Underlying Geology' map is documented there. The categories applied in each map are also outlined there, and are listed in Tables 1 and 2 for the two base maps.

TABLE 1. Categories used in the 'Simplified Geology' map.

SIMPLIFIED GEOLOGY TYPES

Sedimentary rocks:
Alluvial & lake deposits
Unconsolidated sands
Sandstones
Tertiary limestones + marls & chalks
Mesozoic limestones (incl. 'Tsingy') + marls

Metamorphic and igneous rocks:
Basement rocks (Metamorphic & Igneous)
Ultrabasics
Quartzites
Marbles (Cipolin)
Lavas (incl. Basalts & Gabbros)

Other categories:
Mangrove Swamps

TABLE 2. Categories used in the 'Remaining Primary Vegetation' map.

REMAINING PRIMARY VEGETATION 'ZONES'

Evergreen formations (east and centre):
Coastal forest (eastern)
Evergreen, humid forest: low altitude (0–800 m)
Evergreen, humid forest: mid altitude (800–1,800 m)
Evergreen, humid forest: lower montane (1,800–2,000 m)
Montane (*Philippia*) scrubland (> 1,800 m)
Evergreen, sclerophyllous (*Uapaca*) woodland (800–1,800m)

Deciduous formations (west and south):
Coastal forest (western)
Deciduous, seasonally dry, western forest (0–800 m)
Deciduous, dry, southern forest and scrubland (0–300 m)

Other categories:
Mangrove
Marshland

* Copies of this paper and the three maps are available from the authors: a small copy of the 'Remaining Primary Vegetation' map is reproduced here (Map 1). These maps are also available on the Internet at the addresses given at the end of this paper.

Extent of remaining primary vegetation, and the degree of protection provided by the current system of protected areas

Histograms showing the remaining area (in km²) of primary vegetation in each vegetation zone (Fig. 2) and vegetation type (Figs 3 to 5) have been produced. Overlaying a map of the Protected Areas (COEFOR/CI, 1993) on the vegetation maps allows the amounts of protection for each vegetation zone and type to be shown on the histograms, and immediately demonstrates which zones and types are poorly represented within the current system of protected areas. The maps can then be re-examined to show where intact areas of primary vegetation suitable for conservation still exist (see Maps 3–5). If reserves were set up on each significant vegetation type, then the system of reserves would include as wide a range of vegetation types as possible, and therefore the greatest possible diversity of species. This may be regarded as a form of rapid phytodiversity assessment, which gives a measure of plant diversity that can be used in the identification of conservation priorities.

A preliminary histogram has already been published (Fig. 2, and Du Puy & Moat, 1996) showing the areas of primary vegetation remaining within the main vegetation zones (derived from the map of 'Remaining Primary Vegetation', Map 1), and also indicating the areas which are included within the current system of protected areas. The 'deciduous, dry, southern forest and scrubland' was highlighted as being inadequately protected. In the present paper we are publishing further analyses of the maps, based on the 'Remaining Primary Vegetation classified by the Underlying Geology' map in particular, and are presenting the resulting implications for conservation planning and biodiversity management. The vegetation types which are shown in the histograms (Figs 2 to 5) to be least well represented within the current system of protected areas are discussed below, are illustrated in Maps 1 to 5, and are listed in the conclusion to the paper.

Species restricted to a specific rock type or group of rock types

It has been assumed in this study that different rock types will support vegetation containing different species, a widespread phenomenon in temperate vegetation. This implies that if a set of protected areas were set up which covered all the main vegetation types (i.e. all vegetation zones on the major rock types on which they occur), then as large a sample as possible of the phytodiversity would be included in the system of reserves. In the tropics the effect of rock type on species distribution may be obscured by deep weathering, such as occurs on the 'basement rocks' which form the backbone of upland Madagascar, where there are often deep beds of laterite, and also where sandstones are covered by eroded sand. We have used a database of the Papilionoid Legumes, containing about 7,000 collection points in c. 350 species, to look for evidence to support this assumption. This research continues, but preliminary results confirm that many species show distributions which coincide with the geological substrate: examples of species confined to one, or to a subset of the rock types are presented (Table 3). The strongest influence of rock type on species distribution appears to concern occurrence on limestone. Some species are confined to limestone, sometimes distributed uniquely on either 'Mesozoic' or 'Tertiary limestones', while widespread species often appear on a range of rock types but not on limestone. The sister species *Dicraeopetalum mahafaliensis* and *D. capuronianum*, for example, have been demonstrated to occur on Tertiary limestones and Unconsolidated sands respectively, both in a restricted area of south-west Madagascar. The 'lavas (including basalts and gabbros)' also often support their own exclusive species, while others occur around but not on them (for instance those species with a broad distribution on the 'basement rocks' of the central plateaux which avoid the basalts of the Ankaratra Massif, and

TABLE 3. Native papilionoid legume species with distribution patterns in Madagascar which coincide with the underlying geology.

NATIVE SPECIES SHOWING A DISTINCT PREFERENCE FOR SEDIMENTARY ROCKS

Unconsolidated Sands	*Cadia commersoniana, Canavalia rosea, Crotalaria edmundi-bakeri, Crotalaria androyensis, Dalbergia lemurica, Dicraeopetalum capuronianum, Galactia tenuiflora, Sakoanala madagascariensis*
Sandstones	*Crotalaria pervillei, Indigofera blaiseae, Kotschya perrieri, Pyranthus tullearensis, Tephrosia parvifolia, Tephrosia phylloxylon, Tephrosia isaloensis, Vaughania cerighellii, Vaughania dionaeifolia*
Sands (Unconsolidated Sands, Sandstones & Alluviums)	*Crotalaria anomala, Derris trifoliata, Erythrina madagascariensis, Indigofera compressa, Mundulea micrantha, Tephrosia pumila*
Tertiary Limestones	*Crotalaria humbertiana, Crotalaria poissonii, Dicraeopetalum mahafaliensis, Ormocarpopsis tulearensis, Pearsonia madagascariensis, Vaughania mahafalensis, Vaughania humbertiana*
Mesozoic Limestones (incl. 'Tsingy')	*Crotalaria capuronii, Dalbergia glaberrima, Dalbergia humbertii, Dalbergia neoperrieri, Indigofera bemarahaensis, Mucuna gigantea, Neoharmsia madagascariensis, Pongamiopsis viguieri, Rhynchosia viscosa, Stylosanthes fruticosa, Tephrosia bibracteolata*
Limestones (both Mesozoic and Tertiary)	*Tephrosia perrieri*

NATIVE SPECIES SHOWING A DISTINCT PREFERENCE FOR METAMORPHIC & IGNEOUS ROCK TYPES

Basement Rocks (Metamorphic & Igneous Rocks, including Granites, Migmatites, Schists and Gneiss)	*Aeschynomene heurckeana, Argyrolobium pedunculare, Cadia ellisiana, Cadia pubescens, Cordyla haraka, Crotalaria diosmifolia, Crotalaria incana, Crotalaria uncinella, Crotalaria tanety, Crotalaria craspedocarpa, Dalbergia monticola, Dalbergia orientalis, Decorsea meridionalis, Eriosema procumbens, Eriosema parviflorum, Indigofera mangokyensis, Indigofera nummulariifolia, Indigofera arrecta, Kotschya strigosa, Leptodesmia congesta, Mundulea barclayi, Ormocarpopsis mandrarensis, Phylloxylon xylophylloides, Pyranthus pauciflora, Rhynchosia versicolor, Strongylodon madagascariensis, Vigna parkeri, Vigna angivensis, Zornia puberula*
Quartzites often associated with Marbles (Cipolin)	*Argyrolobium itremoensis, Crotalaria ibityensis, Indigofera lyallii, Mundulea anceps, Tephrosia betsileensis, Indigofera itremoensis, Pyranthus ambatoana*
Lavas (incl. Basalts & Gabbros)	*Crotalaria ankaratrana, Dalbergia pseudobaroni, Indigofera thymoides, Indigofera pinifolia, Leptodesmia bojeriana, Leptodesmia perrieri, Pyranthus monantha, Tephrosia retamoides, Trifolium ankaratrense*

TABLE 3 continued.

NATIVE SPECIES SHOWING NO ROCK TYPE PREFERENCE BUT A PRONOUNCED DISLIKE FOR A PARTICULAR GROUP OF ROCK TYPES

Avoiding Basement Rocks	*Dalbergia peltieri, Dalbergia xerophila, Dalbergia trichocarpa, Dalbergia abrahamii, Erythrina variegata, Indigofera longeracemosa, Millettia richardiana, Millettia aurea, Ormocarpum drakei, Ormocarpum bernierianum, Rhynchosia baukea, Sakoanala villosa, Stylosanthes erecta*
Avoiding Lavas and Limestones	*Alistilus jumellei, Crotalaria mandrarensis, Crotalaria grevei, Ophrestia lyallii*
Avoiding Limestones	*Crotalaria fiherenensis, Crotalaria cornu-ammonis, Desmodium repandum, Dumasia villosa, Indigofera bojeri, Indigofera imerinensis, Mundulea laxiflora*

similarly for the areas of ancient lavas in the Mandrare River basin in southern Madagascar). The sandstone and unconsolidated sand categories often have species which are distributed on both categories, although certain species are confined to one or the other. Evidence is accumulating that other species occur exclusively on particular rock types, to varying degrees.

These maps have already helped to explain species distributions and centres of diversity and microendemism in the Papilionoid Legumes, and also in other groups, including both flora and fauna: they may well mirror the distribution patterns of biodiversity as a whole.

Remaining areas of primary vegetation within the major vegetation zones, and their current protection

A histogram showing the areas (in km²) of remaining primary vegetation within each vegetation zone (Fig. 2, and Du Puy & Moat, 1996) was produced by superimposing the map of 'Protected Areas' (COEFOR/CI, 1993) on the map of 'Remaining Primary Vegetation' (Map 1, and Du Puy & Moat, 1996). The categories of protection were divided into two levels, a higher 'well protected' category (including Réserves Naturelles Intégrales (RNI), Réserves Spéciales (RS) and Parcs Nationaux (PN), all now under the auspices of ANGAP), and a lower 'poorly protected' category (including Réserves Forestières (RF) and Forêts Classées (FC), under the auspices of DEF) which are, in general, forestry and forest exploitation areas, and have 'suffered from enormous human pressures with the practice of 'tavy' (shifting agriculture) and an often illegal and abusive forest exploitation' (translated from COEFOR/CI, 1993). The 'protected areas' referred to in this paper therefore only include the top three categories (RNI, RS, PN) which offer the best available protection to the vegetation within their boundaries, although lack of adequate funding, staffing and management means that the protection offered is often nominal and still very incomplete.

The map of 'Remaining Primary Vegetation' (Map 1) shows that only about 18% of the surface area of Madagascar is covered by primary vegetation (see Fig. 1). It should be noted, however, that the satellite images used to produce this map were taken during the 1970s (Faramalala, 1988), and that the present native vegetation cover is probably substantially reduced from this already low percentage. Further examination

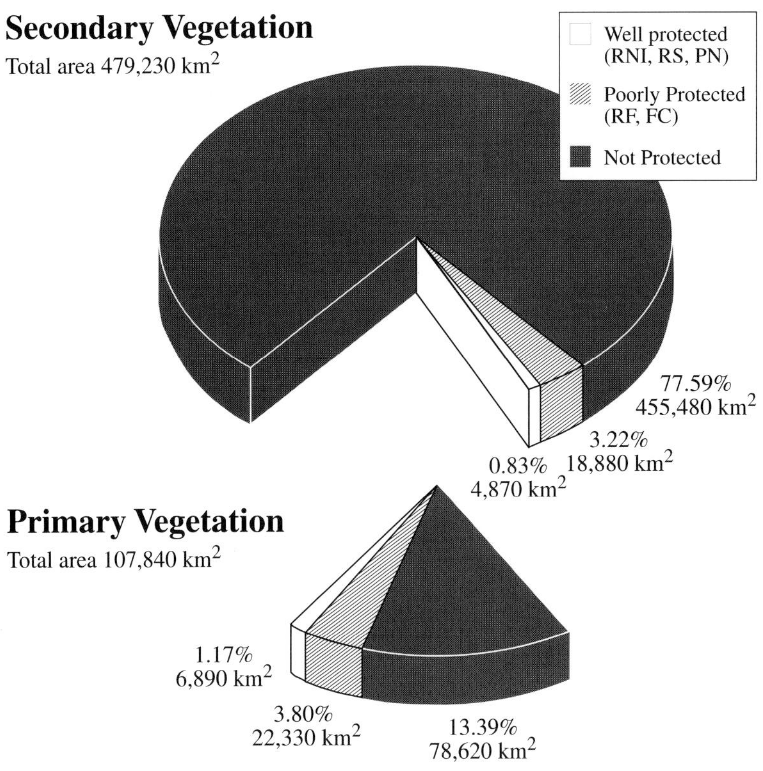

FIG. 1. Proportion of remaining primary vegetation to secondary vegetation in Madagascar and the percentages of these within protected areas.

shows that only 6% of this remaining primary vegetation falls within the current system of protected areas (about 1.17% of the total surface area of Madagascar). Moreover, at least 40% of the vegetation within the current system of protected areas is indicated as secondary vegetation.

Extensive vegetation zones

The histogram showing the remaining areas of primary vegetation within the major vegetation zones (Fig. 2) illustrates certain imbalances in the protection offered by the current system of protected areas. There are four vegetation zones which still have substantial areas of primary vegetation cover. Of these, the 'deciduous, dry, southern forest and scrubland' is the most outstanding example of inadequate protection; it has a far smaller proportion of primary vegetation within protected areas than the other major vegetation zones. There is clearly a strong case to be made for further protected areas to be placed within this southern zone, particularly since it contains a varied geology (see the discussion below) and a large number of endemic species and genera.

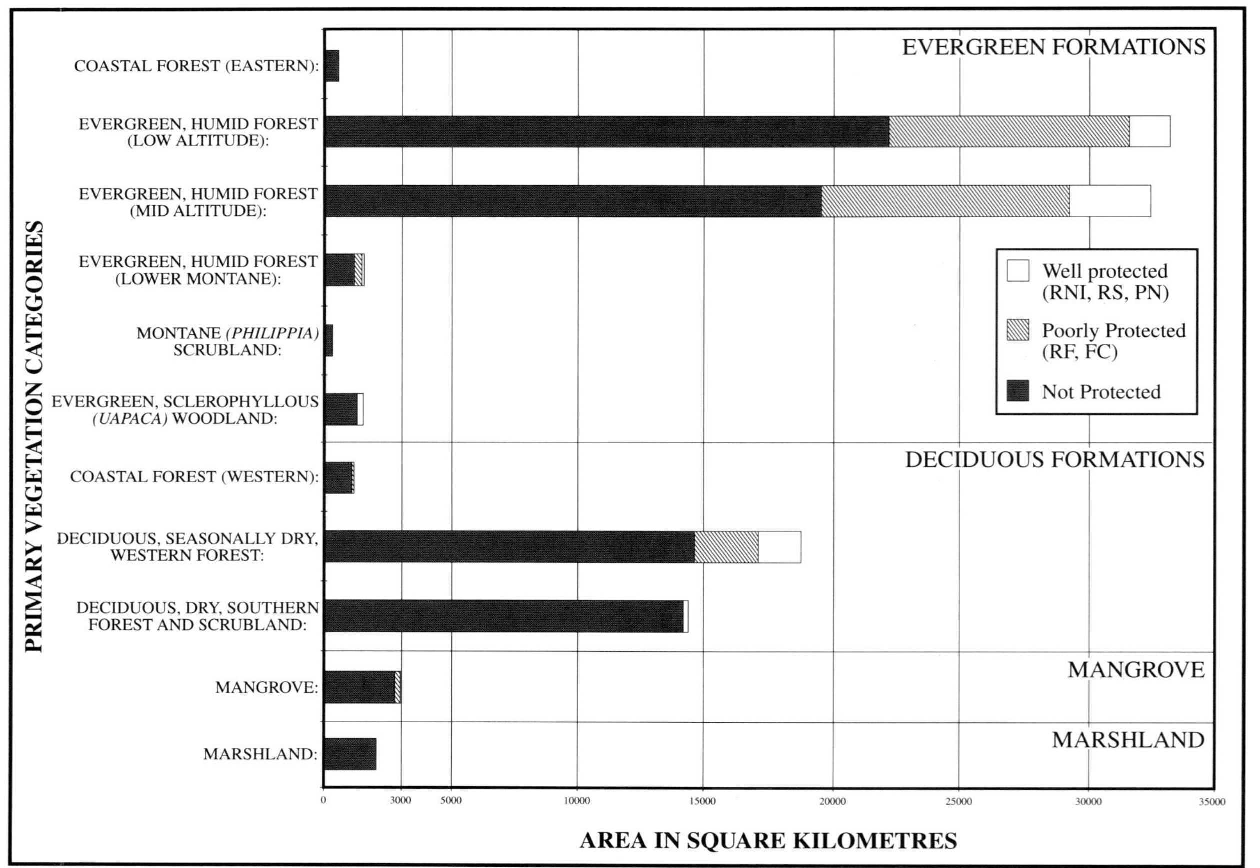

FIG. 2. Areas of remaining primary vegetation in Madagascar: showing their degrees of protection.

The other three major zones, 'deciduous, seasonally dry, western forest', 'evergreen, humid forest: low altitude' and 'evergreen, humid forest: mid altitude', all appear in this histogram (Fig. 2) to have a larger and more adequate proportion of their current area within protected areas. However, only the 'evergreen, humid forest: mid altitude' has as much as 10% of its remaining primary vegetation within protected areas.

The amount of protected vegetation within the 'evergreen, humid forest: low altitude' will increase when the new reserve on the Masoala Peninsula** is taken into account. Nevertheless, it is important to note that the area of 'evergreen, humid forest: low altitude' within protected areas is least at lower altitudes, and that from near sea level up to about 400 metres altitude the forest is both the most degraded and the least well protected. Forest destruction for charcoal and cultivation is proceeding extremely rapidly in these more accessible, and often more densely populated, lowland areas. In fact, very little forest remains at the lowest altitudes, particularly in the southern half of Madagascar. A strong case can certainly be made for ensuring further conservation of 'evergreen, humid forest' at low altitudes, and particularly ensuring that forests on basalts and sandstones are adequately included (Fig. 3). The lowland forests from Vohemar to the Masoala Peninsula, around the Baie d'Antongil and near Mananara, and between Tôlañaro (Fort Dauphin) and Vangaindrano in the south, contain important stands of this rapidly diminishing forest type. The forest of Manombo (to the south of Farafangana) has been singled out in the IUCN Palm Conservation Action Plan (Beentje & Dransfield, 1996) as an area of high value, and it is certainly an important and rare example of lowland 'evergreen, humid forest' in the southern half of the island (see also the discussion of 'coastal forest' below).

Similarly, although the 'deciduous, seasonally dry, western forest' appears to have a relatively large degree of protection, the majority of it is focused on the vegetation on 'Mesozoic limestone'. These limestone areas are of outstanding beauty containing the deeply eroded limestone karst, known locally as 'tsingy' (see discussion below).

Restricted vegetation zones

The remaining vegetation zones (Fig. 2), which are now restricted to small areas of primary vegetation (all less than 3,000 km^2), are of concern and should be considered as priority areas. Amongst these, the 'coastal forest (eastern)' and the 'evergreen, sclerophyllous woodland' dominated by *Uapaca bojeri* (Euphorbiaceae) merit particular mention for the need for conservation (see also Fig. 4).

The 'coastal forest (eastern)' zone has historically been under intense pressure as it is confined to the coastal plains of the east coast, where there are numerous towns and settlements which have used this forest to provide construction wood and charcoal, and it has been easily accessible by sea and along the canal which extends along the coastal plains (the 'Pangalan'). There are very few remaining patches of this forest (Map 1), mainly in scattered stands along the northern stretch of coast from Vohemar to the Bay of Antongil, and at other scattered localities further south, including Fenoarivo Atn. (Tampolo), Isle Sainte Marie and Pointe à Larée, Ambila-Lemaitsu, Manakara and around Tôlañaro (Fort Dauphin) at Cap Sainte Luce and Ampetrika (Petriky). The areas around Tôlañaro are soon to be largely destroyed through mining: the coincidence of sands rich in titanium and the remaining forest remnants is remarkably

---

** Since the submission of this article the creation of the new Masaola reserve has been formalised. This extremely large area (the largest protected area in Madagascar) provides a very significant contribution to the protection of the 'evergreen, humid forest: low altitude'.

This new reserve covers over 2,000 km^2 of primary forest, which is predominantly (80%) 'evergreen humid forest: low altitude': nearly all of it occurs on basement rocks, with very small areas on quartzites and lavas. The rest (20%) is of 'evergreen humid forest: mid altitude' on basement rocks.

high. The few remaining stands of forest are rich in locally endemic species, and have a very high diversity. Orchids, for example, are plentiful near sea level, and do not again form a large element of the flora until around the 600 m contour in the 'evergreen, humid forest' (see Bosser *et al.*, 1996). The threat of mining will remove a substantial proportion of the remaining primary forest in this zone and every effort is required to protect as much as possible. These forests in the south-east are of particular interest as they span the rapid transition zone from humid evergreen to dry deciduous vegetation, and there are many highly localised species in the individual forest remnants.

The 'evergreen, sclerophyllous (*Uapaca*) woodland' (Map 1) is a distinctive vegetation which is almost the only remnant of the forest which is previously thought to have covered large areas of the southern half of the central plateaux. It has been modified by the annual grassland fires, and only persists due to the resistance of the *Uapaca* trees to these fires. It is generally species-poor, and is maintained for various ethnic reasons including native silk production (see Gade, 1985). Figure 4 shows that the majority occurs on sandstones, and that a substantial area is protected within the Isalo National Park (PN). However, the small cluster of remnants which occur on 'quartzites' in the Itremo Massif (Map 2) have a substantially different character, and form mixed stands with *Sarcolaena oblongifolia*. They are associated with a flora on the surrounding exposed rocks containing many endangered orchids (see Bosser *et al.*, 1996) and succulents (such as miniature *Aloe* and *Pachypodium* species and the last major stands known of the palm *Dypsis (Chryslidocarpus) decipiens*, all listed on Appendix I of CITES). Similarly, the 'marbles (Cipolins)' on the eastern flanks of the Massif contain their own unique flora which is highly threatened due to its very restricted nature and the concessions of marble mining rights in the region. The whole of this area (Map 2) should be a priority for inventory and accurate mapping. The vegetation on the 'quartzites' and 'marbles (Cipolin)' in the Itremo Massif is of extremely high value and importance: this region, which is also of great beauty, certainly merits adequate protection.

Remaining areas of primary vegetation in southern and western Madagascar, subdivided according to the rock types on which they occur: implications for the protection of diversity

Deciduous, dry, southern forest and scrubland

According to the histogram showing the remaining areas of primary vegetation within the major vegetation zones (Fig. 2, discussed above), of the four vegetation zones which still contain substantial areas of primary vegetation, the 'deciduous, dry, southern forest and scrubland' (Map 1) is of greatest concern, as there is only an extremely small proportion within current protected areas. Furthermore, if this zone is subdivided into different vegetation types according to the underlying geology (Fig. 5), it becomes apparent that three major rock types still support large areas of primary vegetation: 'unconsolidated sands', 'Tertiary limestones' and 'basement rocks'. The distributions of these vegetation types are shown in Map 3, and are obviously major candidates for protection within reserves. The most appropriate localities for conservation can be chosen from amongst the areas indicated for each vegetation type.

The most poorly protected, and probably the most threatened vegetation type in this zone occurs on the unconsolidated sands (Map 3). The area accessible by the road to the north of Toliara (Tulear) has already been felled for charcoal, leaving only the Mikea Forest, between Toliara (Tulear) and Morombe as the only remaining extensive area. The Mikea Forest is of great importance if the diversity of the southern zone is to be preserved. Another very localised area on sand, of high interest although restricted in size, occurs around Itampolo: the proximity of the Mahafaly Plateau and its escarpment in this region provides the potential for a reserve of great interest and diversity enclosing vegetation on two of the main rock types.

The few remaining areas of this vegetation zone on sandstone are represented in the Beza-Mahafaly reserve (Fig. 5, Map 3). This area is, nevertheless, extremely small and should be substantially extended.

The 'Tertiary limestones' are represented in the 'deciduous, dry, southern forest and scrubland' zone, by the vegetation on the Mahafaly Plateau (Map 3): the Tsimanampetsotsa Reserve (RNI) provides an absolute minimum of protection. It is strongly recommended that other areas of the Mahafaly Plateau should be protected, and the area behind Tulear, including the escarpment edge towards Saint Augustin, offers an outstanding example which could be linked to a profitable tourist economy.

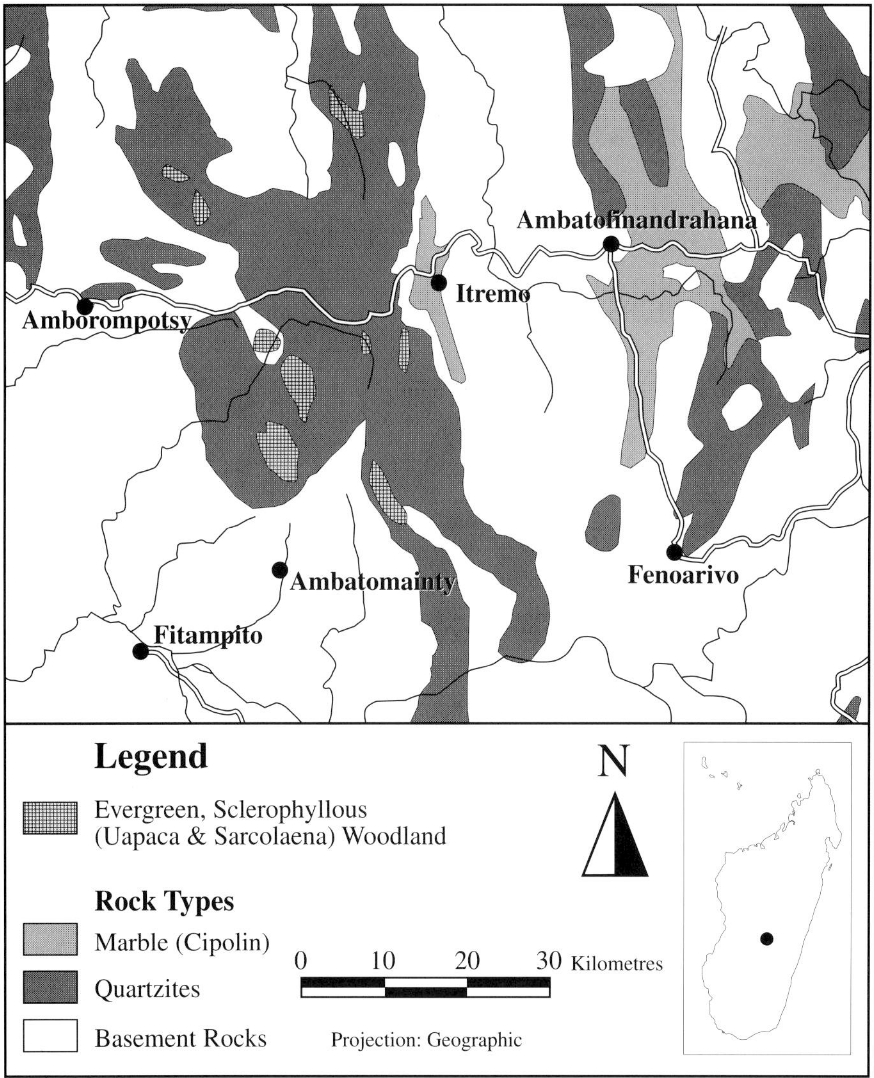

MAP 2. Itremo Massif area, showing the remaining areas of 'Evergreen, Sclerophyllous Woodland', and the occurrence of quartzite and marble ('Cipolin') rock types.

107

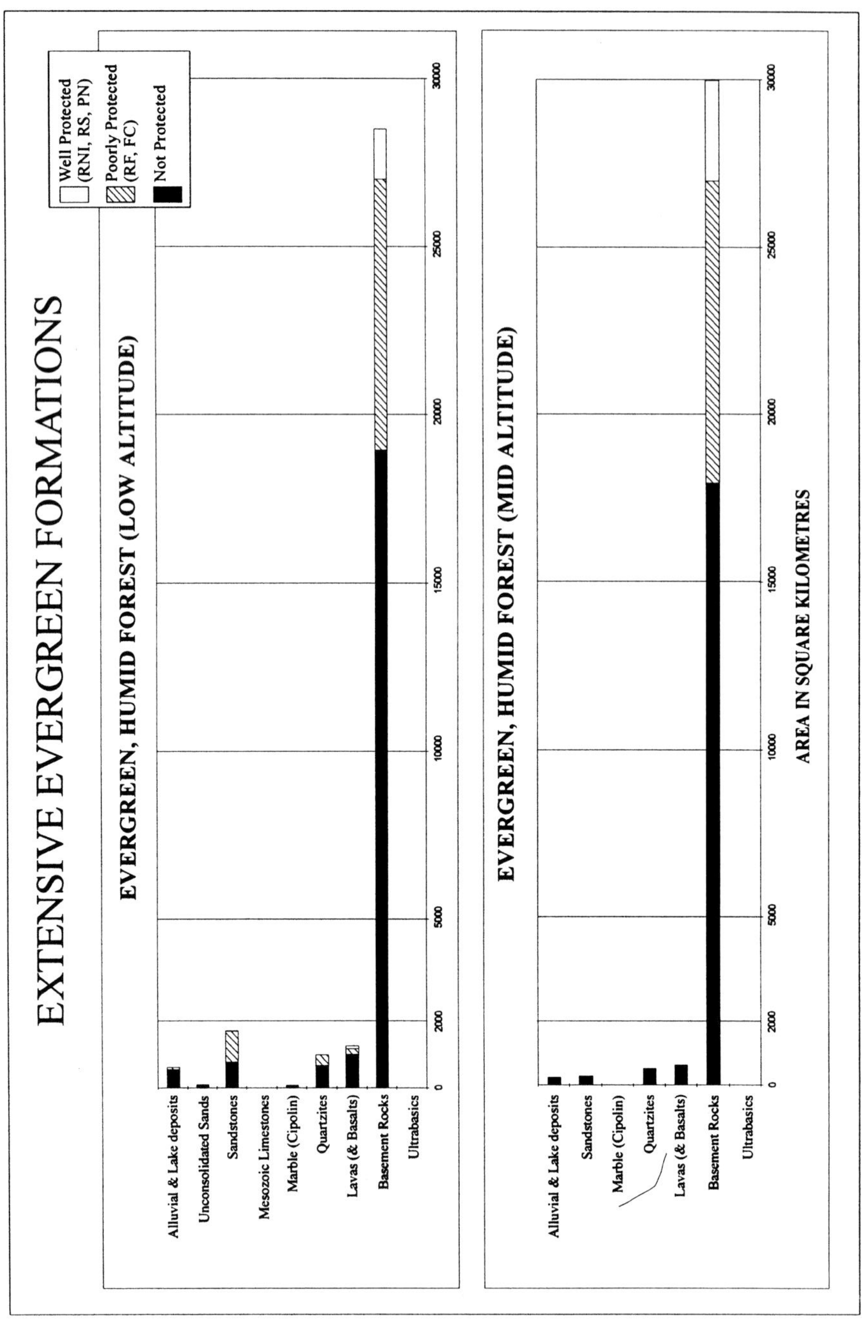

Fig. 3. Areas of extensive primary evergreen formations in Madagascar: classified by the underlying geology and showing their degrees of protection.

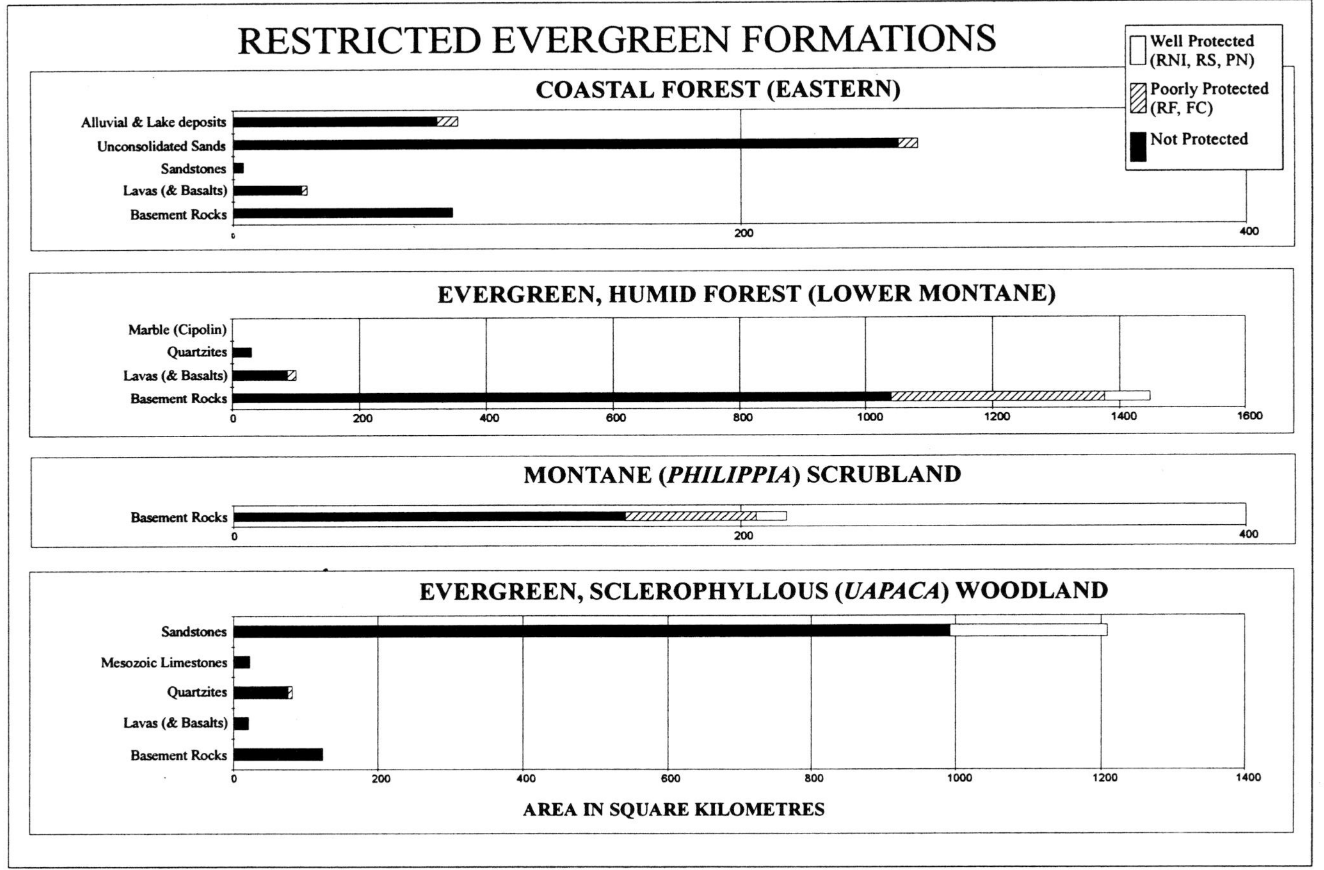

FIG. 4. Areas of restricted primary evergreen formations in Madagascar: classified by the underlying geology and showing their degrees of protection.

109

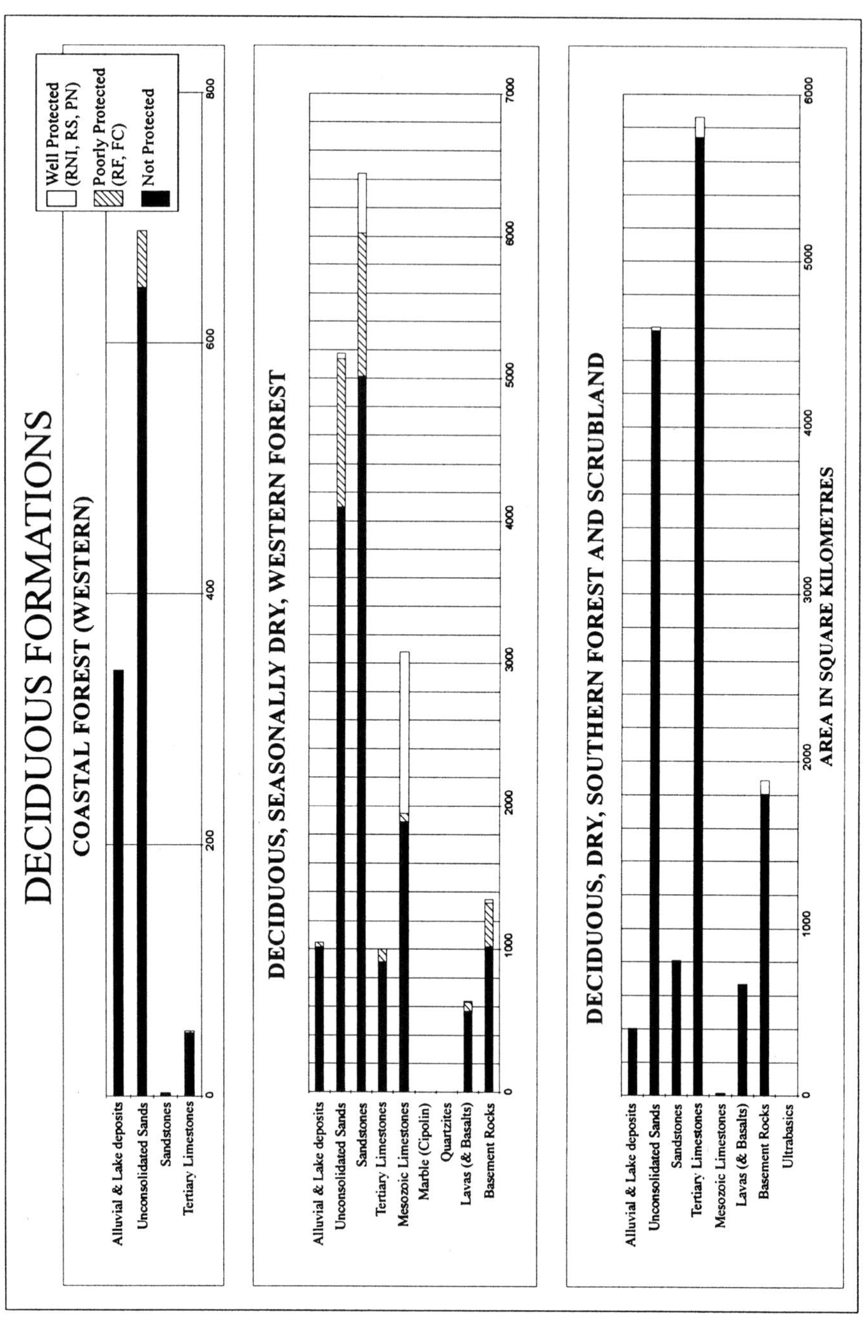

Fig. 5. Areas of primary deciduous formations in Madagascar: classified by the underlying geology and showing their degrees of protection.

Similarly, protection of the vegetation on the 'basement rocks' and the ancient 'lavas (basalts & gabbros)' in the large crater-like basin of the Mandrare River should be extended: the Andohahela Reserve (RNI, parcelles 2 & 3) is in this area and provides protection for only a very small proportion of the remaining vegetation. Furthermore, this reserve does not include any vegetation on the ancient lavas and basalts in this area.

Deciduous, seasonally dry, western forest

Although the histogram in Figure 2 shows a relatively large amount of protection within the 'deciduous, seasonally dry, western forest' (Map 1), it is evident, when this zone is subdivided by rock type (Maps 4, 5 and Fig. 5), that the large majority of this protection is of vegetation on the 'Mesozoic limestones'. These are ancient Jurassic limestones, which have often become deeply eroded into spectacular 'tsingy' or limestone pinnacles traversed by canyons formed through the collapse of underground river systems. In fact, the vegetation in these areas is one of the least in need of active protection as the nature of the habitat renders it of little use for grazing, in little danger from grassland fires, and inaccessible to human exploitation due to the extreme difficulty of access, particularly where bare feet cannot tolerate the sharp rocks. However, within this western zone there still exists a large area of 'deciduous, seasonally dry, western forest' on 'unconsolidated sands' which has very little protection at present (Fig. 5), and which must be considered as a major candidate for the establishment of a protected area. There are substantial areas remaining of this vegetation type along the western coastal plains between Maintirano and Ambohibe (north of Morombe) (see Map 4), including the areas to the west of the Tsingy de Bemaraha, and the forests to the north of the Mangoky River. These forests are inaccessible and little known: they are key areas for which inventory work and botanical exploration are particularly required. The upgrading of the Forêt Classée of Marofihitra (north of Ambohibe), the extension of the Réserve Spéciale d'Andranomena (north east of Morondava), and the protection of the forest to the west of the Bemaraha Massif could provide ideal opportunities for increased conservation of this habitat.

A secondary category in this region, which is little known but also a strong candidate for protection is the forest on 'basement rocks (igneous and metamorphic)' which occurs to the south-west of the Tsaratanana Massif in NW Madagascar (Bora, Map 5).

The remnant forest of Zombitsy (NW of Sakaraha, currently also a Forêt Classée), on sand over sandstone is also a unique area which, due to its proximity to the main road, is undergoing perhaps unparalleled rates of clearance, and is of immediate concern.

Conclusions

The maps presented here, and the histograms derived from them, can be interpreted as reflecting patterns of biodiversity distribution, and in particular plant diversity. If reserves were established in as many different vegetation types as possible (the vegetation types indicated by the vegetation zones subdivided by geological substrate), then the greatest possible diversity of species would be included in the protected areas. Species distributions have already been shown to coincide with the geological categories used in this study, and it is probable that many more follow the same patterns.

The highlighted examples of the areas of importance for conservation in Madagascar are in no way exhaustive, but for the first time recommendations concerning conservation of plant diversity have been supported by statistical evidence. They have also been mapped to assist the selection of appropriate areas for conservation.The main vegetation types identified within this study which are largely excluded from or are inadequately protected by the current system of protected areas are summarised below (not in order of importance), and some potential sites for conservation measures are indicated:

- Deciduous, dry, southern vegetation (Map 3) on:
    1. Unconsolidated sands (Mikea Forest, Itampolo)
    2. Mesozoic limestones (Mahafaly Plateau)
    3. Sandstones (extension of the Beza-Mahafaly reserve)
    4. Basement rocks and ancient Lavas and basalts (Mandrare River basin, north and west of Ifotaka)
- Deciduous, seasonally dry, western forest (Maps 4 & 5) on:
    5. Unconsolidated sand (coastal plains between Ambohibe and Maintirano, Map 4)
    6. Basement Rocks (SW of the Tsaratanana Massif, Bora, Map 5)
    7. Sandstones (Forest of Zombitsy)
- Evergreen, sclerophyllous (*Uapaca*) woodland on:
    8. Quartzites and marbles ('Cipolins'), with associated succulent flora on exposed, non-forested areas (Itremo Massif, Map 2)
- Evergreen, humid forest at low altitude (Map 1):
    9. Particularly at the lowest altitudes (Vohemar to the Masoala Peninsula and around the Bay of Antongil, Manombo and other areas described in the text) .
- Eastern coastal forest (Map 1):
    10. All remaining remnants (described in the text, especially ensuring some protection for remnants in the south-east near Tôlañaro (Fort Dauphin)).

Traditional reserve boundaries, although often transgressed, are generally recognised locally. It is therefore recommended that, wherever possible, existing protected areas with a lower degree of protection should be upgraded and given more adequate protection rather than creating entirely new protected areas, for which it may be more difficult to gain local acceptance. Furthermore, protected areas can only offer true protection if adequate funds and personnel are made available and are assured for the future, and that the will to ensure continued protection is cultivated at a local and a national level.

We hope that this work will contribute to the planning of priorities for the conservation of biodiversity in Madagascar, particularly in the selection of areas suitable for the establishment of new reserves. The inclusion in reserves of habitats not currently covered by the existing series of protected areas will ensure the conservation of as much of the island's phytodiversity as possible. We would also like to encourage a broader use of these maps to assist in planning balanced regional development strategies, which include biodiversity conservation, for rural Madagascar.

Acknowledgements

We would like to thank the Weston Foundation for supporting this research. We would also like to thank ESRI for the donation of computer software, Conservation International, ANGAP, the Ministère des Eaux et Forêts (Madagascar) and FTM for allowing the use of their maps, and the Royal Society and the National Geographic Society for funding research in Paris and field work in Madagascar. We would also like to thank the Parc de Tsimbazaza and the University of Antananarivo for collaboration and assistance particularly with field work. We are grateful to the many individuals who have contributed to our understanding of the vegetation in Madagascar.

References

Beentje, H. and Dransfield, J. (1996). Priorities in Madagascar. In: IUCN/SSC Palm Specialist Group; Status Survey and Conservation Action Plan — Palms, their Conservation and Sustained Utilization. Pp. 66–68. IUCN, Switzerland and U.K.

Besairie, H. (1964). Carte Géologique de Madagascar, au 1:1,000,000ᵉ, trois feuilles en couleur. Service Géologique, Antananarivo.

Bosser, J.M., Du Puy, D.J. and Phillipson, P. (1996). Madagascar and Surrounding Islands. In: IUCN/SSC Orchid Specialist Group; Status Survey and Conservation Action Plan — Orchids. Pp. 103–107. IUCN Switzerland and U.K.

COEFOR/CI (1993). Répertoire et Carte de Distribution : Domaine Forestier de Madagascar. Direction des Eaux et Forêts, Service des Ressources Forestières, Projet COEFOR (Contribution à l'étude des Forêts Classées), et Conservation International, 20 p. + 1 map.

Du Puy, D.J. and Moat, J. (1996). A refined classification of the primary vegetation of Madagascar based on the underlying geology: using GIS to map its distribution and to assess its conservation status. In: W.R. Lourenço (editor). Proceedings of the International Symposium on the Biogeography of Madagascar. Pp. 205–218, + 3 maps. Editions de l'ORSTOM, Paris.

Faramalala, M.H. (1988). Étude de la Végétation de Madagascar à l'aide des Données Spaciales. Doctoral Thesis, Univ. Paul Sabatier de Toulouse, 167 p. + map at 1:1,000,000.

Faramalala, M.H. (1995). Formations Végétales et Domaine Forestier National de Madagascar. Conservation International (*et al.*), 1 map.

Gade, D.W. (1985). Savanna woodland, fire, protein and silk in highland Madagascar. *J. Ethnobiol.* **5**: 109–122.

Humbert, H. (1955). Les Territoires Phytogéographiques de Madagascar. Leur Cartographie. Colloque sur les Régions Ecologiques du Globe, Paris 1954. *Ann. Biol.* (*Paris*) **31**: 195–204, + map.

Internet Addresses – General

General Index:
http://www.rbgkew.org.uk/herbarium/madagascar/mad_index.html
Main Page (Madagascar GIS):
http://www.rbgkew.org.uk/herbarium/madagascar/mad_gis.html
Madagascar Vegetation mapping and biodiversity conservation - overview:
http://www.rbgkew.org.uk/herbarium/madagascar/veg_mapping.html
DU PUY, D.J. and MOAT, J.F. (1997):
http://www.rbgkew.org.uk/herbarium/madagascar/bio_paper.html

Internet Addresses – Maps

Map of the Simplified Geology:
http://www.rbgkew.org.uk/herbarium/madagascar/simp_geol.html
Map of the Remaining Primary Vegetation:
http://www.rbgkew.org.uk/herbarium/madagascar/primary_veg.html
Map of the Remaining Primary Vegetation classified by the underlying Geology:
http://www.rbgkew.org.uk/herbarium/madagascar/veg_geol.html
Map of the Protected areas in Madagascar:
http://www.rbgkew.org.uk/herbarium/madagascar/mad_parks.html
Download data and maps (in ArcView Shape File format):
http://www.rbgkew.org.uk/herbarium/madagascar/download.html

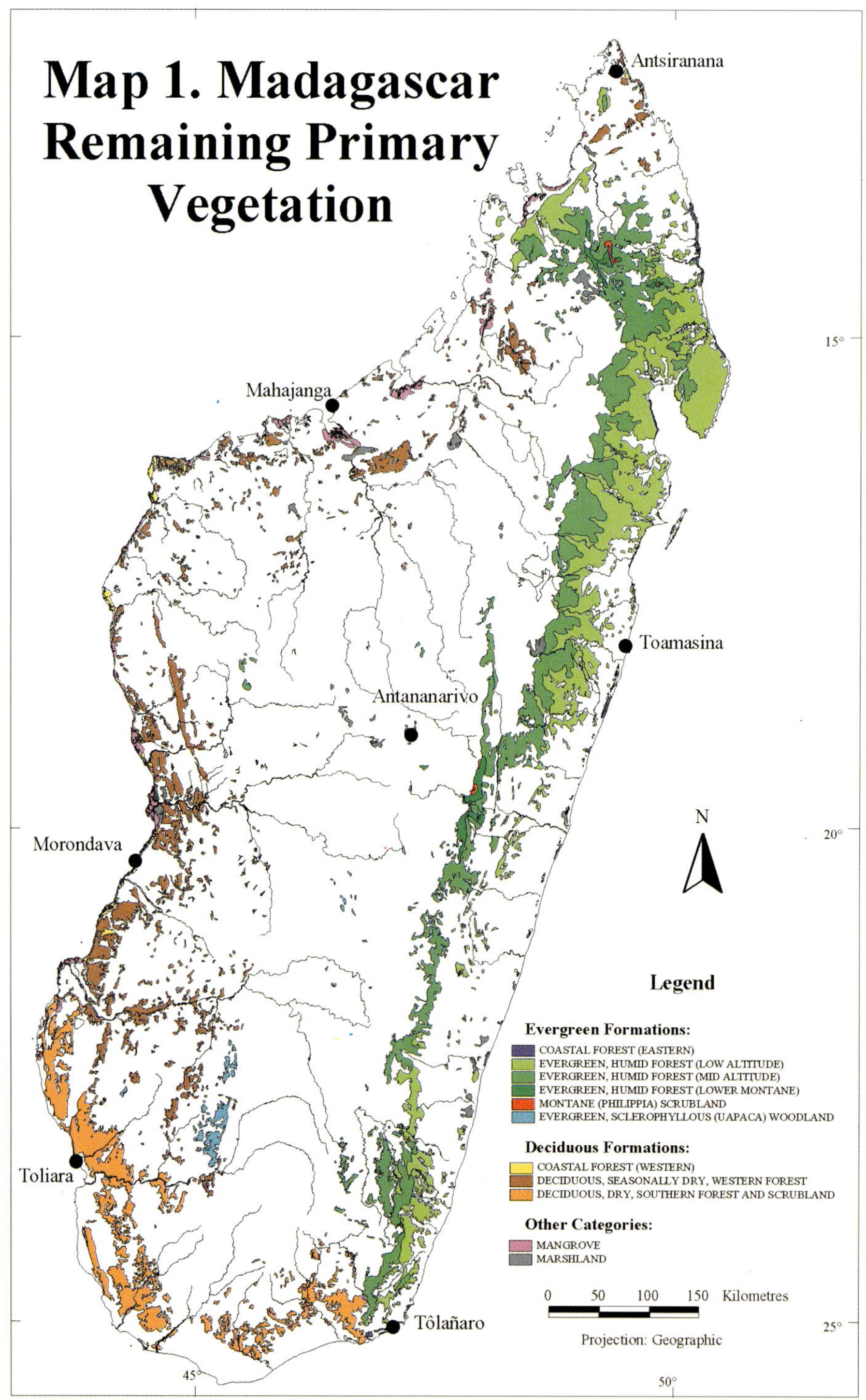

Map 1. Madagascar Remaining Primary Vegetation
Antsiranana
Mahajanga
Toamasina
Antananarivo
Morondava
Toliara
Tôlañaro
N
15°
20°
25°
45°
50°
Legend
Evergreen Formations:
COASTAL FOREST (EASTERN)
EVERGREEN, HUMID FOREST (LOW ALTITUDE)
EVERGREEN, HUMID FOREST (MID ALTITUDE)
EVERGREEN, HUMID FOREST (LOWER MONTANE)
MONTANE (PHILIPPIA) SCRUBLAND
EVERGREEN, SCLEROPHYLLOUS (UAPACA) WOODLAND
Deciduous Formations:
COASTAL FOREST (WESTERN)
DECIDUOUS, SEASONALLY DRY, WESTERN FOREST
DECIDUOUS, DRY, SOUTHERN FOREST AND SCRUBLAND
Other Categories:
MANGROVE
MARSHLAND
0   50   100   150   Kilometres
Projection: Geographic

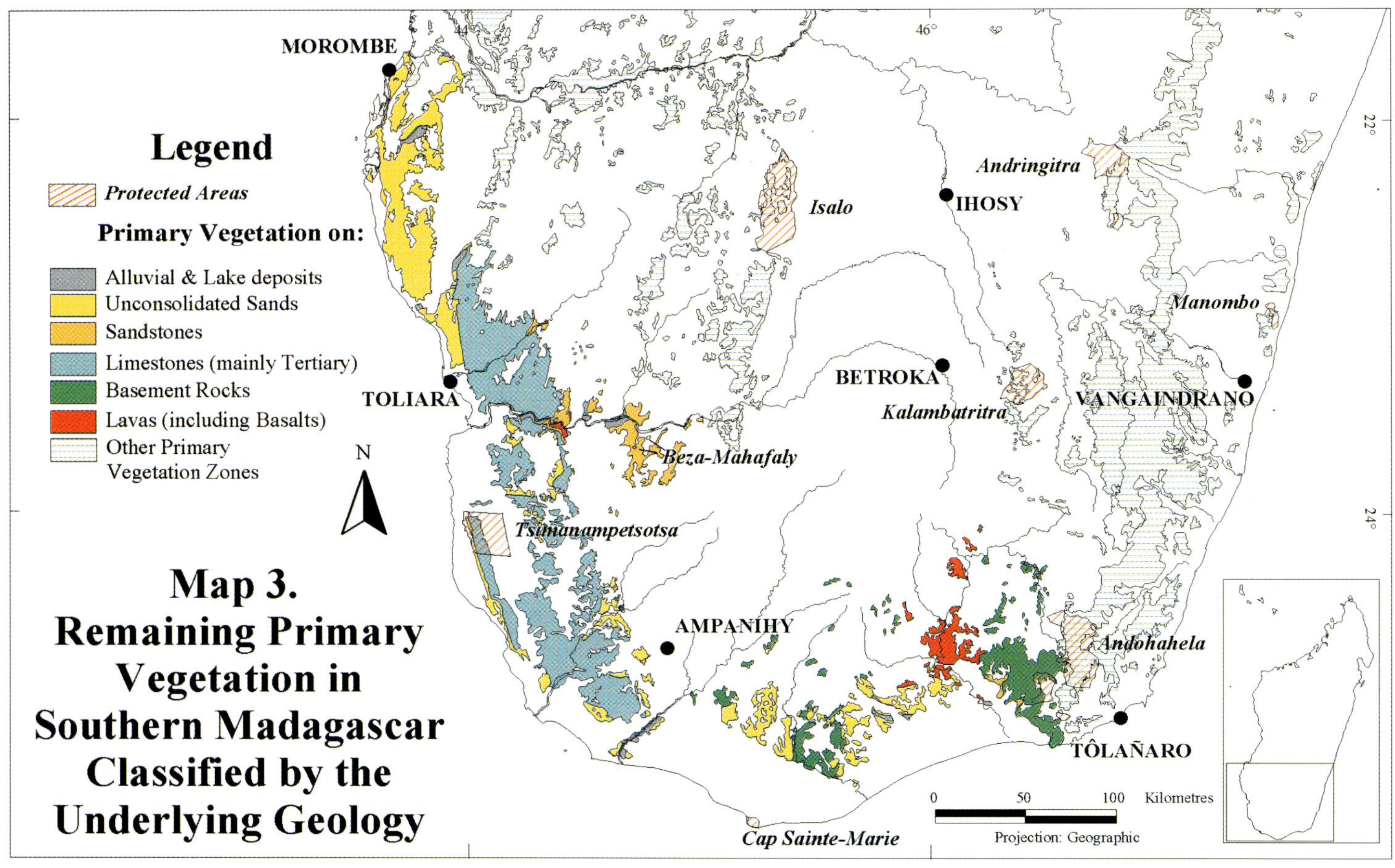
MOROMBE
44°
46°
22°
Andringitra
Isalo
IHOSY
Manombo
BETROKA
Kalambatritra
VANGAINDRANO
TOLIARA
24°
N
Beza-Mahafaly
Tsimanampetsotsa
AMPANIHY
Andohahela
TÔLAÑARO
Cap Sainte-Marie
0   50   100   Kilometres
Projection: Geographic
Legend
Protected Areas
Primary Vegetation on:
Alluvial & Lake deposits
Unconsolidated Sands
Sandstones
Limestones (mainly Tertiary)
Basement Rocks
Lavas (including Basalts)
Other Primary
Vegetation Zones
Map 3.
Remaining Primary
Vegetation in
Southern Madagascar
Classified by the
Underlying Geology

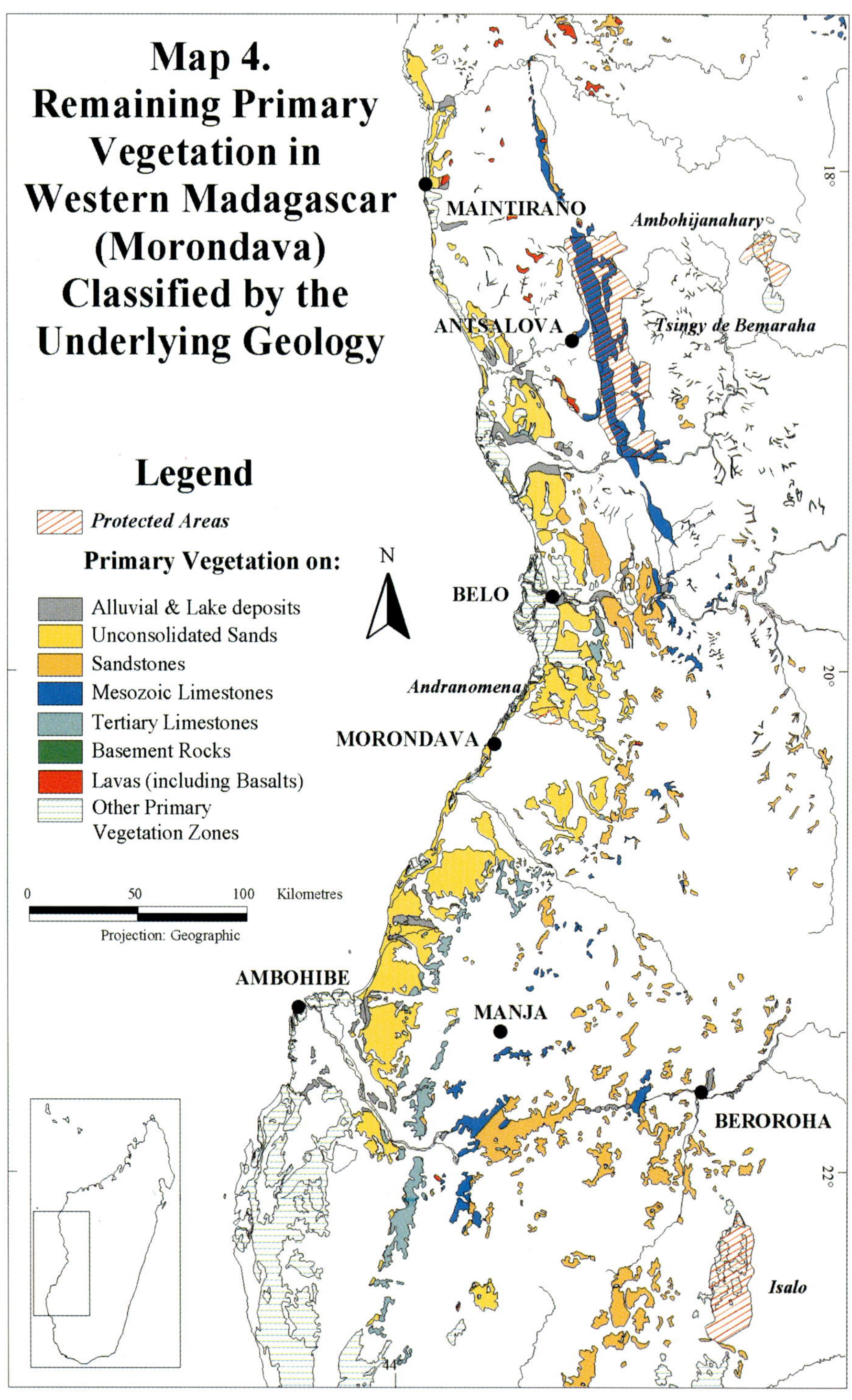

Map 4.
Remaining Primary
Vegetation in
Western Madagascar
(Morondava)
Classified by the
Underlying Geology
Legend
Protected Areas
Primary Vegetation on:
Alluvial & Lake deposits
Unconsolidated Sands
Sandstones
Mesozoic Limestones
Tertiary Limestones
Basement Rocks
Lavas (including Basalts)
Other Primary
Vegetation Zones
N
0
50
100
Kilometres
Projection: Geographic
MAINTIRANO
Ambohijanahary
ANTSALOVA
Tsingy de Bemaraha
BELO
Andranomena
MORONDAVA
AMBOHIBE
MANJA
BEROROHA
Isalo
18°
20°
22°

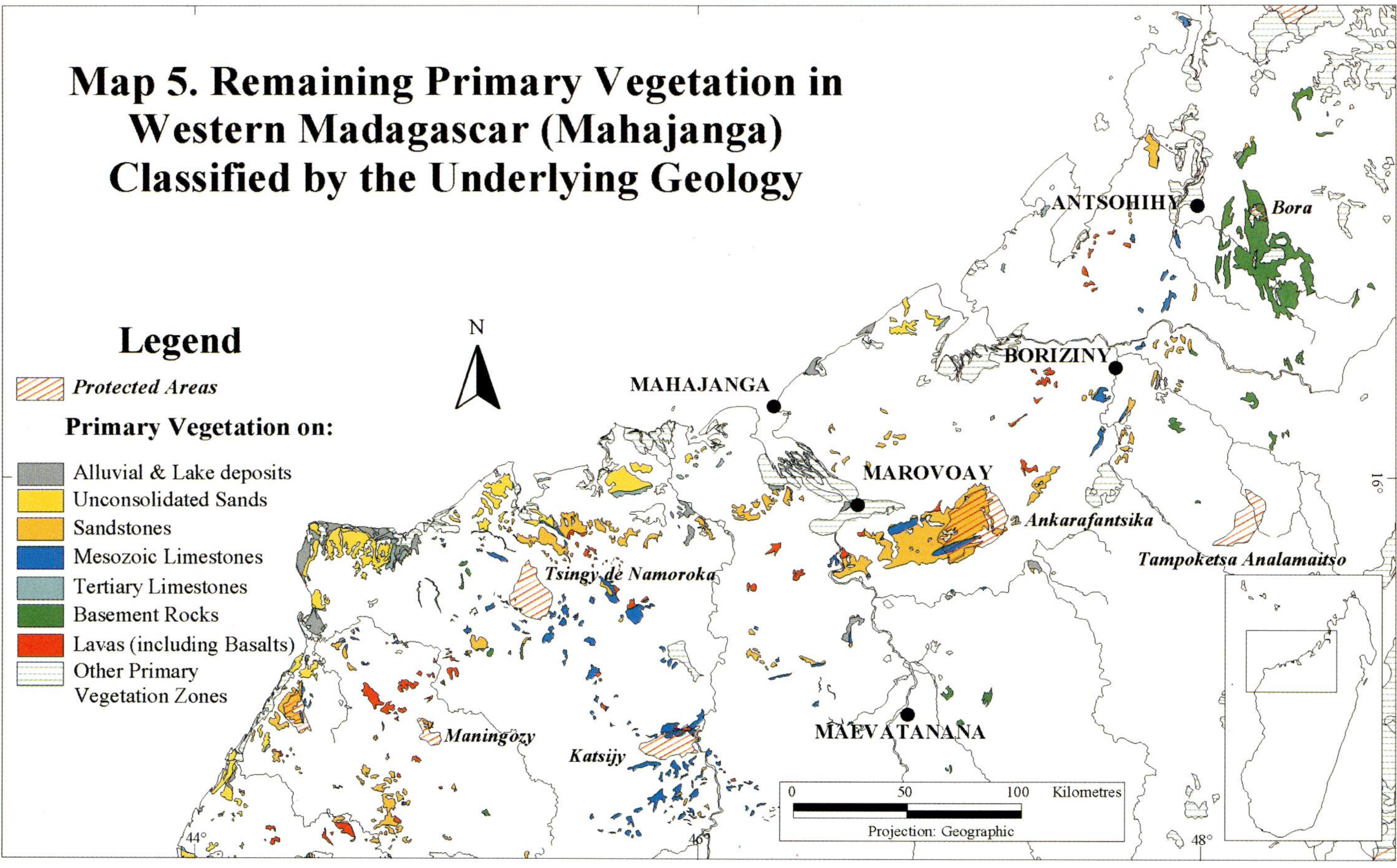

Map 5. Remaining Primary Vegetation in Western Madagascar (Mahajanga) Classified by the Underlying Geology
Legend
Protected Areas
Primary Vegetation on:
Alluvial & Lake deposits
Unconsolidated Sands
Sandstones
Mesozoic Limestones
Tertiary Limestones
Basement Rocks
Lavas (including Basalts)
Other Primary Vegetation Zones
N
Kilometres
0
50
100
Projection: Geographic
16°
48°
44°
Bora
ANTSOHIHY
BORIZINY
Tampoketsa Analamaitso
Ankarafantsika
MAROVOAY
MAHAJANGA
MAEVATANANA
Tsingy de Namoroka
Maningozy
Katsijy

Barthlott, W. & Porembski, S. (1998). Diversity and phytogeographical affinities of inselberg vegetation in tropical Africa and Madagascar. In: C.R. Huxley, J.M. Lock and D.F. Cutler (editors). Chorology, Taxonomy and Ecology of the Floras of Africa and Madagascar. Pp. 119–129. Royal Botanic Gardens, Kew.

8. DIVERSITY AND PHYTOGEOGRAPHICAL AFFINITIES OF INSELBERG VEGETATION IN TROPICAL AFRICA AND MADAGASCAR

WILHELM BARTHLOTT & STEFAN POREMBSKI

Botanisches Institut der Universität Bonn, Meckenheimer Allee 170, 53115 Bonn, Germany

Abstract

Inselbergs occur frequently as dome-shaped rock outcrops in all climatic and vegetational zones of the tropics. Consisting of Precambrian rocks, they form ancient and stable landscape elements. In regard to their edaphic and microclimatic conditions, inselbergs are completely differentiated from their surroundings. It is, therefore, no surprise that their vegetation is likewise very distinct. Well defined inselberg habitats (e.g. cryptogamic crusts, monocotyledonous mats, rock pools, ephemeral flush, wet flush) can be classified on the basis of physiognomy. In tropical Africa, there are considerable regional differences in floristic composition, life-forms and species diversity both in the vegetation of whole inselbergs and in that of individual habitats. The inselberg vegetation of West Africa is floristically relatively uniform and the number of endemics comparatively low. Monocotyledonous mats are almost monospecific stands of the poikilohydric *Afrotrilepis pilosa* (Cyperaceae). The major part of the West African rock outcrop flora consists of widely distributed taxa (e.g. *Cyanotis lanata, Hymenodictyon floribundum*). Inselbergs in the Zambezian Region are floristically richer with many succulents and poikilohydric species, and harbour more endemics than those in West Africa. There is a high degree of floristic divergence with vicariant taxa replacing each other (e.g. *Coleochloa setifera, Xerophyta* spp. as mat-forming elements). The Zambezian inselberg flora shows close relationships with those of Madagascan rock outcrops, which are very rich both in number of species and in endemics. Prominent in this regard are the genera *Xerophyta, Myrothamnus* and *Coleochloa*. For certain taxa (e.g. *Aloe, Kalanchoe, Pachypodium*), Madagascan inselbergs form diversity hot spots. The diversity of African and Madagascan inselbergs is discussed, in the context of global inselberg research.

Introduction

The term "inselberg" which was coined by the German geologist Wilhelm Bornhardt (1900), is widely accepted in international literature. Inselbergs are isolated rock outcrops that stand out abruptly from surrounding plains. There are several categories of inselbergs, such as tabular hills (e.g. the "tepuis" of southern Venezuela consisting of sandstone), castellated hills (e.g. kopjes) and domed hills (so-called "bornhardts", Plate 1A), which characteristically consist of granite or gneiss. Detailed surveys on the geomorphology of inselbergs were provided by Krebs (1942), Bremer & Jennings (1978) and Thomas (1994). Granitic and gneissic rock outcrops are geomorphologically relatively ancient (frequently more than 50 million years old). They are fairly widespread on the old crystalline shields and are particularly frequent in the tropics and subtropics. However, they can also be found in temperate zones (e.g. southeastern USA, southwestern Australia). The size of inselbergs varies considerably from several square kilometres to only a few square metres. Large inselbergs may attain a height of more than 600 m above the surroundings; small outcrops rise only a few metres. So-called "inselberg landscapes" are known from several areas (e.g. the Nigerian Jos Plateau), with numerous rock outcrops in close proximity, sometimes even forming more or less continuous ranges. On the other hand, there are regions with isolated inselbergs hundreds of kilometres apart.

Comparative floristic and vegetational studies of granitic and gneissic inselbergs in Africa, Madagascar, the Seychelles, South America and Australia have been the focal point of our current research project, which started in 1991. During this project, the main emphasis has been placed on island ecological aspects and, in particular, on the identification of regulating factors of species richness in plant communities. Inselbergs possess numerous ecological peculiarities which render them highly suitable subjects for island ecological research just like true oceanic islands.

Whereas temperate inselbergs have attracted some biological interest (for surveys see e.g. Australia: Ornduff, 1987 and Hopper, 1992; USA: Quarterman *et al.*, 1993), their tropical counterparts have largely been neglected in this respect. There are several regional studies available (e.g. Adjanohoun, 1964; Bonardi, 1966; Fleischmann *et al.*, 1996; de Granville, 1978; Hambler, 1964; Ibisch *et al.*, 1995; Porembski, 1996; Porembski *et al.*, 1994; 1995; Richards, 1957; Reitsma *et al.*, 1992; Sarthou, 1992; Villiers, 1981). However, comparative analyses of their vegetation have rarely been published. The aim of the present work is to provide a comparative survey of the vegetation of African and Madagascan inselbergs and to discuss their phytogeographic affinities.

Inselberg vegetation

Due to harsh edaphic (i.e. more or less devoid of soil cover) and microclimatic (i.e. high insolation and evaporation rates) conditions, the vegetation of inselbergs differs markedly from that of their surroundings, underlining their character as azonal continental islands. Inselbergs are ecosystems that comprise clearly distinguished and physiognomically defined plant communities. Despite fundamental differences in floristic composition between distant geographic and climatic regions, these occur almost identically on inselbergs throughout the world. Particularly characteristic are cryptogamic crusts, shallow soil-filled depressions, seasonally water-filled rock pools, monocotyledonous mats and ephemeral and wet flush communities. In the following, the characteristics of these habitats are outlined.

Cryptogamic crusts: exposed rocks are usually completely covered by specialized cyanobacterial lichens (typically *Peltula* spp.) or cyanobacteria (frequently *Gloeocapsa* spp., *Stigonema* spp. and *Scytonema* spp.; Büdel *et al.*, 1994), which are responsible for the typical brownish or greyish colour of inselbergs. The coloration of tropical rock

PLATE 1. **A** Granitic inselberg in West Africa (Ivory Coast). The rocky slopes are covered by mats of the poikilohydric *Afrotrilepis pilosa* (Cyperaceae). **B** *Afrotrilepis pilosa*-mat on Ivorian inselberg. Characterised by a low diversity, these mats offer establishment sites for a limited number of other species, such as *Euphorbia poissoni*. **C** Mats formed by the poikilohydric Cyperaceae *Coleochloa setifera* are typical for East African and Madagascan rock outcrops. Frequently *Myrothamnus flabellifolius* (Africa) and *M. moschatus* (Madagascar) occur as mat colonizers. **D** *Coleochloa setifera*-mats on Madagascan rock outcrops are particularly diverse. Remarkably high is the number of infraspecific variants and cryptic species complexes, for instance within the *Euphorbia milii*-complex.

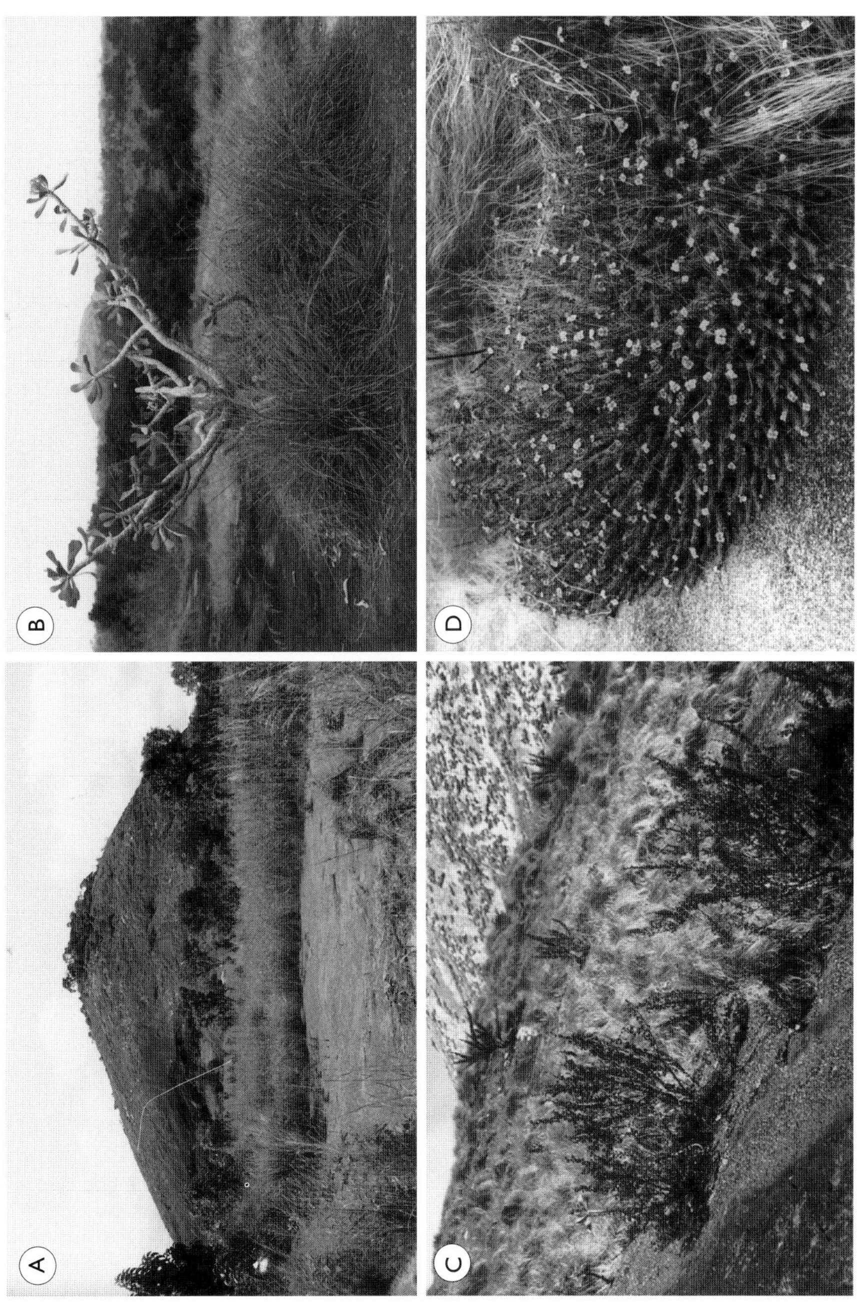

outcrops was noted by Alexander von Humboldt (1819) while traveling in the Orinoco area of what is today southern Venezuela. However, he interpreted the source of the darkish colour of the rocks as a cover of iron and manganese oxide. Cyanobacterial lichens dominate on granite outcrops in seasonally dry climates whereas in rainforest regions the rocky slopes are dominated by cyanobacteria. Particularly where seepage water is available, moss cushions (frequently *Bryum arachnoideum*) may establish (Frahm & Porembski, 1994). A continuous cryptogamic crust is lacking on inselbergs situated in very arid regions, such as in the central Sahara (own unpubl. observations).

Shallow soil-filled depressions: soil cover is relatively thin (usually under 5 cm in depth). Throughout tropical Africa, frequent colonizers of bare soil are liverworts (*Riccia* spp.) and cyanobacteria (e.g. *Schizothrix* spp.). In East Africa, poikilohydric Scrophulariaceae (e.g. *Lindernia wilmsii, L. pulchella, Craterostigma* spp.) and small succulents (e.g. *Portulaca rhodesiana*) are characteristic. In West Africa, therophytes (e.g. *Lindernia exilis, Cyanotis lanata*) dominate. In the latter region, perennials are represented by geophytes (e.g. *Ophioglossum costatum, Pancratium trianthum*). In Madagascar, annuals (e.g. *Exacum* spp.) are widespread in this habitat.

Seasonally water-filled rock pools: temporal habitats that are water-filled after rainfall but dry out soon if not replenished by subsequent rain. Short-lived herbs predominate. Besides typical water plants (e.g. Lemnaceae) and species which are otherwise widespread on marshy ground (e.g. *Cyperus* spp., *Ludwigia* spp., *Rotala* spp.), there are specialists which are restricted to this habitat. Prominent among these are certain Scrophulariaceae, such as the Namibian *Chamaegigas intrepidus* and species belonging to the genera *Lindernia* (e.g. *L. monroi* and *L. conferta* from Zimbabwe) and *Dopatrium* (e.g. *D. longidens*, West Africa). Widespread on East African inselbergs are geophytic water plants of the genus *Aponogeton* (e.g. *A. stuhlmannii*). Characteristic but frequently overlooked are geophytic species of *Isoetes* (e.g. *I. nigritiana*, West Africa), which have also been recorded from extra-tropical inselbergs.

Monocotyledonous mats: Carpet-like mats formed by Cyperaceae and Velloziaceae that cover even steep slopes are widespread on inselbergs throughout the tropics. On African and Madagascan inselbergs, Cyperaceae are dominant. In East Africa and Madagascar, the Velloziaceae attain the status of co-dominants.

The Cyperaceae occur with three apparently closely related mat-forming genera: *Afrotrilepis* is distributed from Senegal southwards to Gabon and comprises two species (*A. pilosa* and *A. jaegeri*) that are largely restricted to rock outcrops. *Coleochloa* comprises eight species which show strong affinities to rock outcrops. *Coleochloa setifera* is the dominant mat element on both East African and Madagascan inselbergs. *Microdracoides squamosus* has a disjunct distribution in Guinea/Sierra Leone and Nigeria/Cameroon. The phytogeographical links of these genera are directed towards South American inselbergs, where the genus *Trilepis* occurs with several mat-forming species (Gilly, 1943).

The genus *Xerophyta* (Velloziaceae) is well represented on inselbergs in Madagascar and in southern and eastern Africa. The representatives of this genus are characterized by a high degree of morphological variation between geographically distinct populations which leads to considerable taxonomic problems. With *Xerophyta schnitzleinia* var. *occidentalis*, one geographical outlier occurs in northern Nigeria (Ayo Owoseye & Sanford, 1972; Kershaw, 1968).

Mat-forming Cyperaceae and Velloziaceae possess convergently developed morphological (treelet-like habit), anatomical (roots possessing a velamen radicum; Porembski & Barthlott, 1995) and physiological (poikilohydry) adaptations that help the plants to withstand the harsh ecological conditions on inselbergs. Most mat-forming species host a highly specific set of epiphytic orchids (e.g. *Polystachya microbambusa* on *Afrotrilepis pilosa* in West Africa).

Apart from monocotyledons, only a few other groups of vascular plants occur as mat formers on inselbergs; such are the poikilohydrous shrub *Myrothamnus* (*M. flabellifolius* in East Africa, *M. moschatus* on Madagascar) and a number of similarly poikilohydric species of the pteridophyte genus *Selaginella* (e.g. *S. niamniamensis*, *S. dregei* in East Africa).

Ephemeral flush vegetation: develops at the foot of steep slopes over thin soil into which water seeps continuously during the rainy season. Poaceae (in particular *Panicum* spp., *Sacciolepis* spp. and *Sporobolus* spp.) and Cyperaceae (frequently *Ascolepis* spp., *Isolepis* spp., *Kyllinga* spp. and *Scleria* spp.) make up the bulk of the vegetation. Most striking are ephemerals including members of Eriocaulaceae (*Eriocaulon* spp.), Xyridaceae (*Xyris* spp.), Burmanniaceae (*Burmannia* spp.), Gentianaceae (*Sebaea* spp.) and carnivorous plants (Droseraceae and Lentibulariaceae: *Utricularia*, *Genlisea*). In tropical Africa, this community is especially well developed (i.e. most rich in species) on inselbergs in savanna zones (Dörrstock *et al.*, 1996a), whereas inselbergs situated in rainforests harbour depauperate ephemeral flush communities.

Wet flush vegetation: Usually forming narrow vertical belts, this very peculiar habitat occurs on inclined bare rocky slopes where water flows steadily during the rainy season. Typical are small-sized annuals, particularly widespread are *Xyris* spp. and *Utricularia* spp. which are attached to cyanobacterial crusts. This community seems to reach its best development in a rainforest climate and is obviously lacking in semi-arid and arid regions. Due to its inaccessibility, our knowledge of this community is still very limited.

Species richness and phytogeography of African and Madagascan inselberg vegetation

Granitic and gneissic rock outcrops occur throughout all tropical and subtropical zonobiomes of Africa and Madagascar (Fig. 1). In the following, a concise survey will be provided concerning species richness and endemics of both African and Madagascan inselbergs. Monocotyledonous mats form a highly specific habitat type, which in many respects is representative of the whole inselberg ecosystem. For this reason monocotyledonous mats will be considered here in more detail than other inselberg habitats.

The vegetation of tropical African inselbergs shows a considerable degree of regional differentiation in floristic composition and number of species, as well as in the number of endemics. However, since the extent of our knowledge of inselberg vegetation is patchy in some regions (e.g. Angola, Mozambique, Somalia, Sudan), it is not possible to present a detailed account for the whole continent.

West Africa

A large number of regional studies (e.g. Adjanohoun, 1964; Bonardi, 1966; Hambler, 1964; Porembski & Barthlott, 1993; Porembski *et al.*, 1994; Porembski & Brown, 1995; Reitsma *et al.*, 1992; Richards, 1957; Villiers, 1981) have contributed to our knowledge of West African inselberg vegetation. Inselbergs in this region, between Senegal and Gabon, are floristically relatively uniform with low alpha- and beta-diversity. Likewise, the number of endemics is comparatively low (e.g. *Afrotrilepis jaegeri*, *Djaloniella ypsilostyla*, *Genlisea barthlottii*, *G. stapfii*, *Lindernia yaundensis*). This may be the result of drastic climatic fluctuations during the Pleistocene which appear to have caused major shifts of the West African vegetation belts.

The major part of the West African rock outcrop flora consists of widely-distributed African taxa (e.g. *Afrotrilepis pilosa*, *Cyanotis lanata*, *Hymenodictyon floribundum*). Moreover a considerable number of species are paleotropical (e.g.

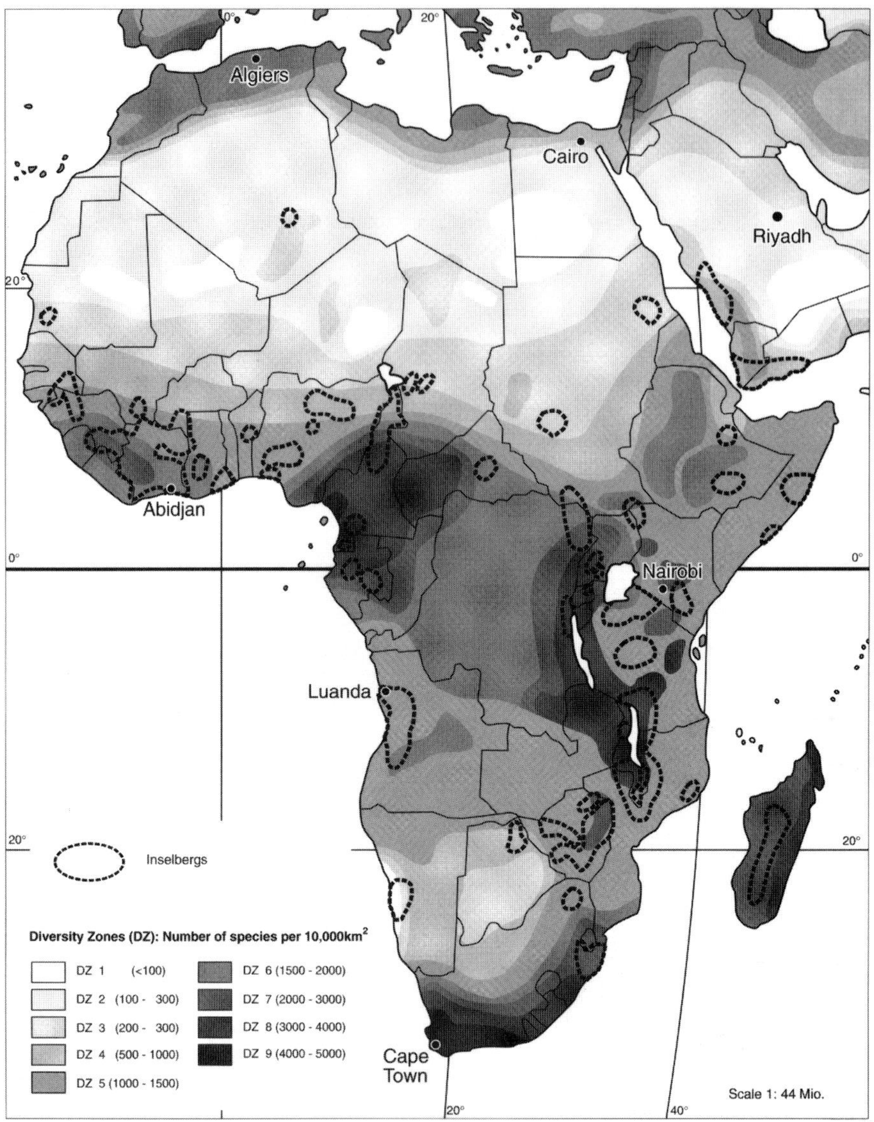

FIG. 1. Inselberg distribution and Diversity Zones (DZ = number of species per 10,000 km²) for Africa, Madagascar and the Arabian Peninsula. Map based on Barthlott *et al.* (1996).

Mariscus dubius) and pantropical (e.g. *Fimbristylis dichotoma*). Many species on West African inselbergs are continuously distributed over the whole region. However, certain species (e.g. *Microdracoides squamosus, Asplenium jaundeense*) are disjunct in their distribution and occur on rock outcrops both in Guinea/Sierra Leone and to the east of the Dahomey gap in the border region between Nigeria and Cameroon. Possibly this is a relict distribution pattern caused by the Pleistocene climatic perturbations mentioned above.

The occurrence of several American-African disjuncts on West African rock outcrops is of phytogeographical interest. Most striking is the presence of *Pitcairnia feliciana*, the only Old World bromeliad on sandstone outcrops in the Fouta Djalon highlands of Guinea. *Neurotheca loeselioides* (Gentianaceae) is characteristic of ephemeral flush communities both in the Neotropics and in West Africa (Dörrstock *et al.*, 1996a) and recently, another American-African disjunct *Utricularia juncea*, has been found for the first time in Africa on inselbergs in the Ivory Coast (Dörrstock *et al.*, 1996b). Not surprisingly, the percentage of Sudano-Zambezian elements is higher on inselbergs in the savanna of the Sudanian Region of West Africa than in the Guineo-Congolian rainforests. Remarkably, plant species richness is much higher on inselbergs located in the savanna zone than on their counterparts in the rainforests of the Upper Guinea Region (Porembski *et al.*, 1995). This inverted gradient of plant diversity is probably due to source-sink effects and metapopulation dynamics which influence extinction and colonisation rates on isolated habitat fragments, like rock outcrops. Since inselbergs are more frequent in the savanna zone, larger metapopulations exist, reducing local extinction rates and resulting in higher species richness.

On West African inselbergs, monocotyledonous mats are almost monospecific stands formed by the poikilohydric Cyperaceae *Afrotrilepis pilosa* (Porembski *et al.*, 1996). *Afrotrilepis* mats (Plate 1B) are the least speciose mat community compared with those in other tropical regions. The occurrence of a particular set of epiphytic orchids is typical of *Afrotrilepis* mats. Throughout the range of *Afrotrilepis pilosa*, the genus *Polystachya* is represented by a number of species (e.g. *P. microbambusa*, *P. pseudo-disa*) which usually are only found in *Afrotrilepis* mats. The taxonomic status of the second species of the genus *Afrotrilepis*, *A. jaegeri*, endemic to the Loma Mts. and Tingi Hills, Sierra Leone, is somewhat obscure. Our own analyses (unpublished data) indicate a close relationship between *A. jaegeri* and the South American genus *Trilepis*.

Central and East Africa

Our knowledge of the vegetation of inselbergs in Central and East Africa is less comprehensive than that of their West African counterparts. There are few studies available dealing with the inselberg vegetation of this region (e.g. Lisowski, 1992; Porembski, 1996; Seine *et al.*, 1998; Sillans, 1958; Taton, 1949). However, despite the lack of data, certain characteristics of Central and East African inselbergs are easily observed. For instance, the inselbergs in the Zambezian Region are floristically richer and harbour more endemics (e.g. in the genera *Begonia* and *Streptocarpus*) than those in West Africa. Many genera, such as *Aloe*, *Brachystelma*, *Encephalartos* and *Xerophyta*, are more speciose on Central and East African inselbergs than in West Africa. Moreover, the numbers of both succulents and poikilohydric vascular plants are considerably higher. In particular, eastern and southeastern Africa possess a large diversity of succulents. Most prominent are Aloaceae (*Aloe* spp.), Asclepiadaceae (e.g., *Brachystelma* spp., *Caralluma* spp., *Ceropegia* spp.), Crassulaceae (e.g., *Cotyledon* spp., *Kalanchoe* spp.), Euphorbiaceae (*Euphorbia* spp.), Lamiaceae (*Aeollanthus* spp.) and Moraceae (*Dorstenia* spp.).

The number of poikilohydric vascular plants is remarkably high. Scrophulariaceae, Gesneriaceae, Velloziaceae, Poaceae, Cyperaceae and ferns (e.g. *Actiniopteris*, *Pellaea*) are richly represented by 'resurrection' plants making East African rock outcrops a global hot-spot for this highly adapted plant group. The small hamamelidaceous family Myrothamnaceae occurs throughout eastern and southern Africa with the poikilohydric shrub *Myrothamnus flabellifolius*. The presence of *Myrothamnus moschatus* on Madagascan inselbergs emphasizes their close phytogeographic relationship to East Africa. Wind-pollinated dioecious shrubs, like *Myrothamnus* are absent from inselbergs in other tropical regions. The rubiaceous genus *Anthospermum* is a remarkable and characteristic element on East African and Madagascan inselbergs and includes several wind-pollinated, shrubby and sometimes dioecious species.

In general, the phytogeographical affinities between Madagascan and East African inselbergs are stronger than those between the latter and West African rock outcrops. This fact is also expressed by the monocotyledonous mats which are formed by *Coleochloa setifera* both on East African and Madagascan inselbergs (Plate 1C). Although physiognomically very similar to *Afrotrilepis*-mats, they are richer in accompanying species (Porembski, 1996). Important colonizers of *Coleochloa*-mats are *Myrothamnus flabellifolius* (respectively *M. moschatus* in Madagascar), several species of *Xerophyta* and a number of succulents. Typical elements of the Zambezian/Madagascan inselberg flora extend far into southern Africa and the Karroo-Namib region. For example, a significant number of succulents (e.g. *Euphorbia, Kalanchoe*) and poikilohydrics (*Myrothamnus flabellifolius, Chamaegigas intrepidus*) with affinities to East Africa are present on inselbergs in Namibia (Nordenstam, 1974; Giess, 1969).

The inselberg floras of West Africa on one hand and East Africa/Madagascar on the other are very different, with vicariant taxa replacing each other (e.g. *Afrotrilepis pilosa, Coleochloa setifera, Xerophyta* spp. as mat-forming elements). Obviously the lowland rainforest of the Zaïre basin has very effectively prevented floristic exchange between inselbergs in West and East Africa.

Madagascar

Despite being fairly conspicuous landscape elements, Madagascan inselbergs have only rarely been the subject of botanical studies. Not a single work has been solely devoted to their unique vegetation. Some of their floristic and ecological characteristics have been mentioned in general accounts on Madagascan vegetation (e.g. Koechlin *et al.*, 1974; Rauh, 1973). A general characterisation of inselberg vegetation on the Central Plateau was given by Nilsson & Rabakonandrianina (1988) in a study concerned with pollination biology. Granitic and gneissic outcrops are particularly frequent on the Central Plateau where they are known to form "a succulent paradise" (Rauh, 1995), whereas the surrounding vegetation has been largely cleared.

About 80% of the Madagascan flora is endemic (White, 1983). Although there is no detailed information about numbers of species and endemics on inselbergs, the preliminary data available clearly suggest that the Madagascan rock outcrop flora also includes a high percentage of endemics. Several endemic genera (e.g. the orchid *Sobennikoffia* and the asclepiad *Stapelianthus*) possess close ecological affinities to inselbergs. Phytogeographically, the Madagascan inselberg vegetation shows strong ties with eastern and southern Africa. This is not only true for vascular plants but also for bryophytes (Pócs, 1975). Apart from examples already mentioned above (e.g. *Myrothamnus*) there is a considerable number of genera which obviously have radiated both on African and Madagascan inselbergs. Of particular interest in this respect are certain genera (e.g. *Aloe, Cynanchum, Euphorbia, Kalanchoe, Pachypodium, Senecio*) which are represented by various leaf and stem succulent species. Remarkably the Madagascan rock outcrops show a higher morphological differentiation and species richness of most succulent taxa when compared with the African mainland. Madagascar is similarly rich in poikilohydric vascular plant species. In a few cases, the same species as in East Africa (e.g. *Coleochloa setifera*) or, more frequently, closely related poikilohydric species occur.

Almost no phytogeographic links exist between Madagascan inselbergs and those of the granitic Seychelles islands situated more than 1,000 km to the north (Fleischmann *et al.*, 1996). At the generic level (e.g. *Lophoschoenus, Nepenthes, Pandanus*) the Seychelles inselbergs show instead more pronounced phytogeographic affinities towards Asia.

Although our knowledge of the vegetation of inselbergs in Sri Lanka and India is far from complete, the few works available (e.g. Bharucha & Ansari, 1962; Willis, 1906) indicate floristic relationships towards Madagascar and East Africa. African-Asian relationships via Madagascan inselbergs are demonstrated for example by succulents (e.g. *Cyanotis, Impatiens*) and poikilohydrics (e.g. the fern genus *Actiniopteris*).

There are some interesting phytogeographic links between East African/ Madagascan inselbergs and those of South America. A remarkable connecting link is formed by the Velloziaceae. Possessing their centre of diversity in southeastern Brazil, they are widespread on rock outcrops throughout East Africa and Madagascar (with the genus *Xerophyta*) and even reach the southern tip of the Arabian peninsula. It can be postulated that the present distribution of the Velloziaceae in the Old World either dates back to Gondwanian times or, alternatively, is the result of later immigration to Africa and subsequent "rock outcrop hopping", using inselbergs as stepping stones. Madagascan inselbergs form a secondary centre of diversity for the New World genus *Rhipsalis* (Cactaceae). Neotenic forms (e.g. *R. baccifera* ssp. *horrida*) of this largely epiphytic genus occur epilithically in Madagascar (Barthlott, 1983). Another example of an American-African-Madagascan relationship is demonstrated by the carnivorous genus *Genlisea* (Lentibulariaceae), which frequently occurs on inselbergs (and other azonal localities) in all three regions (10 species in America, 9 species in tropical Africa and Madagascar).

As in East Africa, *Coleochloa setifera* is the main mat-forming component on Madagascan inselbergs (Plate 1D). Occasionally (in particular at altitudes >2,000 m asl.), *Xerophyta dasylirioides* becomes the dominant mat-former. There is a large proportion of accompanying species amongst which orchids are richly represented. The Madagascan *Coleochloa*-mats seem to represent the most speciose mat community on Old World inselbergs.

Judging from present knowledge, it becomes clear that both the Madagascan and the East African inselberg vegetations are richer in numbers of species and endemics in comparison to West African rock outcrops. The reasons for this striking geographic differentiation are not yet completely understood. However, it can be speculated that this difference is due to the longer continuous development of rock outcrop habitats (i.e. no drastic climatic changes) and the generally larger number of available azonal sites (e.g. the Precambrian crystalline chains in East Africa, such as Usambara and Uluguru). Both factors should reduce local extinction rates in parallel whereas speciation in isolated populations is promoted.

Acknowledgements

Financial support by the Deutsche Forschungsgemeinschaft (Ba 605/4-3) is gratefully acknowledged. The authorities of several countries (Bénin, Côte d'Ivoire, Guinea, Madagascar, Malawi, Zimbabwe) are thanked for permission to conduct research. We would like to express our thanks to L. Ake Assi (Abidjan), N. Biedinger (Bonn), E. Fischer (Bonn), K. E. Linsenmair (Würzburg), R. Seine (Bonn), D. Supthut (Zürich) and I. Theisen (Bonn) for support during fieldwork and valuable comments on the manuscript. In addition we thank M. Gref (Bonn) for skillful technical assistance in producing the phytodiversity map of Africa.

References

Adjanohoun, E. (1964). Végétation des savanes et des rochers découverts en Côte d'Ivoire centrale. *Mém. ORSTOM* **7**: 1–178.

Ayo Owoseye, J. and Sanford, W.W. (1972). An ecological study of *Vellozia schnitzleinia*, a drought-enduring plant of northern Nigeria. *J. Ecol.* **60**: 807–817.

Barthlott, W. (1983). Biogeography and Evolution in Neo- and Paleotropical *Rhipsalinae* (Cactaceae). *Sonderb. Naturwiss. Vereins Hamburg* **7**: 241–248.

Barthlott, W., Lauer, W. and Placke, A. (1996). Global distribution of species diversity in vascular plants: towards a world map of phytodiversity. *Erdkunde* **50**: 317–327.

Bharucha, F.R. and Ansari, M.Y. (1962). Studies on the plant associations of slopes and screes of the Western Ghats, India. *Vegetatio* **61**: 141–154.

Bonardi, D. (1966). Contribution à l'étude botanique des inselbergs de Côte d'Ivoire forestière. Diplome d'études supérieures de sciences biologiques, Université d'Abidjan.

Bornhardt, W. (1900). Zur Oberflächengestaltung und Geologie Deutsch-Ostafrikas. Reimer, Berlin.

Bremer, H. and Jennings, J. (editors). (1978). Inselbergs/Inselberge. *Z. Geomorph., N.F.,* *Suppl.* **31**.

Büdel, B., Lüttge, U., Stelzer, R., Huber, O. and Medina, E. (1994). Cyanobacteria of Rocks and Soils of the Orinoco Lowlands and the Guayana Uplands, Venezuela. *Bot. Acta* **107**: 422–431.

Dörrstock, S., Porembski, S. and Barthlott, W. (1996a). Ephemeral flush vegetation on inselbergs in the Ivory Coast (West Africa). *Candollea* **51**: 407–419.

Dörrstock, S., Seine, R., Porembski, S. and Barthlott, W. (1996b). First record of *Utricularia juncea* (Lentibulariaceae) for tropical Africa. Kew Bull. **51**: 579–583.

Fleischmann, K., Porembski, S., Biedinger, N. and Barthlott, W. (1996). Inselbergs in the sea: vegetation of granite outcrops of Mahé and Silhouette, Seychelles. *Bull. Geobot. Inst. ETH* **62**: 61–74.

Frahm, J.-P. and Porembski, S. (1994). Moose von Inselbergen aus Westafrika. *Trop. Bryol.* **9**: 59–68.

Giess, W. (1969). Die Verbreitung von *Lindernia intrepidus* (Dinter) Oberm. (*Chamaegigas intrepidus* Dinter) in Südwestafrika. *Dinteria* **2**: 23–28.

Gilly, C.L. (1943). An Afro-South-American cyperaceous complex. *Brittonia* **5**: 1–20.

Granville, J.J. de (1978). Recherches sur la flore et la végétation guyanaises. Thèse de Doctorat d'Etat, USTL, Montpellier.

Hambler, D.J. (1964). The vegetation of granitic outcrops in Western Nigeria. *J. Ecol.* **52**: 573–594.

Hopper, S.D. (1992). Patterns of plant diversity at the population and species level in south-west Australian mediterranean ecosystems. In: R.J. Hobbs (editor). Biodiversity of mediterranean ecosystems in Australia.Pp. 27–46. Surrey Beatty & Sons, Chipping, Norton, New South Wales.

Humboldt, A.v. (1819). Relation historique du Voyage aux régions équinoxiales du Nouveau Continent, fait en 1799–1804 par A. de Humboldt et A. Bonpland. Vol. II, N. Maze, Paris.

Ibisch, P.L., Rauer, G., Rudolph, D. and Barthlott, W. (1995). Floristic, biogeographical, and vegetational aspects of Pre-cambrian rock outcrops (inselbergs) in eastern Bolivia. *Flora* **190**: 299–314.

Kershaw, K.A. (1968). A survey of the vegetation in Zaria Province, Nigeria. *Vegetatio* **15**: 244–268.

Koechlin, J., Guillaumet, J.L. and Morat, P. (1974). Flore et végétation de Madagascar. Cramer, Vaduz.

Krebs, N. (1942). Über das Wesen und Verbreitung der tropischen Inselberge. *Abh. Preuss. Akad. Wiss., Math.-Naturwiss. Kl.* **6**.

Lisowski, S. (1992). Notes floristiques de l'Ituri (Haut-Zaïre). *Fragm. Florist. Geobot.* **37**: 215–220.

Nilsson, L.A. and Rabakonandrianina, E. (1988). Hawk-moth scale analysis and pollination specialization in the epilithic Malagasy endemic *Aerangis ellisii* (Reichenb. fil.) Schltr. (Orchidaceae). *Bot. J. Linn. Soc.* **97**: 49–61.

Nordenstam, B. (1974). The Flora of the Brandberg. *Dinteria* **11**: 3–67.

Ornduff, R. (1987). Islands on Islands: Plant Life on the Granite Outcrops of Western Australia. University of Hawaii, H. L. Lyon Arboretum Lecture No. 15. University Press of Hawaii, Honolulu.

Pócs, T. (1975). Affinities between the bryoflora of East Africa and Madagascar. *Boissiera* **24a**: 125–128.

Porembski, S. (1996). Notes on the Vegetation of Inselbergs in Malawi. *Flora* **191**: 1–8.

Porembski, S. and Barthlott, W. (1993). Ökogeographische Differenzierung und Diversität der Vegetation von Inselbergen in der Elfenbeinküste. In: W. Barthlott, C.M. Naumann, K. Schmidt-Loske and K.L. Schuchmann (editors). Animal-plant interactions in tropical environments. Pp. 149–158. Results of the Annual Meeting of the German Society for Tropical Ecology, Bonn 1992.

Porembski, S. and Barthlott, W. (1995). On the occurrence of a velamen radicum in tree-like Cyperaceae and Velloziaceae. *Nordic J. Bot.* **15**: 625–629.

Porembski, S., Barthlott, W., Dörrstock, S. and Biedinger, N. (1994). Vegetation of rock outcrops in Guinea: granite inselbergs, sandstone table mountains, and ferricretes — remarks on species numbers and endemism. *Flora* **189**: 315–326.

Porembski, S. and Brown, G. (1995). The vegetation of inselbergs in the Comoé-National Park (Ivory Coast). *Candollea* **50**: 351–365.

Porembski, S., Brown, G. and Barthlott, W. (1995). An inverted latitudinal gradient of plant diversity in shallow depressions on Ivorian inselbergs. *Vegetatio* **117**: 151–163.

Porembski, S., Brown, G. and Barthlott, W. (1996). A species-poor tropical sedge community: *Afrotrilepis pilosa* mats on inselbergs in West Africa. *Nordic J. Bot.* **16**: 239–245.

Quarterman. E., Burbanck, M.P. and Shure, D.J. (1993). Rock outcrop communities: Limestone, Sandstone, and Granite. In: W.H. Martins, S.G. Boyce & A.C. Echternach (editors). Biodiversity of the southeastern United States. Pp. 35–85. John Wiley & Sons, New York.

Rauh, W. (1973). Über die Zonierung und Differenzierung der Vegetation Madagaskars. *Trop. Subtrop. Pflanzenwelt* **1**.

Rauh, W. (1995). Succulent and Xerophytic Plants of Madagascar. Vol. 1. Strawberry Press, Mill Valley.

Reitsma, J.M., Louis, A.M. and Floret, J.J. (1992). Flore et végétation des inselbergs et dalles rocheuses: première étude au Gabon. *Bull. Mus. Natl. Hist. Nat., B, Adansonia* **14**: 73–97.

Richards, P.W. (1957). Ecological notes on West African Vegetation I. The plant communities of the Idanre Hills, Nigeria. *J. Ecol.* **45**: 563–577.

Sarthou, C. (1992). Dynamique de la végétation pionniere sur un inselberg en Guyane Française. Thèse de Doctorat d'Etat, Université Paris 6.

Seine, R., Becker, U., Porembski, S., Follmann, G. and Barthlott, W. (1998). Vegetation of inselbergs in Zimbabwe. *Edinburgh J. Bot.* **55**: 267–293.

Sillans, R. (1958). Les savanes de l'Afrique centrale. Lechavalier, Paris.

Taton, A. (1949). La colonisation des roches granitiques de la région de Nioka (Haut-Ituri-Congo belge). *Vegetatio* **1**: 317–332.

Thomas, M.F. (1994). Geomorphology in the tropics. A study of weathering and denudation in low latitudes. John Wiley & Sons, New York.

Villiers, J.F. (1981). Formations climaciques et relictuelles d'un inselberg inclus dans la forêt dense camerounaise. Thèse de Doctorat d'Etat, Université Paris 6.

White, F. (1983). The Vegetation of Africa: a descriptive memoir to accompany the UNESCO/AETFAT/UNSO vegetation map of Africa. *Natural Resources Research* **20**. UNESCO, Paris.

Willis, J.C. (1906). The Flora of Ritigala, an isolated mountain in the North-Central Province of Ceylon; a study in Endemism. *Ann. Roy. Bot. Gard. (Peradeniya)* **8**: 271–302.

9. PHYTOGEOGRAPHY, SYSTEMATICS AND DIVERSIFICATION OF AFRICAN MORACEAE COMPARED WITH THOSE OF OTHER TROPICAL AREAS

C.C. BERG

**Botanical Institute, University of Bergen
The Norwegian Arboretum, N-5067 Store Milde, Norway**

Abstract

The moraceous flora of Africa comprises a South American-African element well represented in the Guineo-Congolian subregion and an Asian-Australasian element better represented in the eastern part of the African region (in particular the Madagascan subregion) than in the western part. At the generic level the African moraceous flora equals the two other major phytogeographic regions, but in Africa the number of species (185) is less than in the Neotropics (c. 270) and in Asia-Australasia (c. 600); in particular terrestrial tree species are poorly represented in the African region. The most speciose taxonomic entities in Africa, *Dorstenia*, *Ficus* sect. *Galoglychia* and *Ficus* 'subsect.' *Sycomorus*, are distinctive for the African flora and can be related to second 'waves' of diversification after pantropical establishment of the family. These taxa are well represented both in humid forest and savanna woodland. The role of Moraceae in the composition and structure of lowland rain forest is less significant in Africa (and Asia-Australasia) than in the Neotropics. Features of the diaspores of most Moraceae appear not to favour long-distance dispersal, whereas in *Ficus* the pollination system limits remote establishment. Therefore, world distribution patterns have to be connected with coherence of land masses. The nature of pollination systems in the family allowed early radiation and dispersal of the family.

Introduction

Through a long process of accumulation of knowledge and designing of classifications of the family Moraceae (Trécul, 1847; Miquel, 1853, 1867; Baillon, 1875; Bentham & Hooker 1880; Eichler, 1875; Engler, 1889; Corner, 1962; Berg, 1989b) the present concept of its delimitation and subdivision appear to provide a satisfactory basis for the comparison of the moraceous floras of smaller and larger phytogeographic regions in relation to recognizable evolutionary trends. In the present study these aspects are focused on the African representatives of the family. Field experience of the present author in the various parts of the tropics, fair for the Neotropics, poor for Africa, and none for Asia-Australasia, may influence conclusions. For the present study it implies a somewhat unbalanced approach with regard to the African and neotropical Moraceae on one side and the Asian-Australasian on the other.

131

The two major phytogeographic elements of the Moraceous flora

For the comparison of floras of major phytogeographic regions not only the presence and absence of genera and families count, but even more important are their delimitations and subdivisions at the various supra-specific levels. Redefinition of genera has resulted in the only two amphi-atlantic moraceous genera listed by Thorne (1973), *Chlorophora* and *Trymatococcus*, being made obsolete.

For the African flora region, in the larger part of the continent and Madagascar (plus adjacent Indian Ocean islands) two phytogeographic elements can be recognized for Moraceae, as well as for most other families of the Urticales: (1) the South American-African (West Gondwanan) element and (2) the Asian-Australasian-African element also represented in the northwestern part of the Neotropics.

1. The South American-African element

This element shows more or less strong southern transatlantic taxonomic connections. The taxa involved are distinctly centred in South America: East Brazil, Amazonia, and the Guiana (or Guayana) region. The group also occurs in the Pacific lowlands of South America and extends from there into Central America. A secondary centre can be recognized in northern Central America (extending to the Greater Antilles) very probably established prior to the rise of the present Panama landbridge (as early as in the Late Cretaceous?; cf. Gentry, 1982). In Africa this element shows a concentration of taxa in the Guineo-Congolian subregion, in particular in Cameroun and Gabon.

The transatlantic connection is clearest in the tribe Dorstenieae. This tribe comprises 129 species, of which 105 belong to the genus *Dorstenia* (see Table 1). In contrast to many other genera with southern transatlantic connections, this genus is about equally represented on both sides of the Atlantic: 46 species occur in the Neotropics with a distinct concentration of species in east Brazil (as well as in the secondary Mesoamerican-Antillean centre referred to above). In the African flora region the genus is represented by 58 species and has one of its centres in the Cameroun-Gabon(-Congo) subregion and comprises the most primitive, including the woody, members of the genus. Another, more diffuse, centre is found in Eastern Africa, to which the only Asian species, *D. indica* (in West India and Sri Lanka) is linked. The transatlantic connection is so strong that the continents even share a section of the genus. In that section two groups of species can be distinguished, the neotropical *D. urceolata* group and the neotropical *D. turnerifolia* group, both represented by a single species in Africa: *D. picta* and *D. subdentata* respectively.

The tribe Dorstenieae comprises, in addition to *Dorstenia*, seven much smaller genera, four of them in Africa, namely *Bosqueiopsis*, *Scyphosyce*, *Trilepisum* and *Utsetela*; two of these genera are concentrated in the Cameroun-Gabon region, the others are more widespread. Three genera are American: *Brosimum*, *Helianthostylis*, and *Trymatococcus*. For this group of genera the northern Hylea region can be indicated as a somewhat diffuse centre.

Key for Table 1

( ) number of species; [ ] endemic genera; * incl. two North-American species (*Maclura pomifera* and *Morus rubra*); ○ the same species (*A. toxicaria*); # also in Asia (*F. exasperata* and *F. palmata*); % comprising sect. *Sycidium* (Corner, 1965); § also comprising sect. *Adenosperma*, sect. *Neomorphe* and sect. *Sycocarpus* (Corner, 1965); @ comprising sect. *Kalosyce* and sect. *Rhizocladus* (Corner, 1965); <- America: associated with NW Neotropics, connections with Asian-Australasian taxa; <- Africa: entirely or largely continental; <- Asia-Australasia: ± concentrated in the western part of the region; -> America: associated with the South-American continent; -> Africa: restricted to or predominantly in the Madagascan region; -> Asia-Australasia: ± concentrated in the eastern part of the region; A: tribe Moreae; B: tribe Artocarpeae; C: tribe Dorstenieae; D: tribe Castilleae; E: tribe Ficeae.

TABLE 1. Comparison of moraceous genera and species in the three major tropical phytogeographic regions.

America	Africa	Asia-Australasia	
Maclura (2)<-	*Maclura* (1)<>	*Maclura* (7)<-	
Morus (2)<-	*Morus* (1)<-	*Morus* (c. 8)<-	
Trophis (5)<-	*Trophis* (2)->	*Trophis* (2)<>	
	Bleekrodea (1)->	*Bleekrodea* (1)<-	
	Broussonetia (1)->	*Broussonetia* (7)<-	
	Fatoua (1)->	*Fatoua* (1)<>	
	Streblus (3)->	*Streblus* (20)<>	
	Milicia (2)<-		
A (11*)	**(12)**	**(46)**	**(69*)**
	Treculia (3)<-	*Artocarpus* (51)<-	
		Hullettia (2)<-	
		Parartocarpus (2)<-	
		Prainea (4)<-	
	(3)	**(59)**	**(62)**
Bagassa (1)->			
Batocarpus (3)->			
Clarisia (3)->			
Poulsenia (2)->			
Sorocea (14)->			
(23)			**(23)**
		Antiaropsis (2)->	
		Sparattosyce (1)->	
B (23)	**(3)**	**(3)**	**(88)**
Dorstenia (46)->	*Dorstenia* (58)<-	*Dorstenia* (1)<-	
	Bosqueiopsis (1)<-		
	Scyphosyce (2)<-		
	Trilepisium (1)<>		
	Utsetela (1)<		
Brosimum (15)->			
Helianthostylis (2)->			
Trymatococcus (2)->			
C (65)	**(63)**	**(1)**	**(129)**
	Antiaris (1)<>	*Antiaris* (1)<>¤	
	Mesogyne (1)<-		
Castilla (3)->			
Helicostylis (7)->			
Maquira (4)->			
Naucleopsis (22)->			
Perebea (9)->			
Pseudolmedia (9)->			
D (54)	**(2)**	**(1)**	**(57)**
	Galoglychia (71)<>	*Urostigma* (c. 80)<-	
	Urostigma (8)-	*Oreosycea* (c. 50)->	
	Oreosycea (4)->		
Americana (c. 100)<-			
Pharmacosycea (c. 20)<-		*Malvanthera* (22)->	
(c. 120)	**(83)**	**(c. 150)**	**(c. 350)**
	Ficus (1)->#	*Ficus* (c. 60)<-	
	Sycidium (9)->#	*Sycidium* (c. 90)->%	
	Sycomorus (12)->	*Sycomorus* (c. 120)->§	
		Synoecia (c. 80)<-@	
E (c. 120)	**(83+22)**	**(c. 150+350)**	**(c. 720)**
19 [14] gen. (c. 270)	**17 [7] gen. (185)**	**16 [6] gen. (c. 600)**	

The tribe Castilleae shows the same pattern, although less evidently than the Dorstenieae. The tribe comprises 54 species, of which all but two are neotropical, being members of six genera, and showing close affinity to the Amazon region. Two monotypic genera occur in Africa, the rare *Mesogyne* and the common *Antiaris*, the latter occurring throughout the African region as well as in Asia-Australasia, from India and Sri Lanka to Tonga Island.

The southern transatlantic connection is also found in the related family Cecropiaceae, with c. 165 species in three genera, *Cecropia* (c. 65 spp.), *Coussapoa* (47 spp.), and *Pourouma* (25 spp.) in the Neotropics (and there associated with the South American continent) and with two genera, *Musanga* (2 spp.) and *Myrianthus* (7 spp.), mainly in western and central tropical Africa. The sixth genus of the family, *Poikilospermum* (20 spp.), is Malesian, and taxonomically remotely related to the other five genera (Berg, 1988, 1989b).

2. The Asian-Australasian-African element

The taxa belonging to this element are centred in Asia, some largely mainland Asia, others West Malesian and others Australasian. This element shows connections with the northwestern part of tropical America: Central America, the northern Andean region and the Caribbean. In the African region this element is better represented in the Madagascan subregion than on the continent, and on the continent with a tendency to better representation in the eastern part than in the west.

This element is most clearly represented by the tribe Moreae, comprising eight genera and c. 70 species. All genera are represented in the African region, however, by only twelve species. The only endemic genus is *Milicia*, the others, *Bleekrodea*, *Broussonetia*, *Fatoua*, *Maclura*, *Morus*, *Streblus*, and *Trophis*, are represented by one, two or three species and have in most cases more species in Asia-Australasia, and in *Maclura*, *Morus*, and *Trophis*, even more in the northwestern Neotropics.

The tribe Artocarpeae has a remarkable composition. It lacks the homogeneity of the other tribes and comprises three distinct groups (to be regarded at least as subtribes): a group of five genera with in total 23 species, being associated with South American and showing superficial similarities to the Moreae, a group of two monotypic genera in Australasia showing superficial similarities to the Castilleae, and a group of five genera, of which four are largely associated with the Malesian region: *Artocarpus* and allied genera, *Hullettia*, *Parartocarpus*, and *Prainea*. The fifth genus, *Treculia*, is African. This group strengthens the Asian-Australasian element in the African flora, but as two of the species are confined to the Cameroun-Gabon subregion, it differs in distribution from the tribe Moreae.

The genus *Ficus* (the only genus of the tribe Ficeae) also supports the link between the Asian-Australasian and the African moraceous flora. This large genus of c. 720 species can be subdivided in two major groups, one comprising the subgenera *Urostigma* and *Pharmacosycea*, with together about 50% of the species. Both are pantropical and all species are monoecious. The other group, comprising (the 'subgenera') *Ficus*, *Synoecia* (incl. *Kalosyce* and *Rhizocladus*), *Sycidium* and *Sycomorus* (incl. *Adenosperma*, *Neomorphe* and *Sycocarpus*), is entirely palaeotropical and the majority of the species are (gyno)dioecious. In woody life forms and growth habits it is more diverse than the rest of the Moraceae (see below).

The subdivisions of *Ficus* used are not fully in accordance with the classification proposed by Corner (1965), but a looser and more practical application of the major subdivisions (Berg 1989a, 1990a). Corner's subgenera *Ficus* and *Sycomorus* are combined and four major groups ("subg.") distinguished in the largely (gyno)dioecious group of *Ficus*.

Subg. *Urostigma* (c. 280 spp.) is the largest subdivision of *Ficus*. The greater part of the species are hemi-epiphytic. Four subdivisions can be recognized: a neotropical one (sect. *Americana*) — a high number of them in Central America and the northern

Andes, a small and rather uniform and largely Australasian one (sect. *Malvanthera*), and a largely Asian one (sect. *Urostigma*), also rather uniform. The latter section extends into the African flora region with eight species: four on the continent and four in the Madagascan subregion. These species are more or less clearly linked to species from the western part of the section range and more or less adapted to relatively dry conditions.

The entirely African section *Galoglychia* is morphologically and ecologically the most diverse of the subgenus. Sixty-six species are continental; three extend to the Comoro Islands and Madagascar. Six species are endemic to the Madagascan subregion. Morphological features and relationships of pollinators (Wiebes, 1994) suggest that *Galoglychia* is most closely related to the Australasian(-Australian) section *Malvanthera*.

Subg. *Pharmacosycea* consists of the neotropical section *Pharmacosycea*, well represented in the northern Andean region and Central America, and the largely Australasian-Asian section *Oreosycea*, two of its species occurring in Madagascar and two on the African continent.

Considering the morphological differences, the sections of the subgenera *Urostigma* and *Pharmacosycea* could be treated as subgenera, and in comparisons within the family as genera, as indicated in Table 1.

One of the four "subgenera" of the largely (gyno)dioecious group of species, *Synoecia*, with species of root-climbers, is confined to the Asian-Australasian region, but the other three are represented in Africa: *Ficus* with one species, *F. palmata*, in NE Africa, *Sycidium* with five species in the Madagascan subregion and four species on the continent, and *Sycomorus* with eight species in the Madagascan subregion, one of them also occurring on the continent, where in total five species are found. These twelve African-Madagascan species are among the 20 belonging to one of the subdivisions of ('subg.') *Sycomorus*. All African-Madagascan species of this subdivision are monoecious, whereas the Asian-Australasian ones are (gyno)dioecious, except for one species, *F. racemosa*, which is related to the African *F. sur*. This group of monoecious species can be regarded as an African element. Monoecy is found in only two species of another subdivision of ('subg.') *Sycomorus*, both in New Guinea.

In general, the African *Ficus* flora largely links up with infrageneric entities more or less clearly centred in the easten part of the Asian-Australasian region. However, some, in particular elements associated with relatively dry areas in Africa, link with entities more or less clearly centred in the Asian mainland and West Malesia. These may represent a relatively recent immigration into Africa.

The families Urticaceae and Ulmaceae, in particular subfam. Celtidae, show more or less clearly the same phytogeographic patterns as the Asia-centred groups of the Moraceae.

Subdivisions of the African region

Table 2, based on Berg (1977c, 1978, 1982, 1986a, 1986b, 1988), Berg *et al.* (1984, 1985), Berg & van Heusden (1985) and Berg & Hijman (1989), gives a survey of floristic composition of three areas of Africa and Madagascar and indicates degrees of endemism.

The percentage of endemic moraceous species is relatively high in the Madagascan subregion. This subregion contains a mixture of Asian-Australasian and essentially continental African elements. The former category is represented by endemic species of the tribe Moreae, and some Asian-Australasian centred subdivisions of *Ficus*. The latter category is represented by some widespread species: *Antiaris toxicaria, Dorstenia cuspidata, Ficus lutea, F. polita, F. sycomorus, F. trichopoda, Treculia africana* and *Trilepisium madagascariense*. Only six species of *Ficus* section *Galoglychia* are endemic to the Madagascan subregion. Only 14 out of the 37 species of Moraceae of this subregion are associated with the African continent.

The Asian-Australasian element is less well represented on the continent, where *Ficus* sect. *Galoglychia* and *Dorstenia* are the dominant groups, both well-represented in forest and savanna woodland. If one regards *Ficus* sect. *Galoglychia* as typically African, only 23 out of the 160 continental species of Moraceae are associated with groups centred in Asia-Australasia.

The differences in the degree of endemism between eastern and central Africa are small. The percentages of endemic species would hardly be changed by including the whole Guineo-Congolian region and adjacent transitional zones, and by extending the Flora of Tropical East Africa area to northeastern and to southern Africa.

In East Africa the Asian-Australasian element is represented by a few species of some subdivisions of *Ficus,* mostly related to taxa of the westernmost part of the Asian-Australasian range of the genus and showing adaptations to relatively dry habitats, as *F. ingens, F. cordata, F. palmata* and *F. verruculosa,* by the subendemic *Maclura africana,* and by the widespread species *Milicia excelsa, Streblus usambarensis* and *Treculia africana.* In addition to the genus *Dorstenia,* the transatlantic element is represented by the disjunctly more or less continuously widespread *Bosqueiopsis gilletii, Mesogyne insignis* and *Trilepisium madagascariense.*

TABLE 2. Comparison of generic composition of the Moraceae of three areas in the African and Madagascan region.

	Ficus	*Dorstenia*	Other genera
Madagascan region	26(22)	1(0)	10(7)
East Africa (Flora Trop. E. Afr.)	50(10)	24(16)	9(0)
Cameroun (Flore de Cameroun)	59(9)*	23(17)*	10(3)
West Africa (Ivory Coast)	41(2)	5(3)	7(1)

( ) endemic; * endemic to the western part of the eastern section of the Guineo-Congolian region, thus including Gabon, SE Nigeria, Congo, part of Zaïre and part of the Central African Republic.

In the forested parts of Central Africa (the eastern part of the Guineo-Congolian region with adjacent transitional regions) the Asian-Australasian element is represented by species of *Treculia* and a few *Ficus* species, such as *F. asperifolia* and *F. variifolia.* The transatlantic element while more apparent than in the other continental subregions, is represented by only a few species, such as those of *Scyphosyce* and *Utsetela,* and by several shrubby species of *Dorstenia.*

The forested part of West Africa (the western part of the Guineo-Congolian region and the adjacent transitional zone) is largely an extension of the previous region but with fewer species, in particular those of the transatlantic element. The Asian-Australasian element of the previous region is extended by the endemic *Milicia regia.*

For a discussion about details of distribution patterns of African representatives of *Dorstenia* and *Ficus* see Berg & Hijman (in press) and Berg (1990b) respectively.

Numerical aspects

Table 3 shows that the African Flora region holds 17 genera (185 spp.), the Asian-Australasian 16 genera (c. 600 spp.) and the Neotropics 19 genera (c. 270 spp.) of Moraceae. The composition of the moraceous floras of these three main regions is more equal if one counts the major subdivisions of *Ficus* as genera, as shown in Table 3. The African moraceous flora is not less diverse than that of the other two main regions, either in the representation of genera and "subgenera", or, if one excludes the *Ficus-Synoecia-Sycidium-Sycomorus*-group, of diversity of the range of morphological and ecological features.

The majority of the 47 genera of the Moraceae are small: six are monotypic, eight have two species, five have three species, two have four species, six have 5–10 species, four have 10–20 species, three have 20–51, *Dorstenia* has 105 species, and *Ficus* c. 720 species with its major subdivsions between 15 and c. 130 species. The category of small entities, with 1–3 species, increases from 19 to about 40 if one also takes into account infrageneric taxonomic entities (subgenera and sections).

In the African region 15 (out of the 17) genera, including the endemic ones, are represented by one, two or three species. In the Neotropics 9 (out of 19) genera have more than three species and in Asia-Australasia 7 (out of 16).

The poor representation of the Asia-Australasia-centred genera in Africa could be referred to occurrence in the periphery of generic (or in *Ficus* subgeneric) ranges of distribution. In the remaining group of genera, eight comprise one to three species and only two, *Dorstenia* and *Ficus*, are speciose.

Taking into account the general tendency towards small genera (subgenera and sections), it could be a matter of chance that Africa contains relatively many small genera.

TABLE 3. Survey of numbers of genera, subdivisions of *Ficus*, species, and categories of species.

	America	Africa	Asia-Australasia
Genera	19	17	16
Genera + major subdivsions ("subgenera") of *Ficus*	20	22	22
Species	270	185	600
Species, excl. *Ficus* (*Ficus-Synoecia-Sycidium-Sycomorus*-group	270	165	265
Species, excluding *Ficus*	150	80	110
Species, excl. *Ficus* and herbs	105	35	110
Species, terrestrial tree species only, excluding *Ficus*	90	15	75
Species, only macrospermous and non-anemophilous trees	80	6*	60

Except for those indicated with *, all numbers for species are approximate.

It is clear that the African moraceous flora is less speciose than the other two floras, even more so if only terrestrial forest tree species are taken into account.

Speciation in Africa

A contrast in species richness of forest trees between Africa and the Neotropics is also found in the related tree family of Cecropiaceae. The African pioneer tree genus *Musanga* has only two species, one lowland and the other montane, whereas its neotropical ecological equivalent of similar habit, *Cecropia*, comprises about 65 species, c. 60% lowland and c. 40% (sub)montane. The African genus *Myrianthus* comprises seven species, but its neotropical ecological equivalent *Pourouma*, resembling the former in habit and leaf variation, has 25 species. Even if one excludes speciation related to the Andean region, the difference in species numbers and, in *Cecropia*, the ecological diversification in the neotropical lowlands, the differences in the numbers of species are striking. Judging from infrageneric morphological continuities one could conclude that these differences in numbers can be ascribed to speciation rate rather than to extinction.

In view of the number and diversity of species in *Dorstenia* and *Ficus* sect. *Galoglychia* and 'subg.' *Sycomorus* 'sect.' *Sycomorus* (discussed below), a low rate of speciation and radiation is not a general aspect of the African moraceous flora.

Speciation in these groups is to a large extent associated with rain forest conditions, but these units have radiated into drier types of vegetation such as savanna woodland and, in *Dorstenia* and *Galoglychia*, to a limited extent even into arid zones; this development has contributed considerably to the diversity of the family in Africa. Adaptations to dry types of vegetation distinguish these entities from their neotropical and Asian-Australasian counterparts.

However, the fact remains that the number of species of most genera or their major subdivisions are small in Africa. If that can be referred to the effects of occurrence at the periphery of present (and ancient?) ranges of distribution it seems to imply that most terrestrial tree species are of considerable age. Distribution patterns in and outside Africa appear to support this supposition. In contrast to herbacous and (suf)frutescent taxa the ranges of terrestrial lowland tree species tend to be large, continuous or disjunct, and appear not to be susceptible to extinction.

It is often stated (e.g. Richards, 1973) that the relatively species-poor state of the tropical African flora can be related to the effects of climatic changes. This species-poor state is only evident for the continental African (non-*Ficus*) tree genera, most of them endemic. This might be the result of an accumulation of factors other than drastic climatic changes, such as chance, rate of speciation, occurrence at the periphery of ranges of distribution, ecological substitution by other taxa, and forest structure.

The morphological diversity of the rain forest component of *Dorstenia* is greater in Africa than that in the Neotropics. The same applies to *Galoglychia* in comparison with the other sections of subgenus *Urostigma* and to the African segment of 'sect.' *Sycomorus* in comparison with the Asian-Australasian segment.

Habit and habitat occupation

If one excludes *Ficus* and herbaceous and succulent life forms, the remainder of the Moraceae comprises c. 240 species of terrestrial trees, shrubs and some climbers and subshrubs: in Africa 32 species, in both the Neotropics and Asia-Australasia a few more than 100, and the North American continent two species. The majority are components of lowland humid tropical forest as canopy trees, undergrowth treelets or (sub)shrubs or as emergents, most of them sclerophyllous. Many of the members of the anemophilous Moreae are typically forest margin components of deciduous or semi-deciduous forest and mesophyllous. *Morus* and *Broussonetia* have features such as shedding of shoot apices which associate them with (warm-)temperate conditions. A few species of the (non-*Ficus*) group of lowland tropical genera have become (sub)montane.

The essentially herbaceous life forms are confined to 94 species of *Dorstenia*, and to *Fatoua*. Two of them occur in Asia-Australasia and the others in about equal numbers in Africa and the Neotropics, being most diverse in Africa.

In *Ficus* about 300 species (subg. *Urostigma* and one of the sections of 'subg.' *Sycidium*) are basically hemi-epiphytic, although some facultatively and some others being largely hemi-epilithic. About 80 species are root-climbers, all of them confined to Asia-Australasia. The remainder (c. 340 spp.) are various types of trees and shrubs, being light-demanding for germination, normally establishing in disturbed places, thus often in riverine forest and secondary forest. Several of them can be long-lived forest trees, such as *F. mucuso* and *F. variifolia* in Africa.

The few African moraceous forest tree species, such as *Antiaris toxicaria, Ficus mucuso, F. variifolia, Milicia excelsa, M. regia, Morus mesozygia, Treculia africana* and *Trilepisium madagascariense*, are listed as important or even (co-)dominant components of some types of forest, in particular those marginal to humid lowland rain forest: peripheral semi-evergreen lowland rain forest, forest transitional to woodland and secondary forest (Beentje, 1990; Gentry, 1988; Lubini, 1990; Richards, 1952; White, 1983). For some of the species the association with these more or less open types of forest can be explained by microspermy, and the need of light for germination and/or by anemophily. Only the three macrospermous (non-anemophilous) species, *Antiaris toxicaria, Treculia africana* and *Trilepisium madagascariense* can be reckoned as potential components of mature evergreen forest. Even if other macrospermous and non-anemophilous, rarer or less widespread African tree species (*Bosqueiopsis gilletii, Mesogyne insignis* and *Treculia obovoidea*) are added, the six species in this category is a surprisingly small number, in comparison with the Asian-Australasian region with c. 60 of such species, and the Neotropics with c. 80. This seems to be the most profound difference between the moraceous floras of the three main tropical regions.

Not only the small number of forest tree species but also their role in forest ecosystems distinguishes the African and neotropical moraceous terrestrial tree floras.

In the Neotropics Moraceae often rank among the top three families with regard to species representation in various types of lowland rain forest (Balslev & Renner, 1989; Boom, 1986; Gentry, 1988). In censuses, Cecropiaceae are often included in Moraceae, but even if one subtracts Cecropiaceae, the Moraceae retain their position. Representation of Moraceae in such forests is often not just by single specimens but often by numerous individuals (Boom, 1986), so that the family plays an important role in the structure of these forests. The number of lowland rain forest species involved is about 90, many of them as small to medium-sized trees, with only a few species (of *Brosimum, Ficus* and *Maquira*) as emergents, and the majority macrospermous and non-anemophilous. However, Moraceae can be rare in lowland rain forest, apparently in relation to soil conditions. They are then often replaced by Annonaceae, or less markedly by Myristicaceae, both showing similarities in tree habit and/or architecture (the model of Roux and the closely related model of Cook, cf. Berg 1977a; Hallé *et al.*, 1978). The less important role of at least the Moraceae, other than *Ficus,* in the composition and construction of lowland rain forest in the Old World might be caused by substitution by members of other families, such as Dipterocarpaceae in the Asian-Malesian region, where non-*Ficus* moraceous tree species seem to play a minor role in composition and structure of rain forest.

Moraceae are also scarce in neotropical lowland rain forest dominated by emergents. The fact that moraceous forest tree species are rare in African forest dominated by *Parinari excelsa* (White, 1983: 178) seems to be in accordance with experience in the Neotropics and could indicate that Moraceae can be largely excluded by the structure of the forest.

Both the assumption of substitution and of exclusion appears to be supported by data compiled by Gentry (1988, 1993) in his study of floristic composition in several parts of the world.

Differentiation of floral structures and pollination

The flowers of Moraceae are unisexual, small and basically 4-merous with a simple perianth. Reductions are frequent and are related to increased complexity of inflorescences (cf. Berg, 1977b, 1989b). Maintenance of the basic floral structure of staminate flowers appears to be essential for the mode of anemophily as occurring in the Moreae, as it is in Urticaceae and some Ulmaceae.

The diversity of inflorescences in Moraceae can hardly be matched by that in other angiosperm families (cf. Berg, 1977b, 1989b; Corner, 1962). Inflorescences vary from simple in structure, mostly with single flowers as units of pollination (anemophily) to condensed and more or less complex with the inflorescences as units of pollination (entomophily, geitonogamous autogamy). Bisexual inflorescences (in Brosimeae and Ficeae) tend to be more complex than unisexual ones. Much of the differentiation of flowers and inflorescences can be explained by protection of developing stamens until anthesis, and developing ovules and ovaries, until maturity of the seeds.

As differentiation of flowers and inflorescences appears to be strongly related to function, such as pollination and protection, as well as to the sexuality of inflorescences, it appears not to be justified to relate in general degrees of complexity and floral reduction to the age of taxa. Secondary simplification of inflorescences has also occurred.

Three modes of pollination can be distinguished: anemophily, entomophily based on breeding insects, and autochory. In most anemophilous taxa (belonging to the tribe Moreae) the flowers have the urticaceous construction of the staminate flower: complete perianth, stamens inflexed in the bud and a distinct pistillode. The pollen is released explosively by sudden opening of the flower and sudden and elastic bending of the stamens, throwing pollen out in the air. This method of pollen release can be associated with forest habitat, allowing herbs, shrubs and treelets occurring along streamlets to disperse the pollen by weak air movement caused by streaming water. Most species with this type of pollination are, however, forest margin plants or components of semi-decidous forest. This mode of pollination is the prevailing one in the Moreae. A more common mode of anemophily may occur in a few other species of Moraceae (such as in some species of *Artocarpus*, cf. Jarrett, 1959, and *Brosimum alicastrum*, cf. Peters, 1991).

More common is entomophily (documented or presumed) based on breeding by insects. In *Ficus* there is a unique system in which fig wasps (Hymenoptera) are involved. In other genera with condensed unisexual inflorescences pollination is probably carried out in the same way as found in *Artocarpus* species (van der Pijl, 1953): the staminate inflorescences are used as breeding sites for larvae of beetles and/or flies and the pistillate ones protected against penetration by these larvae (cf. Berg, 1990c). As many pollination studies revealed, e.g. in palms (cf. Henderson, 1986), pollination based on breeding as a primary attraction, often in combination with the offer of pollen, thermogenesis and/or mating conditions, is common for components of evergreen tropical forest which are neither adapted to wind pollination nor to Post-Cretaceous pollinators. It is likely that this mode of pollination was in operation in the early history of angiosperms as it could have been inherited from gymnosperms.

Autogamy might be prevailing in many representatives of the Dorstenieae with bisexual inflorescences. There are indications (as seed-production in pistillate inflorescences of *D. cayapia* under greenhouse conditions) that agamospermy occurs in *Dorstenia*. In some species of this genus vegetative reproduction by small tubers is found.

These types of pollination could not have been limiting factors in range expansion for most Moraceae. However, the species-specific pollination system of *Ficus* seems to be a limiting factor in this respect.

Differentiation of fruits and dispersal

The fruits of Moraceae are drupe(let)s or achenes (enclosed in fleshy structures). Fruits are often parts of compound fleshy structures. Dehiscent drupe(lets) squeezing out or ejection endocarp bodies (autochory) are peculiar to the Moraceae. The majority of the genera are macrospermous, the seeds remaining viable for only a short time and are not light-demanding with regard to germination. *Dorstenia* comprises both macrospermous and microspermous species, but small seeds seem to behave like large ones in relation to longevity. *Ficus* and some genera of the Moreae are microspermous, with seeds generally remaining viable for a long period, but mostly requiring light for germination.

Cauliflory occurs in some Old World genera (*Artocarpus, Ficus, Treculia*) but is absent from the Neotropics, where ramiflory is only found in some species of *Clarisa* and *Ficus*. Dehiscent drupe(let)s, often in combination with autochory, are absent from the Neotropics but present in most tribes in the Old World (Artocarpeae, Dorstenieae, Ficeae, and Moreae).

In the majority of the Moraceae dispersal is carried out by animals, in most cases by arboreal ones and mostly endozoochorous. In all these cases the seeds are enclosed in fleshy structures (pericarp, perianth or receptacle). Dry diaspores are the endocarp bodies extruded or ejected from dehiscent drupe(lets). They drop on the soil, or in water which carries them further.

The size and longevity of seeds and/or the modes of dispersal means that long-distance dispersal can not (or hardly) be taken into consideration to explain disjunct distributions. Long-distance dispersal has much a better chance in *Ficus*, but in this genus the chance of establishment of the pollination system is small.

Historical aspects

The present distribution (centres and transatlantic links), morphological and ecological differentiation patterns, and the presumed lack of possibilities for long-distance dispersal strongly suggest that the genus *Dorstenia* was in existence and was diversified to some extent before the separation of South America from Africa. The other genera of Moraceae may also have arrived at such a state in that epoch and the family as a whole became pantropical. The assumption that modern genera already existed in the Cretaceous seems to be supported by data provided by Romero (1993) for South American palaeofloras and in line with findings, such as in Hamamelidaceae (Endress, 1989), but it is rejected by authors such as Smith (1973) and Thorne (1978). Anyhow, data provided by Muller (1981) indicate an early rise and diversification in groups with magnolioid flowers and with amentiferous floral features.

As stated above, patterns of distribution (centres, ranges and disjunctions) in combination with infraspecific variation in several moraceous tree species suggest considerable age of several of them and "conservatism" with regard to morphological differentiation.

The implication of the existence of small, simple, and often reduced flowers in condensed inflorescences in the early history of the Moraceae corresponds with detection of small flowers and condensed inflorescences in Lower Cretaceous fossils (Crane *et al.*, 1986). These features of flowers and inflorescences match the (probably ancient) mode of pollination based on breeding insects.

The commonness of macrospermy in Moraceae and consequently the lack of possibilities for long-distance dispersal appear not to comply with the scenario described by Tiffney (1984) based on fossil diaspores, indicating prevailing microspermy in the early history of the angiosperms and hence a pioneer behaviour of tree species. One may wonder to what extent this scenario applies to tropical taxa. The general tendency in Moraceae is rather a moving out of rain forest conditions than into.

Judging from the present patterns of distribution and morphological similarities and dissimilarities one gets the impression that two 'waves' of diversification (evolution) have taken place in Moraceae. The first one resulted in the establishment of the present genera and major subdivisions of *Ficus*. It appears that this occurred before the segregation of the southern (Gondwana) continent where the moraceous flora consisted of a western element and an eastern element, the latter connected with the southwestern part of the present North American continent through Laurasia (cf. Krutzsch, 1989); as land connections at a suitable latitude and time allowing tropical taxa to migrate from Asia to America have not been demonstrated. At least three of the tribes (Artocarpeae, Castilleae and Dorstenieae) were established in the present major tropical phytogeographic regions, and there as components of humid tropical lowland forest. A few entities (*Broussonetia* and *Morus*) established themselves in a northern warm-temperate environment without, however, losing the possibility of re-establishing in the tropics, as demonstrated by the two African species of these genera.

In the majority of the taxonomic entities further development seems to have been largely an increase of the number of species without profound morphological and ecological differentiation and providing rather uniform, mostly small genera. Speciation has clearly been more profuse in the hemi-epiphytic subdivision of the monoecious group of *Ficus* than in the terrestrial monoecious groups, including sect. *Oreosycea* and sect. *Pharmacosycea*, possibly due to the difference in life form.

Secondary differentiation

The second wave of diversification (evolution) has apparently occurred in some subdivisions of *Ficus* and in *Dorstenia* carrying these taxa beyond the primary morphological and ecological set of traits. The products of these developments are largely confined to each of the major phytogeographical subdivisions.

Subsequent evolution has occurred most dramatically in the present *Ficus-Synoecia-Sycidium-Sycomorus* group resulting in a tremendous diversity of habits, from huge trees to tiny shrubs with different types of architecture and life forms (root-climbers, rheophytes, hemiepiphytes, holoepiphytes), with inflorescences which are axillary, ramiflorous, cauliflorous, trunciflorous or flagelliflorous, with a considerable diversity in the inflorescences, flowers and fruits, and occupying a wide range of habitats. Some of this diversification can be observed in Africa, namely in the group of monoecious species of "subsect." *Sycomorus*. The differentiation of this group has apparently taken place largely in the African region. This group comprises huge trees (*F. mucuso*) and small ones (e.g. *F. vallis-choudae*), has inflorescences axillary, ramiflorous, tronciflorous or flagelliflorous (geocarpic), and shows a wide variation in leaf characters. The species occur in conditions ranging from humid through transitional to relatively dry conditions.

Galoglychia is another group of *Ficus* which has differentiated considerably. It remained confined to the African floral region and is more diverse than any of the other sections of *Urostigma*, in particular in characters of the inflorescences (figs), such as size and position on the tree, and those of the fruitlets. The degree of morphological and ecological differentiation is reflected in the seven pollinator fig wasp genera (partly associated with subsections of *Galoglychia*, cf. Wiebes in Berg & Wiebes, 1992), whereas only one or five genera are associated with the other sections (Wiebes, 1994). In origin the group is associated with humid types of forest but has radiated into drier types of vegetation as savanna woodland and even into arid regions. About 50% of the species are components of humid forest, about 20% occur (often with distinct subspecies) in drier types of forest, and about 30% are more distinctly associated with drier types of vegetation, such as savanna woodland, and some occur in arid habitats.

The third group is the genus *Dorstenia*. For this genus nine sections can be recognized, six of them are African and one is shared with the Neotropics. Among the African sections, two (with in total 11 species) are woody and comprise the most primitive species of the genus. About $^2/_3$ of the species are inhabitants of humid forest and most of them occur in the Guineo-Congolian region. The others are more or less clearly associated with drier types of vegetation, several of them are components of savanna woodland and some occur in arid habitats. This group is largely East African. The genus shows considerable diversity in life and growth forms; the diversity is broader in Africa than in the Neotropics. In Africa it ranges from treelets (*D. oligogyna*) to shrubs, subshrubs to herbs. In the large group of herbaceous species the spectrum ranges from tall to small, caulescent to acaulescent, phanerophytic to geophytic, mesomorphic to succulent (stem, tuber and/or leaf succulence), from perennials to annuals, and from terrestrial to epiphytic. In the Neotropics two sections can be recognized in addition to the one shared with Africa. In these two sections most of the differentiation took place, but it did not result in the diversity found in Africa and only to a limited extent in radiation into drier types of habitat.

References

Baillon, H. (1875 – 1876). Histoire des plantes, Vol. 6. Libairie Hachette and Cie., Paris.

Balslev, H. and Renner, S.S. (1989). Diversity of Ecuadorian forests. In: L.B. Holm-Nielsen, I.C. Nielsen and H. Balslev (editors). Tropical forests — Biological dynamics, speciation and diversity. Pp. 287–295. Academic Press, London, San Diego, New York, Berkeley, Boston, Sydney, Tokyo, Toronto.

Beentje, H.J. (1990). The forests of Kenya. *Mitt. Inst. Allg. Bot. Hamburg* **23a**: 265–286.

Bentham, G. and Hooker, J.D. (1880). Genera Plantarum, Vol. 3(1). L. Reeve and Co., London.

Berg, C.C. (1977a). The Castilleae, a tribe of the Moraceae, renamed and redefined due to the exclusion of the type genus *Olmedia* from the 'Olmedieae'. *Acta Bot. Neerl.* **26**: 73–82.

Berg, C.C. (1977b). Urticales, their differentiation and systematic position. *Pl. Syst. Evol.*, Suppl. **1**: 349–74.

Berg, C.C. (1977c). Revisions of African Moraceae (excluding *Dorstenia, Ficus, Musanga,* and *Myrianthus. Bull. Jard. Bot. Belg.* **47**: 267–407.

Berg, C.C. (1978). Revisions of African Moraceae (excl. *Dorstenia, Ficus, Musanga* and *Myrianthus*); corrections and additions. *Bull. Jard. Bot. Belg.* **48**: 466–468.

Berg, C.C. (1982). The reinstatement of the genus *Milicia* Sim (Moraceae). *Bull. Jard. Bot. Belg.* **52**: 225–229.

Berg, C.C. (1986a). The *Ficus* species (Moraceae) of Madagascar and the Comoro Islands. *Bull. Mus. Natl. Hist. Nat., B, Adansonia,* série 4, **8**: 17–55.

Berg, C.C. (1986b). The delimitation and subdivision of the genus *Maclura* (Moraceae). *Proc. Kon. Ned. Akad. Wetensch. C* **89**: 241–246.

Berg, C.C. (1988). The genera *Trophis* and *Streblus* (Moraceae) remodelled. *Proc. Kon. Ned. Akad. Wetensch. C* **91**: 345–356.

Berg, C.C. (1989a). Classification and distribution of *Ficus*. In: The comparative biology of figs. *Experientia (Basel)* **45**: 605–611.

Berg, C.C. (1989b). Systematics of Urticales. In: P.R. Crane and S. Blackmore (editors). Evolution, systematics, and fossil history of the Hamamelidae. 2, 'Higher' Hamamelidae. Pp. 193–220. Clarendon Press, Oxford.

Berg, C.C. (1990a). Reproduction and evolution in *Ficus* (Moraceae): traits connected with the adequate rearing of pollinators. *Mem. New York Bot. Gard.* **55**: 169–185.

Berg, C.C. (1990b). Distribution of African taxa of *Ficus*. *Mitt. Inst. Allg. Bot. Hamburg* **23a**: 401–405.

Berg, C.C. (1990c). Differentiation of flowers and inflorescences of Urticales in relation to their protection against breeding insects and to pollination. *Sommerfeltia* **11**: 13–34.

Berg, C.C. and Hijman, M.E.E. (1989). Moraceae. In: R.M. Polhill (editor). Flora of Tropical East Africa. A.A. Balkema, Rotterdam, Brookfield.

Berg, C.C. and Hijman, M.E.E. (in press). The genus *Dorstenia* (Moraceae). *Ilicifolia* **2**. Botanical Institute, University of Bergen, Bergen.

Berg, C.C., Hijman, M.E.E. and Weerdenburg, J.C.A. (1984). Flore du Gabon 26, Moracées. Muséum national d'histoire naturelle, Laboratoire de phanérogamie, Paris.

Berg, C.C., Hijman, M.E.E. and Weerdenburg, J.C.A. (1985). Flore du Cameroun 28, Moracées (incl. Cécropiacées). Ministere de l'enseignement superieure et de la recherche scientifique (MESRES), Yaoundé.

Berg, C.C. and van Heusden, E.C.H. (1985). Flore des Mascareignes, Fam. 164, Moracées. L. Carl Achille, Port Louis.

Berg, C.C. and Wiebes, J.T.(1992). Africa fig trees and fig wasps. *Verh. Kon. Ned. Akad. Wetensch., Afd. Natuurk.*, Tweede Sect. **89**: 1–298.

Boom, B.M. (1986). A forest inventory in Amazonian Bolivia. *Biotropica* **18**: 287–294.

Corner, E.J.H. (1962). The classification of Moraceae. *Gard. Bull. Singapore* **19**: 187–252.

Corner, E.J.H. (1965). Check-list of *Ficus* in Asia and Australasia with key to identification. *Gard. Bull. Singapore* **21**: 1–186.

Crane, P.R., Friis, E.M and Pedersen, K.P. (1986). Lower Cretaceous angiosperm flowers: fossil evidence and early radiation of dicotyledons. *Science* **232**: 852–854.

Eichler, A.W. (1875). Blüthendiagramme. Verlag von Wilhelm Engelmann, Leipzig.

Endress, P.K. (1989). Aspects of evolutionary differentiation of Hamamelidaceae and the Lower Hamamelidae. In: F. Ehrendorfer (editor). Woody plants — evolution and distributions since the Tertiary. *Pl. Syst. Evol.* **162**: 193–211.

Engler, G.H.A. (1889). Moraceae. In: G.H.A. Engler and K.A.E. Prantl. Die natürlichen Pflanzenfamilien, Vol. 3(1): 66–98. Verlag von Wilhelm Engelmann, Leipzig.

Gentry, A.H. (1982). Neotropical floristic diversity: phytogeographical connections between Central and South American Pleistocene climatic fluctuations, or an accident of the Andean orogeny? *Ann. Missouri Bot. Gard.* **69**: 557–593.

Gentry, A.H. (1988). Changes in plant community diversity and floristic composition on environmental and geographical gradients. *Ann. Missouri Bot. Gard.* **75**: 1–34.

Gentry, A.H. (1993). Diversity and floristic composition of lowland tropical forest in Africa and South America. In: P. Goldblatt (editor). Biological relationships between Africa and South America. Pp. 500–547. Yale University Press, New Haven, London.

Hallé, F., Oldeman, R.A.A. and Tomlinson, P.B. (1978). Tropical trees and forests — An architectural analysis. Springer-Verlag, Berlin, Heidelberg, New York.

Henderson, A. (1986). A review of pollination studies in the Palmae. *Bot. Rev. (Lancaster)* **52(3):** 221–259.

Jarrett, F.M. (1959–1960). Studies in *Artocarpus* and allied genera, I–V. *J. Arnold Arbor.* **40**: 1–37; 113–155; 298–368; **41**: 73–109; 111–140; 320–340.

Jongkind, C.C.H. (1995). Novitates gabonensis (24). A new species of *Utsetela* (Moraceae). *Bull. Jard. Bot. État.* **64**: 179–181.

Knapp, R. (1973). Die Vegetation von Afrika. Gustav Fischer Verlag, Stuttgart.

Krutzsch, W. (1989). Paleogeography and historical phytogeography (paleochorology) in the Neophyticum. In: F. Ehrendorfer (editor). Woody plants — evolution and distributions since the Teriary. *Pl. Syst. Evol.* **162**: 5–61.

Lubini, A. (1990). La flore de la Reserve Forestier de Luki (Bas-Zaire). *Mitt. Inst. Allg. Bot. Hamburg* **23a**: 135–154.

Miquel, F.A.W. (1853). Urticales. In: C.F.P. von Martius. Flora Brasiliensis 4 (1): 78–218. Leipzig.

Miquel, F.A.W. (1867). Annotations de *Ficus* specibus. *Ann. Mus. Bot. Lugduno-Batavum* **3**: 260–300.

Muller, J. (1981). Fossil pollen records of extant angiosperms. *Bot. Rev. (Lancaster)* **47(1)**: 1–142.

Peters, C.M. (1991). Reproduction, growth and the population dynamics of *Brosimum alicastrum* Sw. in a moist tropical forest in Central Veracruz, Mexico. Dissertation 1989. U.M.I., Ann Arbor.

Pijl, L. van der (1953). On the flower biology of some plants from Java — with general remarks on fly-traps (species of *Annona, Artocarpus, Typhonium, Gnetum, Arisaema* and *Abroma*). *Ann. Bogor.* **1**: 77–99.

Richards, P.W. (1952). The Tropical Rain Forest. Cambridge University Press, Cambridge.

Richards, P.W. (1973). Africa, the "Odd Man Out". In: B.J. Meggers, E.S. Ayensu and W.D. Duckworth (editors). Tropical forest ecosystems in Africa and South America — A comparative review. Pp. 21–26. Smithsonian Institution Press, Washington.

Romero, E.J. (1993). South American paleofloras. In: P. Goldblatt (editor). Biological relationships between Africa and South America. Pp. 62–85. Yale University Press, New Haven, London.

Smith, A.C. (1973). Angiosperm evolution and the relationship of floras of Africa and America. In: B.J. Meggers, E.S. Ayensu and W.D. Duckworth (editors). Tropical forest ecosystems in Africa and South America — A comparative review. Pp. 49–61. Smithsonian Institution Press, Washington.

Thorne, R.F. (1973). Floristic relationships between tropical Africa and tropical America. In: B.J. Meggers, E.S. Ayensu and W.D. Duckworth (editors). Tropical forest ecosystems in Africa and South America — A comparative review. Pp. 27–47. Smithsonian Institution Press, Washington.

Thorne, R.F. (1978). Plate tectonics and angiosperm distribution. *Notes Roy. Bot. Gard. Edinburgh* **36**: 297–315.

Tiffney, B.H. (1984). Seed size, dispersal syndromes, and the rise of angiosperms: evidence and hypothesis. *Ann. Missouri Bot. Gard.* **71**: 551–576.

Trécul, A. (1847). Sur la famille des Artocarpées. *Ann. Sci. Nat. Bot.*, Sér. 3, **8**: 38–157.

White, F. (1983). The Vegetation of Africa: a descriptive memoir to accompany the UNESCO/AETFAT/UNSO vegetation map of Africa. *Natural Resources Research* **20**. UNESCO, Paris.

Wiebes, J.T. (1994). The Indo-Australian Agaoninae (pollinators of figs). *Verh. Kon. Ned. Akad. Wetensch., Afd. Natuurk.*, Tweede Sect. **92**: 1–208.

APPENDIX: List of taxa of Moraceae from the African flora region.

With total number of species for each genus and references to descriptions and applied names for species, subspecies and varieties.

1. ***Antiaris*** Lesch. (1 sp.) (Berg, 1977c; 1978)
   1. *A. toxicaria* Lesch.
      a. subsp. *welwitschii* (Engl.) C.C. Berg
         var. *welwistchii*
         var. *africana* A. Chev.
         var. *usambarensis* (Engl.) C.C. Berg
      b. subsp. *madagascariensis* (H. Perrier) C.C. Berg
      c. susbp. *humbertii* (Léandri) C.C. Berg

2. ***Bleekrodea*** Blume (2 spp.) (Berg, 1977c)
   1. *B. madagascariensis* Blume

3. ***Bosqueiopsis*** De Wild. (1 sp.) (Berg, 1977c)
   1. *B. gilletii* De Wild.

4. ***Broussonetia*** Vent. (8 spp.) (Berg, 1977c)
   1. *B. greveana* (Baill.) C.C. Berg

5. ***Dorstenia*** L. (105 spp.) (Berg & Hijman, in press)
   1. *D. djettii* Guillaumet
   2. *D. oligogyna* (Pellegr.) C.C. Berg
   3. *D. elliptica* Bureau
   4. *D. africana* (Baill.) C.C. Berg
   5. *D. kameruniana* Engl.
   6. *D. dorstenioides* (Engl.) Hijman & C.C. Berg
   7. *D. involuta* Hijman & C.C. Berg
   8. *D. turbinata* Engl.
   9. *D. angusticornis* Engl.
   10. *D. alta* Engl.
   11. *D. scaphigera* Bureau
   12. *D. subdentata* Hijman & C.C. Berg
   13. *D. picta* Bureau
   14. *D. mannii* Hook. f.
       a. var. *mannii*
       b. var. *mougasii* Hijman
       c. var. *alternans* (Engl.) Hijman
       d. var. *mungensis* (Engl.) Hijman
       e. var. *stipulata* (Rendle) Hijman
       f. var. *humilis* (Hijman & C.C. Berg) Hijman
   15. *D. ciliata* Engl.
   16. *D. lujae* De Wild.
       a. var. *lujae*
       b. var. *batesii* (Rendle) Hijman
   17. *D. tenera* Bureau
       a. var. *tenera*
       b. var. *obtusibracteata* (Engl.) Hijman & C.C. Berg
   18. *D. zenkeri* Engl.
   19. *D. barteri* Bureau
       a. var. *barteri*
       b. var. *multiradiata* (Engl.) Hijman & C.C. Berg
       c. var. *subtriangularis* (Engl.) Hijman & C.C. Berg
       d. var. *paucinervis* Hijman & C.C. Berg
   20. *D. yambuyaensis* De Wild.
   21. *D. poinsettiifolia* Engl.
       a. var. *poinsettiifolia*
       b. var. *longicaudata* (Engl.) Hijman & C.C. Berg
       c. var. *angusta* (Engl.) Hijman & C.C. Berg
       d. var. *librevillensis* (De Wild.) Hijman & C.C. Berg
       e. var. *angularis* Hijman & C.C. Berg
       f. var. *staudtii* (Engl.) Hijman & C.C. Berg
       g. var. *glabrescens* Hijman & C.C. Berg
   22. *D. dinklagei* Engl.
       a. var. *dinklagei*
       b. var. *reducta* (De Wild.) Hijman
       c. var. *brieyi* (De Wild.) Hijman
       d. var. *binzaensis* (De Wild.) Hijman
       e. var. *bequaertii* (De Wild.) Hijman
   23. *D. letestui* Pellegr.
   24. *D. prorepens* Engl.
   25. *D. convexa* De Wild.
       a. var. *convexa*
       b. var. *oblonga* (Compère) Hijman
   26. *D. nyungwensis* Troupin
   27. *D. tayloriana* Rendle
       a. var. *tayloriana*
       b. var. *laikipiensis* (Rendle) Hijman
   28. *D. variifolia* Engl.
   29. *D. dionga* Engl.
   30. *D. ulugurensis* Engl.
   31. *D. thikaensis* Hijman
   32. *D. bicaudata* Peter
   33. *D. schliebenii* Mildbr.
   34. *D. holstii* Engl.
       a. var. *holstii*
       b. var. *longestipulata* Hijman
   35. *D. brownii* Rendle
   36. *D. soerensenii* Friis
   37. *D. afromontana* R.E. Fr.
   38. *D. psilurus* Welw.
       a. var. *psilurus*
       b. var. *scabra* Bureau
   39. *D. embergeri* G. Mang.
   40. *D. goetzei* Engl.
   41. *D. annua* Friis & Vollesen
   42. *D. zanzibarica* Oliv.
   43. *D. warneckei* Engl.
   44. *D. tenuiradiata* Mildbr.
   45. *D. bergiana* Hijman
   46. *D. hildebrandtii* Engl.
       a. var. *hildebrandtii*
       b. var. *schlechteri* (Engl.) Hijman
   47. *D. cuspidata* A. Rich.

a. var. *cuspidata*
b. var. *preussii* (Engl.) Hijman
c. var. *brinkmaniana* Hijman
d. var. *humblotiana* (Baill.) Léandri
48. *D. benguellensis* Welw.
49. *D. buchananii* Engl.
    a. var. *buchananii*
    b. var. *longepedunculata* Rendle
50. *D. vivipara* Welw.
51. *D. foetida* (Forssk.) Schweinf.
52. *D. gypsophila* Lavranos
53. *D. gigas* Balf. f.
54. *D. astyanactis* Aké Assi
55. *D. zambesiaca* Hijman
56. *D. barnimiana* Schweinf.
    a. var. *barnimiana*
    b. *tropaeolifolia* (Schweinf.) Rendle
57. *D. ellenbeckiana* Engl.
58. *D. socotrana* A.G. Mill.

6. *Fatoua* Gaudich. (2 spp.) (Berg, 1977c)
1. *F. madagascariensis* Léandri

7. *Ficus* L. (approx. 720 spp.) (Berg & Wiebes, 1992)
1. *F. palmata* Forssk.
2. *F. exasperata* Vahl
3. *F. pachyclada* Baker
    a. subsp. *pachyclada*
    b. subsp. *arborea* (H. Perrier) C.C. Berg
4. *F. lateriflora* Vahl
5. *F. asperifolia* Miq.
6. *F. capreifolia* Delile
7. *F. pygmaea* Hiern
8. *F. bojeri* Baker
9. *F. brachyclada* Baker
10. *F. politoria* Lam.
11. *F. sycomorus* L.
12. *F. mucuso* Ficalho
13. *F. sur* Forssk.
14. *F. vogeliana* (Miq.) Miq.
15. *F. vallis-choudae* Delile
16. *F. mauritiana* Lam.
17. *F. tiliifolia* Baker
18. *F. torrentium* H. Perrier
19. *F. polyphlebia* Baker
20. *F. botryoides* Baker
21. *F. trichoclada* Baker
22. *F. karthalensis* C.C. Berg
23. *F. variifolia* Warb.
24. *F. dicranostyla* Mildbr.
25. *F. assimilis* Baker
26. *F. ampana* C.C. Berg
27. *F. ingens* (Miq.) Miq.
28. *F. cordata* Thunb.
    a. subsp. *cordata*
    b. subsp. *lecardii* (Warb.) C.C. Berg
    c. subsp. *salicifolia* (Vahl) C.C. Berg
29. *F. verruculosa* Warb.
30. *F. madagascariensis* C.C. Berg

31. *F. densifolia* Miq.
32. *F. menabeensis* H. Perrier
33. *F. humbertii* C.C. Berg
34. *F. saussureana* DC.
35. *F. chlamydocarpa* Mildbr. & Burret
    a. subsp. *chlamydocarpa*
    b. subsp. *fernandesiana* (Hutch.) C.C. Berg
    c. subsp. *latifolia* (Hutch.) C.C. Berg
36. *F. lutea* Vahl
37. *F. platyphylla* Delile
38. *F. bussei* Mildbr. & Burret
39. *F. recurvata* De Wild.
40. *F. vasta* Forssk.
41. *F. wakefieldii* Hutch.
42. *F. glumosa* Delile
43. *F. stuhlmannii* Warb.
44. *F. nigropunctata* Mildbr. & Burret
45. *F. tettensis* Hutch.
46. *F. muelleriana* C.C. Berg
47. *F. abutilifolia* (Miq.) Miq.
48. *F. populifolia* Vahl
49. *F. jansii* Boutique
50. *F. trichopoda* Baker
51. *F. grevei* Baill.
52. *F. rubra* Vahl
53. *F. marmorata* Baker
54. *F. bivalvata* H. Perrier
55. *F. calyptrata* Vahl
56. *F. fischeri* Mildbr. & Burret
57. *F. amadiensis* De Wild.
58. *F. craterostoma* Mildbr. & Burret
59. *F. lingua* De Wild. & T. Durand
    a. subsp. *lingua*
    b. subsp. *depauperata* (Sim) C.C. Berg
60. *F. natalensis* Hochst.
    a. subsp. *natalensis*
    b. subsp. *leprieurii* (Miq.) C.C. Berg
61. *F. faulkneriana* C.C. Berg
62. *F. burtt-davyi* Hutch.
63. *F. ilicina* (Sond.) Miq.
64. *F. antandronarum* (H. Perrier) C.C. Berg
    a. subsp. *antandronarum*
    b. subsp. *bernardii* C.C. Berg
65. *F. reflexa* Thunb.
    a. subsp. *reflexa*
    b. subsp. *sechellensis* (Baker) C.C. Berg
    c. subsp. *aldabrensis* (Baker) C.C. Berg
66. *F. thonningii* Blume
67. *F. kamerunensis* Mildbr. & Burret
68. *F. elasticoides* De Wild.
69. *F. burretiana* Hutch.
70. *F. oreodryadum* Mildbr.
71. *F. pseudomangifera* Hutch.
72. *F. usambarensis* Warb.
73. *F. adolfi-friderici* Mildbr.
74. *F. louisii* Boutique & J. Léonard
75. *F. leonensis* Hutch.
76. *F. conraui* Warb.
77. *F. tesselata* Warb.
78. *F. pachyneura* C.C. Berg

79. *F. ardisioides* Warb.
    a. subsp. *ardisioides*
    b. subsp. *camptoneura* (Mildbr.) C.C. Berg
80. *F. preusii* Warb.
81. *F. abscondita* C.C. Berg
82. *F. cyathistipula* Warb.
    a. subsp. *cyathistipula*
    b. subsp. *pringsheimiana* (J. Braun & K. Schum.) C.C. Berg
83. *F. scasselatii* Pamp.
    a. subsp. *scasselatii*
    b. subsp. *thikaensis* C.C. Berg
84. *F. cyathistipuloides* De Wild.
85. *F. scott-elliotii* Mildbr. & Burret
86. *F. densistipulata* De Wild.
87. *F. subcostata* De Wild.
88. *F. barteri* Sprague
89. *F. lyrata* Warb.
90. *F. sagittifolia* Mildbr. & Burret
91. *F. subsagittifolia* C.C. Berg
92. *F. wildemaniana* De Wild. & T. Durand
93. *F. oresbia* C.C. Berg
94. *F. crassicosta* Warb.
95. *F. ottoniifolia* (Miq.) Miq.
    a. subsp. *ottoniifolia*
    b. subsp. *multinervia* C.C. Berg
    c. subsp. *lucanda* (Ficalho) C.C. Berg
    d. subsp. *macrosyce* C.C. Berg
    e. subsp. *ulugurensis* (Mildbr. & Burret) C.C. Berg
96. *F. tremula* Warb.
    a. subsp. *tremula*
    b. subsp. *kimuenzensis* (Warb.) C.C. Berg
    c. subsp. *acuta* (De Wild.) C.C. Berg
97. *F. artocarpoides* Warb.
98. *F. polita* Vahl
    a. subsp. *polita*
    b. subsp. *brevipedunculata* C.C. Berg
99. *F. bizanae* Hutch. & Burtt-Davy
100. *F. chirindensis* C.C. Berg
101. *F. sansibarica* Warb.
    a. subsp. *sansibarica*
    b. subsp. *macrosperma* (Mildbr. & Burret) C.C. Berg
102. *F. dryepondtiana* De Wild.
103. *F. umbellata* Vahl
104. *F. bubu* Warb.
105. *F. ovata* Vahl

8. *Maclura* Nutt. (11 spp.) (Berg, 1977c; 1986a)
    1. *M. africana* (Bureau) Corner

9. *Mesogyne* Engl. (1 sp.) (Berg, 1977c)
    1. *M. insignis* Engl.

10. *Milicia* Sim (2 spp.) (Berg, 1977c 1982).
    1. *M. excelsa* (Welw.) C.C. Berg
    2. *M. regia* (A. Chev.) C.C. Berg

11. *Morus* L. (approx. 12 spp.) (Berg, 1977c)
    1. *M. mesozygia* A. Chev.

12. *Scyphosyce* Baill. (2 spp.) (Berg, 1977c)
    1. *S. manniana* Baill.
    2. *S. pandurata* Hutch.

13. *Streblus* Lour. (23 spp.) (Berg, 1977c; 1988)
    1. *S. usambarensis* (Engl.) C.C. Berg
    2. *S. mauritianus* (Jacq.) Blume
    3. *S. dimepate* (Bureau) C.C. Berg

14. *Treculia* Decne. (3 spp.) (Berg, 1977c)
    1. *T. africana* Decne.
       a. subsp. *africana*
          var. *africana*
          var. *mollis* (Engl.) J. Léonard
       b. subsp. *madagascarica* (N.E. Br.) C.C. Berg
          var. *madagascarica*
          var. *ilicifolia* (Léandri) C.C. Berg
          var. *sambiranensis* (Léandri) C.C. Berg
    2. *T. acuminata* Baill.
    3. *T. obovoidea* N.E. Br.

15. *Trilepisium* Thouars (1 sp.) (Berg, 1977c)
    1. *T. madagascariense* DC.

16. *Trophis* P. Browne (9 spp.) (Berg, 1977c; 1988)
    1. *T. borbonica* (Duch.) C.C. Berg
    2. *T. montana* (Léandri) C.C. Berg

17. *Utsetela* Pellegr. (1 sp.) (Berg, 1977c)
    1. *U. gabonensis* Pellegr. — incl. *U. neglecta* Jongkind (1995).

Hedberg, I & Hedberg, O. (1998). Plant Variation, Phylogeny and Systematics. In: C.R. Huxley, J.M. Lock and D.F. Cutler (editors). Chorology, Taxonomy and Ecology of the Floras of Africa and Madagascar. Pp. 149–154. Royal Botanic Gardens, Kew.

10. PLANT VARIATION, PHYLOGENY AND SYSTEMATICS

INGA AND OLOV HEDBERG

Department of Systematic Botany, Uppsala University, Villavägen 6, S-752 36 Uppsala, Sweden

Abstract

A solid foundation for all work on tropical African vegetation is provided by Frank White's survey of African phytochoria – his vegetation map and vegetation descriptions are invaluable for all work on African botany. Another field, which may have attracted less general attention, is his fundamental contributions to African plant taxonomy. Through his elaborate studies on Ebenaceae, Meliaceae, Chrysobalanaceae and many other groups he showed an impressive knowledge of how populations and taxa vary in nature, and how such variation should be tackled for taxonomic purposes. Through rigorous analysis of, among other things, morphological and ecological variation, and careful documentation of the results, he illustrated the multidimensional diversity found in the tropical groups he studied, and demonstrated that nature is often neither parsimonious nor forms simple hierarchical systems. He also underscored the necessity of sound taxonomic knowledge as a basis for reliable phytogeographical conclusions. Our paper is an attempt to elucidate the importance of these aspects of Frank White's work, and to pass on his message to younger generations.

To botanists interested in the tropics Frank White will be remembered primarily for his innovations in African phytogeography, where he revolutionised vegetation description and mapping. To describe the vegetation he subdivided the African continent according to a successively refined scheme into a number of regional phytochoria, based on the richness of their endemic floras at the species level. Within each phytochorion he distinguished physiognomic vegetation types (cf. White, 1993a). His scheme has been accepted by the majority of botanists currently working on African flora and vegetation. The phytochoria provide the mapping units in his Vegetation Map of Africa (White, 1983a), which he prepared on behalf of AETFAT (Association pour L'Étude Taxonomique de la Flore d'Afrique Tropicale), supported by the vegetation map committee of the same organisation as well as by UNESCO and UNSO (United Nations SudanoSahelian Office). The task was enormous and consequently took a long time. But the resulting map and the comprehensive accompanying book *The Vegetation of Africa* (White, 1983b), both of which were published by UNESCO, provide an invaluable source of information. The author managed to weave together the enormous amount of basic data into a coherent whole, where all existing information is easily accessible. Based on this solid infrastructure and combining evidence from chorology and ecology with taxonomy he convincingly demonstrated that reliable conclusions and hypotheses about Quarternary vegetation changes in Africa can only be achieved if they are based on an intimate knowledge of the present flora and vegetation. Some earlier hypotheses based on the few and partly contradictory early pollen-analytical results available from tropical Africa were found to be premature (White, 1981: 37; 1990: 164).

Another of Frank White's achievements, which may have attracted less general attention, is represented by his innovative taxonomic work. Together with his research students he made versatile and conscientious revisions of a number of critical groups. In order to fully appreciate the implication of this work for systematics in general, and for systematics on tropical material in particular, it is necessary to first briefly overview the evolution of systematic botany from the time of Linnaeus.

The purpose of research in systematic botany has changed with time. To Linnaeus it meant description and classification of the great diversity of the plant kingdom in order to promote its rational use by man — Linnaeus believed all species had been produced by divine creation and for the benefit of man. Since the time of Linnaeus, taxonomic botany has experienced a series of new approaches or waves of fashion: anatomy, embryology, serodiagnostics, cytotaxonomy, biosystematics, palynology, phytochemistry, numerical taxonomy, cladistics and molecular systematics. Some of these were originally claimed by their proponents to be not only more modern but also more scientific than traditional systematic botany (cf. O. Hedberg, 1988; 1995a). Most of those methods did not, however, lead to any drastic taxonomic changes, and have gradually been subsumed into the toolbox of taxonomists, whereas the most recent ones, especially cladistics, have caused intense debates because of their revolutionary aspirations.

Most post-Darwinian taxonomic botanists have regarded the variation pattern found in nature as resulting from evolution and — with variable success — attempted to arrange their taxa in an evolutionary pattern. Although special categories have been devised, e.g., for agricultural purposes and for experimental taxonomy, most botanists have been satisfied with the "general purpose classification" inherited from earlier generations and continuously but carefully revised, and used in regional and local floras. With the explosive development of cladistics the situation has changed — many cladists consider their computer-supported classification to be superior to the traditional one, and hence try to introduce far-reaching taxonomic and nomenclatural changes. This has led to a bifurcation of systematic botany, with one branch producing a general purpose classification serving the need for information storage and retrieval for plants and plant products, the other attempting to trace and document the phylogeny of the plant kingdom. The second branch results in phylogenetic schemes, but, in the absence of fossil records, explicit phylogenies are still just as hypothetical as when Lotsy (1916) formulated his famous criticism (cf. O. Hedberg, 1996). In this connection it is worth recalling the view of Patterson (1982) "Taxonomy has always been a dual-purpose weapon: it organises knowledge and reflects relationships. These two purposes sometimes come into conflict." As emphasised by Schrire and Lewis (1996: 363): "The conflict between the strong desire of achieving a natural classification and the need for stability of names, is likely to escalate as the number of cladistic analyses emerging of major plant groups increases. Many users of classifications express concern about nomenclatural stability, especially when changes to existing classifications are proposed, based on poorly supported cladistic studies". Already, in 1952, Woodger concluded that "the taxonomic system and the evolutionary phylogenetic schemes are quite different jobs and only confusion will result from identifying or mixing them". This statement was recently brought even further by Brummitt (1996: 382): "The practical effects of applying cladistic principles to classification and nomenclature are potentially disastrous".

Many 'general purpose' taxonomists consider that the nearest we can get to a reliable survey of the system is a 'Dahlgrenogram' (Fig. 1), representing a cross section of the hypothetical 'evolutionary tree'. There is no harm in attempts to track evolution below that cross-sectional surface through cladistic studies, and to hypothesise about phylogeny. It is, however, unacceptable to make extensive rearrangements in the general purpose taxonomic system on the basis of such hypothetical generalisations, also because, as emphasised by Hall (1993: 610) the experimentally divisive methods of cladistics can only cope with a limited number of taxa at a time. This forces many cladists to include only one species for each of the genera they compare, which leads

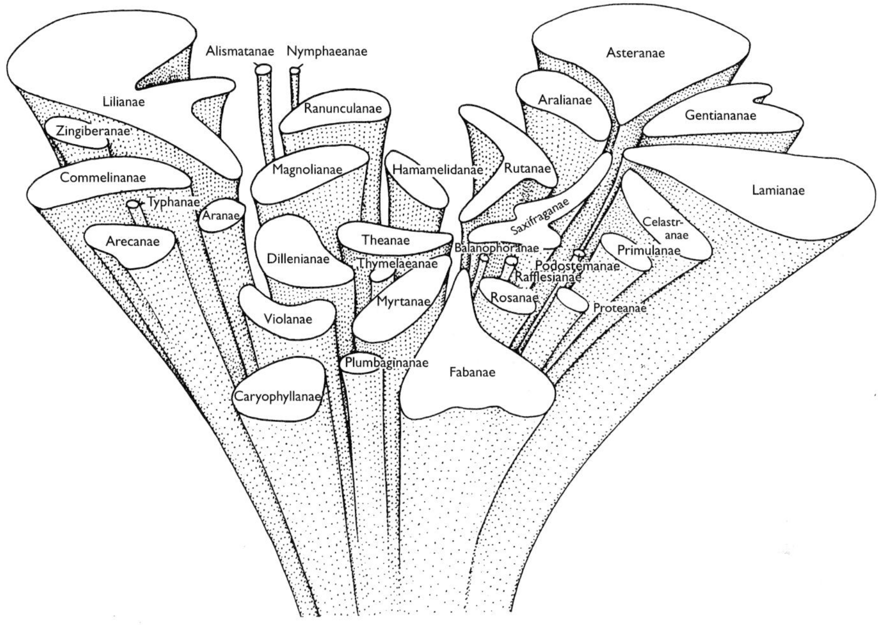

FIG. 1. Cross section of a hypothetical evolutionary tree for the angiosperms ("Dahlgrenogram"), from Dahlgren 1979: 188.

them on treacherous paths. A drastic example of this kind has recently been exposed by Leeuwenberg (1997) in the taxonomic restructuring proposed by Struwe *et al.* (1994) for the order Gentianales.

The latest taxonomic bandwagon is molecular systematics, which has turned out to be an invaluable companion to cladistics. The amount of DNA difference between related taxa was earlier believed to give more or less infallible evidence on the rate of genetic differentiation, but recent studies have led to questioning the reliability of this 'molecular clock' (Lewin, 1990). Harris and Ingram (1991: 407) considered the use of cpDNA to be "*a priori* no better or no worse than any other data source used in biosystematic studies", and stressed that not enough is known about the degree of intra-individual and intraspecific variation and the mode of plastid transmission. According to Dover (1987: 56) "there is considerable controversy over crude 'single-copy' DNA-DNA hybridization studies and the support they are assumed to give to concepts of universal time-dependent clocks". There are also often conflicts between molecular and morphological data (Knox, 1996).

In Frank White's pioneer studies in tropical Africa, a general purpose classification, based on detailed field studies and utilising "rigorous techniques" (White, 1993b), was a necessary prerequisite for his phytogeographic studies, not least for the delimitation of phytochoria. Confronted with the bewildering and sometimes astounding variation of the tropical African tree flora he soon felt the need for adequate methods to analyse and portray the complex variation pattern. Like us (I. Hedberg, 1967; I. & O. Hedberg, 1977; O. Hedberg, 1957; 1985; 1993 etc.) he therefore made use of pictorialized scatter diagrams and developed pictorialized distribution maps. In his work on, among others, the families Ebenaceae (White, 1978, etc.), Meliaceae (White, 1986) and Chrysobalanaceae (Prance & White, 1988), for which he utilised not only characters such as gross morphology, anatomy and embryology, but also palynology, chromosome numbers, phytochemistry and chorology, he demonstrated an impressive knowledge about how populations and taxa vary in nature, and how such variation should be tackled for taxonomic purposes. He paid special attention to chorological and ecological studies of some widespread and ecologically versatile species, where the infraspecific variation might reflect responses to recent climatic changes (White, 1993b). The detailed monographic work on Chrysobalanaceae led to the conclusion that, although the family as well as each of its 17 genera is doubtless monophyletic, the distribution of character states within it is so reticulate as to make cladistic or other phylogenetic groupings unworkable (Prance & White, 1988: 4). This conclusion is very important for other similar studies of tropical groups — it is likely to be matched in many other families when they become adequately studied.

Like Frank White we have long been working on tropical material and learned early in our work that modes of variation often differ from those in temperate areas. In the depauperate flora of Boreal areas, which have been subjected to repeated Pleistocene glaciations, the variation pattern is often relatively simple, inviting belief in constancy of character state difference between populations, species and genera. In the tropics the flora seems to have been less influenced by drastic climatic changes, and so to have maintained much more of the variation acquired during millions of years of evolution. Ignorance of those conditions, together with scarcity of material during the early exploratory phase led to excessive description of new species in many groups. At times there seems to have been almost a race between botanists and different botanical institutes not only to explore the wealth of plant forms occurring in tropical Africa but also to exploit it in terms of numbers of new species. The variation pattern found here is also much less conducive to the construction of dichotomous keys and hierarchical systems than that in the north.

Sometimes absence of clear discontinuities in individual features between closely related species necessitates studies of the integrated variation in several taxonomic features, as in the genus *Polycarpaea* (O. Hedberg, 1995b). A pertinent example of extensive variation in the tropics is given by the variation in leaf shape in African *Myrica* documented by White (1993b), whose revision of the African material reduced the number of recognised species from twenty-six to six.

Occasionally a group which shows discontinuous variation in some areas may display overlapping variation elsewhere, as *Swertia crassiuscula* on the East African mountains (O. Hedberg, 1957: 313). Experience of this kind makes us agree with White (1993b; see also, for example, Gillett, 1958; Stace, 1980; Stuessy, 1993) that, ideally, taxonomic work should attempt to be monographic in outline, being based on as much field and herbarium material as possible.

However, the current upsurge of interest in conservation of biodiversity , triggered by the Rio conference, has called attention to the urgent need for a less definite but workable classification of the plant world, facilitating rapid and simple information storage and retrieval. This is particularly essential for the intensive field surveys of surviving stands of natural vegetation, pioneered and intensely pleaded for by Frank White (1993b and earlier), which are a fundamental prerequisite for future conservation. In this time of painful awakening to the disastrous speed of environmental degradation, and to the incompleteness of the botanical inventory of the world (Prance, 1984), we may have to resort, for the time being, to an α or β classification in the sense of Turrill (1950), but this must not prevent us from advancing towards the optimum monographic work so brilliantly exemplified by Frank White and his collaborators.

References

Brummitt, R.K. (1996). In defence of paraphyletic taxa. In: L.J.G. van der Maesen, X.M. van der Burgt and J.M. van der Medenbach de Rooy (editors). The Biodiversity of African Plants. Pp. 371–394. Proceedings XIVth AETFAT Congress. Kluwer Academic Publishers, Dordrecht, Boston, London.

Dahlgren, R. (1979). Angiospermernes Taxonomi. Bind 1. Ed. 2., Universitetsforlaget, Copenhagen.

Dover, G.A. (1987). DNA Turnover and the Molecular Clock. *J. Molec. Evol.* **26**: 47–58.

Gillett, J.B. (1958). Ways and means of promoting the production of monographs of families and genera. *Mem. Soc. Brot.* **13**: 107–111.

Hall, A.V. (1993). Classification of evolutionary groups: the Uniter program's approach to greater resolving power. *Taxon* **42**: 609–625.

Harris, S.A. and Ingram, R. (1991). Chloroplast DNA and biosystematics. The effects of intraspecific diversity and plastid transmission. *Taxon* **40**: 393–412.

Hedberg, I. (1967). Cytotaxonomic studies on *Anthoxanthum odoratum* L. *s. lat.* II. Investigations of some Swedish and of a few Swiss population samples. *Symb. Bot. Upsal.* **18(5)**: 1–88.

Hedberg, I. and Hedberg, O. (1977). The genus *Dipsacus* in tropical Africa. *Bot. Not.* **129**: 383–389.

Hedberg, O. (1957). Afroalpine Vascular Plants. A taxonomic revision. *Symb. Bot. Upsal.* **15**: 1–411

Hedberg, O. (1985). The genus *Blaeria* (Ericaceae) in Ethiopia. *Nordic J. Bot.* **5**: 463–467.

Hedberg, O. (1988). Classical taxonomy versus modern systematic studies in African botany. *Monogr. Syst. Bot. Missouri Bot. Gard.* **25**: 567–573.

Hedberg, O. (1993). *Rumex nepalensis* (Polygonaceae) — a Himalayan element in the afromontane flora. *Fragm. Florist. Geobot. Suppl.* **2(1)**: 91–96.

Hedberg, O. (1995a). Cladistics in taxonomic botany — master or servant? *Taxon* **44**: 3–11.

Hedberg, O. (1995b). Studies in the genus *Polycarpaea* (Caryophyllaceae) in Ethiopia. *Nordic J. Bot.* **15**: 513–517.

Hedberg, O. (1996). Reply to Neil Snow's "The phylogenetic paradigm of comparative biology". *Taxon* **45**: 91–92.

Knox, E.B. (1996). The problem of generic delimitation when conflict exists between molecular and morphological data. In: L.J.G. van der Maesen, X.M. van der Burgt and J.M. van der Medenbach de Rooy (editors). The Biodiversity of African Plants. Pp. 421–43. Proceedings XIVth AETFAT Congress. Kluwer Academic Publishers, Dordrecht, Boston, London.

Leeuwenberg, A.J.M. (1997). Amusing superficiality of orthodox cladists. *Taxon* **46**: 535–537.

Lewin, R. (1990). Molecular clocks run out of time. *New Sci.* 10 February 1990.

Lotsy, J.P. (1916). Evolution by means of hybridization. M. Nijhoff, The Hague.

Patterson, C. (1982). Cladistics and classification. *New Sci.* 29 April 1982.

Prance, G.T. (1984). Completing the Inventory. In: V.H. Heywood and D.M. Moore (editors). Current Concepts in Plant Taxonomy. Pp. 365–395. The Systematics Association Special Volume No. 25. Academic Press, London.

Prance, G.T. and White, F. (1988). The genera of Chrysobalanaceae: a study in practical and theoretical taxonomy and its relevance to evolutionary biology. *Philos. Trans., Ser. B.* **320**: 1–184.

Schrire, B.D. and Lewis, G.P. (1996). Monophyly: a criterion for generic delimitation, with special relevance to Leguminosae. In: L.J.G. van der Maesen, X.M. van der Burgt and J.M. van der Medenbach de Rooy (editors). The biodiversity of African

Plants. Pp. 421–432. Proceedings XIVth AETFAT Congress. Kluwer Academic Publishers, Dordrecht, Boston, London.

Stace, C.A. (1980). Plant taxonomy and biosystematics. Edward Arnold, London

Struwe, L., Albert, V.A. and Bremer, B. (1994). Cladistics and family level classification of the Gentianales. *Cladistics* **10**: 175–206.

Stuessy, T.F. (1993). The role of creative monography in the biodiversity crisis. *Taxon* **42**: 313–321.

Turrill, W.B. (1950). Modern trends in the classification of plants. *Advancem. Sci.* **26**: 1–16.

White, F. (1978). The taxonomy, ecology and chorology of African Ebenaceae. I. The Guineo-Congolian species. *Bull. Jard. Bot. Belg.* **48**: 245–358.

White, F. (1981). The history of the Afromontane archipelago and the scientific need for its conservation. *African J. Ecol.* **19**: 33–54.

White, F. (1983a). UNESCO/AETFAT/UNSO Vegetation Map of Africa. Scale 1: 5 000 000 (in colour). UNESCO, Paris.

White, F. (1983b). The Vegetation of Africa: a descriptive memoir to accompany the UNESCO/AETFAT/UNSO vegetation map of Africa. *Natural Resources Research* **20**. UNESCO, Paris.

White, F. (1986). The taxonomy, chorology and reproductive biology of southern African Meliaceae and Ptaeroxylaceae. *Bothalia* **16**: 143–168.

White, F. (1990). *Ptaeroxylum obliquum* (Ptaeroxylaceae), some other disjuncts, and the Quaternary history of African vegetation. *Bull. Mus. Natl. Hist. Nat., B, Adansonia* **12(2)**: 139–185.

White, F. (1993a). The AETFAT chorological classification of Africa: history, methods and applications. *Bull. Jard. Bot. Belg.* **62**: 225–281.

White, F. (1993b). African Myricaceae and the history of the Afromontane flora. *Opera Bot.* **121**: 173–188.

Woodger, J.H. (1952). From biology to mathematics. *Brit. J. Philos. Sci.* **3**: 1–21.

Cronk, Q.C.B. (1998). The ochlospecies concept. In: C.R. Huxley, J.M. Lock and D.F. Cutler (editors). Chorology, Taxonomy and Ecology of the Floras of Africa and Madagascar. Pp. 155–170. Royal Botanic Gardens, Kew.

11. THE OCHLOSPECIES CONCEPT

Q.C.B. Cronk

**Royal Botanic Garden, 20A Inverleith Row, Edinburgh, EH3 5LR, UK
and
Institute of Cell and Molecular Biology, University of Edinburgh, Kings Buildings,
Mayfield Road, Edinburgh, EH9 3JH, UK**

Abstract

A review is presented of White's ochlospecies, which represents a particular type of species problem distinct from those resulting from familiar complications of hybridisation, autogamy or apomixis. An ochlospecies is a polymorphic species with chaotic infraspecific variation which is intractable to formal taxonomic treatment; ten criteria are suggested for recognising ochlospecies. Two hypotheses are presented as possible explanations for this type of variation pattern. One, the 'Prance hypothesis', refers to processes of isolation and refusion driven by Pleistocene climatic change. The other, the 'rapid expansion hypothesis', proposes that genetic processes associated with the rapid population expansion of a colonising generalist could result in the observed patterns without an allopatric phase. A test of these hypotheses is suggested.

Species such as these are a nightmare to the taxonomist, though when aspects of their biology other than taxonomy are considered they may have much to offer to general, and, in particular, to evolutionary studies.

White, 1998.

Introduction

It is the common experience of the monographer of a large genus of plants that the entities recognised as species are not homogeneous with respect to their variation patterns, and cannot be made so. They differ both with respect to their infraspecific variation and in their relations to other species. This has been expressed as follows by Hilliard & Burtt (1971) in their monograph of *Streptocarpus*: "Discussions on species often seem to imply that species in, say, orchids are rather different from those of grasses or oaks. Is this true? We think not. Of course, there are special situations that occur in one group but not in another. In the main, however there is more evidence that any group of adequate size will contain species of many different kinds than that different groups will have different sorts of species".

At a meeting of the Systematics Association on Taxonomy and Geography, White (1962) reported on this phenomenon, as revealed by his studies of African Ebenaceae. Of 90 species of *Diospyros*, most were monotypic, seven polytypic and three polymorphic (i.e. ochlospecies, to be discussed here). Forty-eight species were taxonomically isolated: clearly distinguishable from other species by a number of *diagnostic* characters. In contrast 36 species belonged to groups of two to five closely related species, and these could further be divided into allopatric (for which White used the term 'superspecies' of Mayr) or sympatric groups. Species belonging to these species groups could only be distinguished from one another by *differential* characters (White's term).

The three polymorphic species, *Diospyros ferrea, D. natalensis* and *D. mespiliformis,* were quite different from the others in that while the total amount of variation within these species was equivalent to that of a polytypic species, or even a superspecies or species group, the variation pattern proved utterly intractable to strict taxonomic treatment (Fig. 1). Consequently the taxonomist has no recourse other than to recognise a single variable species with no formal variants. For these species White introduced the term 'ochlospecies' from the Greek root, *ochlos,* meaning an irregular crowd or a mob (with an apposite secondary meaning of trouble or annoyance).

Plants have many well documented sources of taxonomic complexity, mainly due to cytological complexity, hybridisation, apomixis or autogamy. These classes of species problem have recently been reviewed by Mayr (1992). White did not intend the term ochlospecies to be used for these cases, for which, in general, other terminologies are available. If putative ochlospecies resolve themselves as hybrid swarms or apomictic groups, then the term ochlospecies is not applicable. When these other types of species problems are separated, the sexual, outbreeding ochlospecies becomes another class of species problem. As such, ochlospecies have a number of characteristic and curious features (see below), but they have never been subject to experimental study, mainly because most of the recognised examples are long-lived tropical trees and large herbs, unsuited to life on the laboratory bench. Nevertheless, as White himself recognised, they still offer interesting opportunities for evolutionary studies.

Definition

A general definition of an ochlospecies would be (after White, 1998):

"A very variable (polymorphic) species, whose variation, though partly correlated with ecology and geography, is of such a complex pattern that it cannot be satisfactorily accommodated within a formal classification".

However, this would be too broad a category, as ochlospecies have some highly characteristic traits and can be rather narrowly defined (as was done by White) to give a rather homogeneous set of problem species. We can recognise four *strong traits* which are essential to the concept, and six *weaker traits* which are usual or common in ochlospecies but not essential.

Strong traits
1. Species which show variation that is strongly polymorphic but only weakly polytypic (i.e. non-hierarchical).
2. Character-state distribution shows only partial correlation with geography and ecology (checkerboard variation).
3. Characters vary independently, and not in a correlated fashion.
4. Complexity of variation is not due to hybridisation between currently recognisable species, or to a specialised breeding system (i.e. an ochlospecies is not a hybrid swarm or an aggregate of apomictic microspecies).

FIG. 1. Pictorialized distribution map of *Diospyros mespiliformis* from F. White (1962) showing the main features of variation. **A** leaf (*Conservator of Forests s.n.*); **B** indumentum of lower leaf surface (*Conservator of Forests* 208); **C** immature fruit (*Andoh* FH 5173); **D** calyx of mature fruit (*Nortey* 103); **E** calyx of mature fruit (Conservator of Forests 208); **F** female flower (*Ledermann* 2958); **G** female flower and mature fruit (*Hoyle* 761); **H** leaf, calyx of female flower and calyx of mature fruit (*Semsei* FH 2900); **I** leaf and calyx of mature fruit (*Santo* 2219); **J** ditto (*White* 2993); **K** ditto (*Lovemore* 76); **L** ditto (*Torre* 2669); **M** indumentum of lower leaf-surface (*Santo* 2219). The continuous line demarcates the distribution of *D. kirkii.* Reproduced with permission from the Systematics Association.

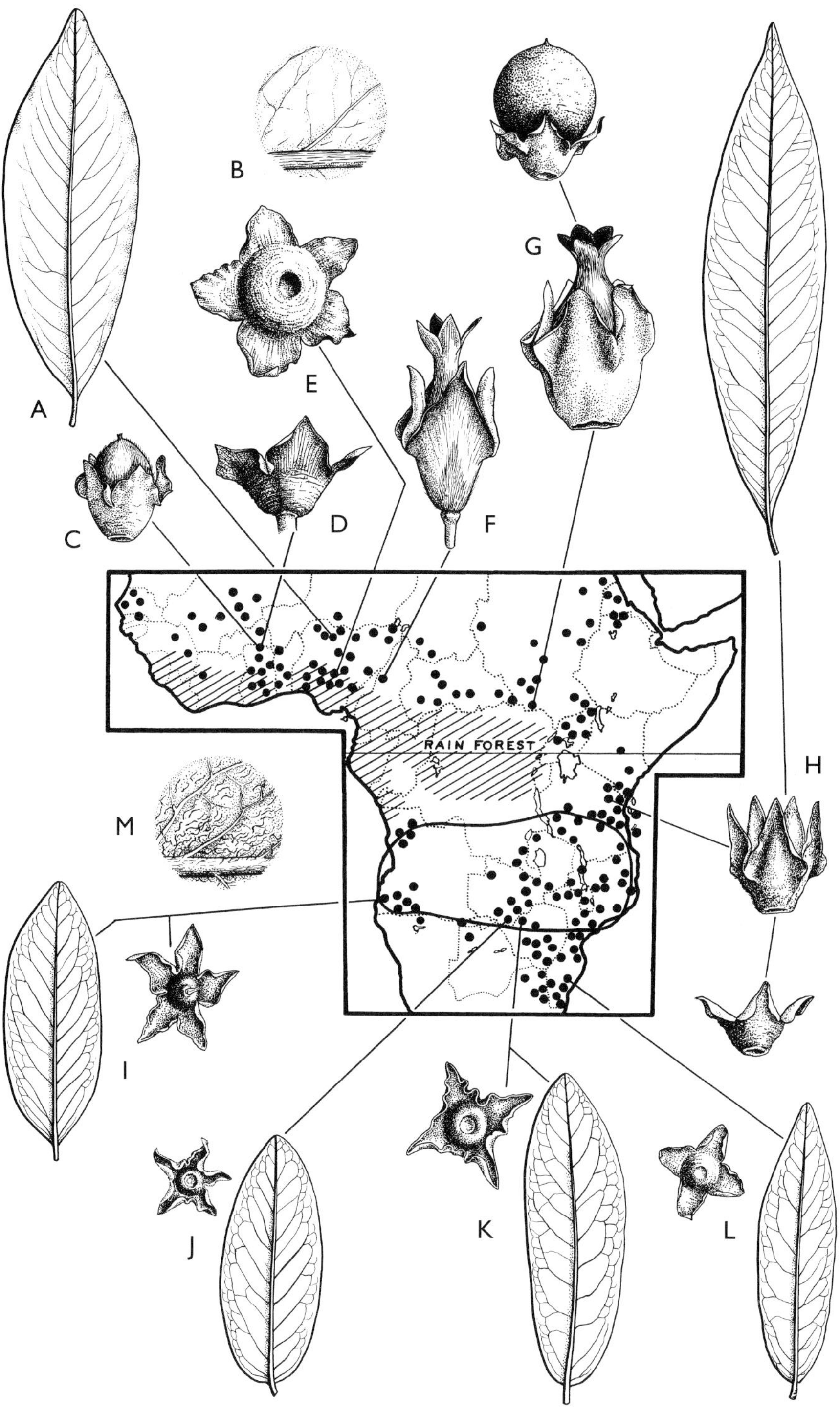

RAIN FOREST

Weak traits

5. Ochlospecies are usually geographically and ecologically widespread, occurring in several climatic and vegetation zones ('ecological and chorological transgressors' of White). An exception is the ochlospecies *Trichilia micrantha* which, while geographically widespread, is not an ecological or chorological transgressor (Pennington, 1981).

6. Distinct variants are recognisable locally but not globally. At a particular locality two distinct and non-intergrading forms may be found, and other forms may be found at other localities, but taken together all the forms intergrade and the classification breaks down.

7. Ochlospecies sometimes have a closely related, but morphologically distinct and monotypic satellite species. For example *Diospyros kirkii* is the satellite of the ochlospecies *D. mespiliformis*, *Ficus kamerunensis* is the satellite of *F. thonningii*, *Heracleum minimum* is the satellite of *H. sphondylium* and *Myrmecodia beccarii* is the satellite of *M. tuberosa*.

8. Similar variants may occur in widely separated localities and appear to be polytopic in origin.

9. An apparently unifying characteristic of ochlospecies is that they tend to occur in medium to large genera usually with more than 50 species, and in such genera tend to occur with a low frequency of 1–8%. No ochlospecies so far definitely known is a member of a monotypic or oligotypic genus.

10. Ochlospecies often have long synonymies: the variation has driven a proliferation of names that eventually prove untenable. Thus ochlospecies are often created by the lumping of a dozen or more species.

Many ochlospecies will display all ten traits, but to qualify as an ochlospecies, a species must have the first four and should have at least the first six.

The tenth criterion is important because it reveals the usual epistemic (rather than evolutionary) route to the formation of ochlospecies. It follows from the incomplete nature of tropical floristics that many ochlospecies may lie unrecognised as a series of poorly collected related 'species' awaiting the attention of a taxonomist. Extreme cases are those of *Vavaea amicorum* where 21 species were reduced to one apparent ochlospecies with no infraspecific variants (Pennington, 1969), and *Allophyllus cobbe* into which a vast amount of variation and names were disposed, somewhat contentiously, by Leenhouts (1994). The three 'Pennington Conditions' used to justify the treatment of *Vavaea* are also, of necessity, ochlospecies conditions, and Mabberley (1979) used the Pennington Conditions to justify treating *Chisocheton lasiocarpus* as an ochlospecies and not recognising formal infraspecific variants. The Pennington Conditions are as follows:

1. None of the variants are well correlated with geography.
2. Different variants may occur in the same locality and in the same habitat.
3. Variants are based on slight vegetative differences, the most striking of which appear to have originated polytopically.

The question of polytopic origins is interesting as they may be characteristic of variation in ochlospecies. White (1998) suggests that extreme variants, particularly the rheophytic and microphyllous forms of *Diospyros natalensis,* have originated more than once. Rheophytes in this ochlospecies occur in widely disjunct places such as Mt. Mulanje and the Dwesa River in South Africa. In these places the rheophytes form part of larger non-rheophytic populations and are likely to be independently locally derived.

A word is appropriate here concerning phenotypic plasticity. There is as yet no evidence that ochlospecies variation is due to exceptional phenotypic plasticity. White himself cultivated ten species of New Caledonian *Diospyros* in Oxford for some 20 years, including the ochlospecies *D. parviflora* (Fig. 1 in Pannell, this volume), and all "retained their species-specific vegetative characters" (White, 1993). Nowhere does he suggest that ochlospecies variation becomes uniform under conditions of uniform cultivation. However, discussing the variation of *D. natalensis* on Mt. Mulanje, White (1988) notes that some of it has been shown to be phenotypic. Whatever the final picture may be, the invocation of phenotypic plasticity does not explain everything about ochlospecies. Phenotypic plasticity is a highly important plant adaptation, under tight genetic control, although it remains a difficult phenomenon to study (Sultan, 1995). There is as yet little clear idea as to why some species have a large phenotypic response and others are invariant with respect to environment, although it has been suggested that evolution in a spatially heterogeneous environment may promote plasticity (Zhivotovsky *et al.*, 1996). It would thus be extremely interesting if all ochlospecies variation was due to phenotypic plasticity, although at present this appears unlikely.

Is the ochlospecies concept really necessary?

In cases where medium to large genera of tropical trees have been recently revised, very often an ochlospecies is proposed. Tables 1 & 2 give the occurences of different species types in 90 species of African Ebenaceae (*Diospyros*) and 67 species of South American Meliaceae (*Trichilia*). Monotypic species, polytypic species and ochlospecies refer to different patterns of infraspecific variation, whereas group species refer to the pattern of variation just above the species level. These two types of variation are largely

TABLE 1. A classification of species types, according to interspecies relationships (group, isolated) and infraspecific variation (monotypic, polytypic, ochlospecies) for 90 African Ebonies of the genus *Diospyros*. Only 86 species are treated as four are too little known (White, 1962).

Species relationship:	*Isolated*	*Group*	Total
Species type:			
Monotypic	46	30	76
Polytypic	2	5	7
Ochlospecies	2	1	3
Total	50	36	86

TABLE 2. A classification of species types, following Table 1, for 67 species of the genus *Trichilia*, after Pennington (1981).

Species relationship:	*Isolated*	*Group*	Total
Species type:			
Monotypic	14	44	58
Polytypic	1	6	7
Ochlospecies	2	0	2
Total	17	50	67

independent, and so most group species are also monotypic. Nevertheless, it is convenient to divide species into these categories, as the species level in one taxon may be equivalent to variation at the subspecies level in another. Even if all the species in particular species groups were lumped, as the variation in ochlospecies is lumped, there would still be definable hierarchical elements amenable to formal classification. This is not the case in ochlospecies. Furthermore, in *Trichilia* the species within groups are largely allopatric whereas the component elements of ochlospecies are significantly sympatric and this appears to be a general phenomenon.

If the same sort of analysis is given to the species of *Streptocarpus* dealt with by Hilliard & Burtt (1971), remarkably similar patterns result if their three 'species aggregates' do duty for ochlospecies (Table 3). These aggregate species are like ochlospecies in that they contain a series of intergrading forms. For instance in the *S. rexii* aggregate, there is a north-south intergrading series of forms: *S. parviflorus* – *S. cyaneus* – *S. primulifolius* – *S. rexii*, in that order (there is even a well-marked satellite species, *S. gardenii*). But here the constituent species can be retained, as the chaos of non-hierarchical variation has not reached up so extensively from where it is natural: at the population level. Nevertheless, it would be perfectly possible to lump the component species of the *rexii* group and even, for a serious lumper, to deny the utility of subspecies.

TABLE 3. A classification of species types, following Table 1, for the genus *Streptocarpus* subgenus *Streptocarpus* in Africa. Data modified after Hilliard & Burtt (1971) with their 'aggregate species' doing duty for 'ochlospecies' for the sake of comparison.

Species relationship:	*Isolated*	*Group*	Total
Species type:			
Monotypic	49	8	57
Polytypic	4	3	7
Aggregate	1	1	2
Total	54	12	66

I have no doubt of the greater complexity of the *Diospyros mespiliformis* ochlospecies than of the *Streptocarpus rexii* aggregate, but we must nevertheless consider the possibility that what stands between the ochlospecies and the aggregate is merely a dash of personal preference and a pinch more intergradation. That would put the ochlospecies in the familiar realm of the age-old dialogue between splitters and lumpers. Against this is the perfectly reasonable contention that where so much complex polymorphism extends over such a wide geographical range and comprises so much variation, as in *D. mespiliformis* (Fig. 1), then some special and unknown process is at work, putting the ochlospecies outside the ordinary splitting/lumping continuum. The splitting/lumping debate is then confined to resolving the difference between polytypic species (a species with numerous subspecies) and the species group (several closely related species). In this we should bear in mind that polymorphic variation, being segregational, is fundamentally different from species differences, which are differences in particular gene combinations, often very complex combinations. The absence of discrete combinatorial variation distinguishes the polymorphic ochlospecies. Nevertheless, we should expect, and indeed do find, intergradation between these two types of variation.

Mention must be made here of Gentry's (1990) criticism of Pennington's taxonomic treatment of entities under the *Guarea glabra* ochlospecies. He noted that at the La Selva field station in Costa Rica six of the entities co-occur and are clearly distinguishable in the field. Here they are recognised as good species, passing locally what Gentry calls a "test of sympatry", which he suggested should take precedence over the complexity of variation patterns elsewhere. The implication of this is that when poorly known ochlospecies (such as little-studied tropical trees) become well known in the field, and characters that are cryptic in the herbarium are revealed, such as flowering and fruiting phenology, then the ochlospecies will resolve itself, like white light passing through a prism, into a spectrum of distinct forms. If there are then still problems of intergradation, it is incumbent on the taxonomist to try to solve them as best as possible, rather than to conceal important variation within the ochlospecies (although White intended ochlospecies variation to be well illustrated by means of pictorialized distribution maps). Further work is needed to clarify how many ochlospecies will be resolved in this way, but Gentry did concede that there appears to be a difference between the neotropics, where he contends that variation patterns are clear-cut, and Africa, where variation patterns appear to be more complex. Three further caveats need to be put on Gentry's critique.

Firstly, it is not clear to me why a morphotype should not behave as a good species in one place (e.g. La Selva) and not elsewhere. Indeed, it would seem an evolutionary axiom that this should occasionally be the case. A good example is that studied by Beeks (1962) in *Diplacus* (Scrophulariaceae). In Southern California, complete intergradation occurs between *D. longiflorus* and *D. puniceus,* and they behave therefore as intergrading races. However, in parts of the Santa Ana Mountains populations of the two entities occur together without intergradation, and behave as sympatric species.

Secondly, it is still disputed whether variation patterns in South America are uniformly clear-cut in comparison to Africa as Gentry suggested from his studies of the Bignoniaceae. Variation patterns do indeed appear to be clear-cut in the geologically old regions of the Guianan and Brazilian shields, but are not always so in the Andes or in Central America (T. Pennington, pers. comm.). This observation is based on repeated subjective inference from identification of large field collections from different parts of South America, and if confirmed would imply a process of 'variation sorting' in geological time.

Thirdly, it could be argued that if there are indeed serious problems with the recognition of taxa, at least over part of the range, then it is incumbent on the taxonomist to admit these problems frankly under the ochlospecies, and not to offer a false dawn of a workable taxonomic treatment where none exists. Thus we come back to our bottom line: whether ochlospecies is part of the age-old debate between splitters and lumpers, or represents variation that is genuinely outside that debate.

Examples of ochlospecies and their patterns

Tropical trees

The abundance of tropical trees among the recorded cases of ochlospecies (Table 4) begs the question of whether this pattern of variation is especially characteristic of tropical woody plants. I suspect, however, that the preponderance is a result of the research interests of those prepared to use the term ochlospecies (White worked almost exclusively on tropical woody plants). It may also reflect the discovery procedure for ochlospecies which involves thorough revision of a relatively large genus over a wide geographical area. Such studies, often associated with the *Flora Malesiana* or *Flora Neotropica* projects, have generally been tropical and, because of the economic, botanical and ecological importance of trees, have often involved woody plants.

TABLE 4. A table of some actual or putative ochlospecies.

Dicotyledons

Impatiens hawkeri	Balsaminaceae	Grey-Wilson (1979)
Licania apetala	Chrysobalanaceae	Prance (1972)
Licania heteromorpha	Chrysobalanaceae	Prance (1972)
Diospyros ferrea	Ebenaceae	White (1962)
Diospyros mespiliformis	Ebenaceae	White (1962)
Diospyros natalensis	Ebenaceae	White (1962, 1998)
Diospyros parviflora	Ebenaceae	White (1993)
Chisocheton lasiocarpus	Meliaceae	Mabberley (1979)
Cedrela odorata	Meliaceae	Pennington (1981)
Guarea glabra	Meliaceae	Pennington (1981)
Trichilia pallida	Meliaceae	Pennington (1981)
Trichilia micrantha	Meliaceae	Pennington (1981)
Vavaea amicorum	Meliaceae	Pennington (1969)
Ficus thonningii	Moraceae	Berg & Wiebes (1992)
		Berg & Hijman (1989)
Cybianthus spicatus	Myrsinaceae	Pipoly (1983)
Triplaris americana	Polygonaceae	Brandbyge (1986)
Myrmecodia tuberosa	Rubiaceae	Huxley (1981)
		Huxley & Jebb (1993)
Drimys piperita	Winteraceae	Vink (1970)

Monocotyledons

Alocasia nicolsonii	Araceae	Hay & Wise (1991)
Cyrtosperma macrotum	Araceae	Hay (1988)
Schizochilus zeyheri	Orchidaceae	Linder (1980)
Pentaschistis pallida	Poaceae	Linder & Ellis (1990)
Vellozia hirsuta	Velloziaceae	Mello-Silva (1990)

Recorded cases of ochlospecies exist from the Myrsinaceae, Chrysobalanaceae and Ebenaceae, but the studies of Pennington (1981) and Mabberley (1979) on the Meliaceae have been particularly fruitful in bringing forward cases of ochlospecies. Pennington prefers to use the term 'complex species' rather than ochlospecies, to avoid imposing a term coined for African ebonies onto other groups in other areas. However, using the criteria suggested here, Pennington's complex species are equivalent to ochlospecies.

Herbaceous and suffruticose ochlospecies

Although there appear to be fewer herbaceous ochlospecies than woody ones, these species are of particular interest as they may provide more tractable experimental material with which to investigate the phenomenon. This is particularly true of the *Impatiens hawkeri* ochlospecies described by Grey-Wilson (1979) in New Guinea. Many of the forms of this ochlospecies are already in cultivation in Europe and North America under a variety of species names.

Another Papuasian species which might form a typical or strong ochlospecies is *Alocasia nicolsonii* as treated by Hay & Wise (1993). However, the local segregation of distinct entities tends to be allopatric rather than sympatric. Another Papuasian species, *Cyrtosperma macrotum,* described by Hay (1988) may also be an ochlospecies (A. Hay pers. comm.).

A case of a strong ochlospecies confirmed by White himself is the widespread ant-inhabited epiphyte of the Rubiaceae, *Myrmecodia tuberosa* (Huxley, 1981; Huxley & Jebb, 1993). In this case 23 names became one, and all the definitional criteria are met. The whole entity forms a continuum of variation from Indochina to Australia and the Philippines to the Solomon Islands (Figs 2 and 3).

Several of the ochlospecies already mentioned come from Papua New Guinea, and there are other examples such as *Chisocheton lasiocarpus* (Mabberley, 1979). Mabberley (1992) implicates the possible exacerbating effect of the geological instability of New Guinea on variation patterns, if those patterns are archaic.

Animal ochlospecies

The only case in which White's term has found its way into zoology is in the case of the polymorphic South American scorpion *Tityus silvestris* (Lourenço, 1988). Lourenço (1994) described a number of variation types to accommodate the major patterns of distribution and differentiation of tropical South American scorpions, defining monomorphic, polytypic, mosaic polymorphic, clinal polymorphic and ochlospecies polymorphic species. His ochlospecies polymorphic type corresponds to the ochlospecies of White. His mosaic polymorphic and clinal polymorphic types have intergradation of forms, but greater ecogeographic patterning to the forms than in the most complex ochlospecies. These two forms may constitute weak ochlospecies of other authors. For instance, *Alocasia nicolsonii*, discussed above, may perhaps be more exactly characterised as a mosaic polymorphic species.

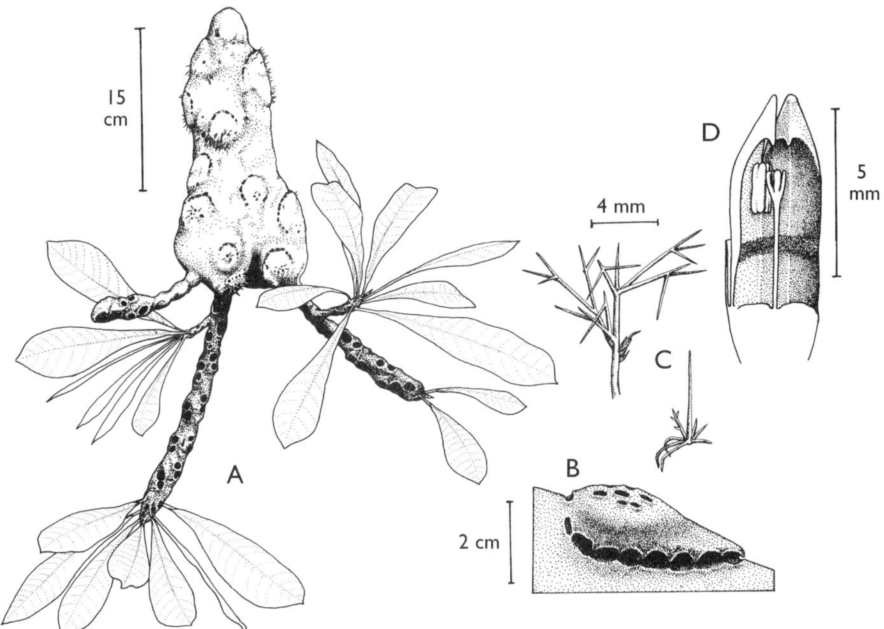

FIG. 2. Example of an ochlospecies from New Guinea, the ant-inhabited epiphyte *Myrmecodia tuberosa* 'versteegii'. **A** habit; **B** tuber surface showing raised area surrounded by entrance holes; **C** spines; **D** section of flower. Note the bract filled alveoli along the stem. Drawn by Eleanor Catherine. Reproduced with permission from Blumea (Huxley & Jebb, 1993).

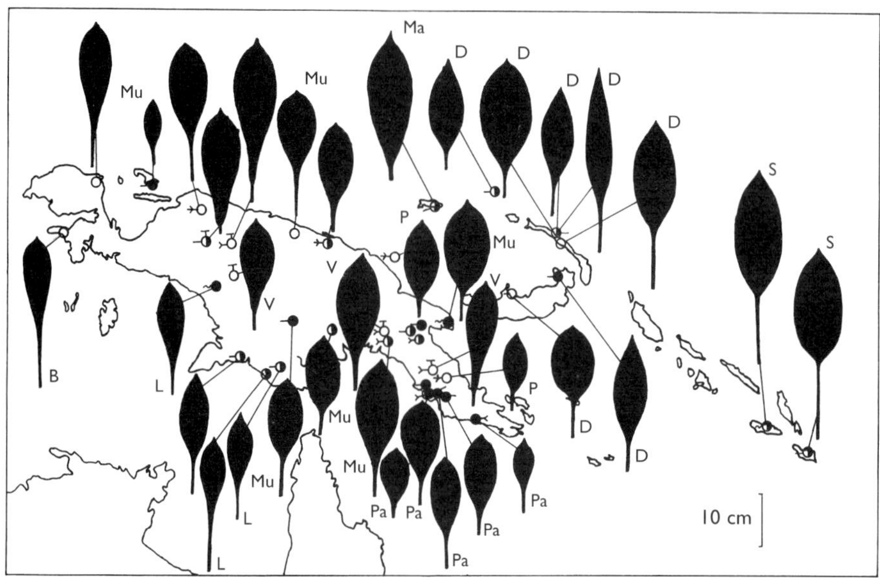

FIG. 3. A pictorialised distribution map of the ant-plant ochlospecies *Myrmecodia tuberosa* in New Guinea and the Solomon Islands. Leaf silhouettes and ideographs depicting spine and bract characters are shown for representative specimens of the eight informally recognized variants. **B** = *M. tuberosa* 'bullosa'; **D** = *M. tuberosa* 'dahlii'; **L** = *M. tuberosa* 'lanceolata'; **Ma** = *M. tuberosa* 'manusensis'; **Mu** = *M. tuberosa* 'muelleri'; **Pa** = *M. tuberosa* 'papuana'; **Pu** = *M. tuberosa* 'pulvinata'; **S** = *M. tuberosa* 'salomonensis'; **V** = *M. tuberosa* 'versteegii'. The ideographs show variation in the following characters: O = bracts not conspicuous; ◑ = bracts filling alveoli in stem; ● = bracts forming cushions protruding from the alveoli; ⌀ = raised areas on tuber surrounded by arcs of entrance holes; O = spines absent or sparse; –O = spines mostly simple; ⊱O = spines mostly branched; +O = spines mostly club-like; ~O = spines mostly sinuate. (The spine symbols are shown at different angles where the ideographs are too close for them to be in the usual position). (From Huxley, 1981).

It is likely that there may be more animal ochlospecies than is immediately suggested by the useful, but simple, prevailing model of species in zoology: the biological species concept. Recent suggestions to replace the zoological concept of the biological species with a phylogenetic species concept, now that the significance of hybridization among differentiated taxa is more adequately acknowledged (Zink & McKitrick, 1995), may lead to increasing reappraisal of complex variation patterns in zoology.

Temperate ochlospecies

All the examples of ochlospecies given here are tropical, which raises the question of whether they are exclusively so. I will argue here that the apparent tropical bias is a historical artefact of the way floristics has developed this century. To explain this idea, an anecdote may be allowable. Several years ago, I put White's ideas of ochlospecies to Max Walters, who had argued cogently that much apparent general taxonomic pattern was eurocentric artefact (Walters, 1963; 1986). I was interested to learn whether Walters considered that ochlospecies could be just such an artefact.

The causes of such artefact can broadly be termed 'psychohistorical' (Cronk, 1990) and stem from two main processes. The first of these is *chaining* whereby new species tend to accrete in higher taxa described a long time ago, usually in Europe by early European botanists. As such entities grow they become more amorphous and easier to place new species inside (Cronk, 1989). This eventually results in colossal, ill-defined and unmanageable higher taxa, such as the genus *Senecio*. The second process is the excessive *splitting* of genera with a long history of eurocentric interest, relative to tropical or less familiar groups. Classic examples are the Umbelliferae (many small genera) versus the Araliaceae (fewer, larger genera) (Walters, 1963) and the papilionoid legumes versus the mimosoid legumes (Cronk, 1990).

Walters did indeed consider that the ochlospecies might be psychohistorical in origin, with the relative undercollection of tropical plants making complex but tractable patterns of variation appear chaotic. This circumstance then might lead to the understandable psychohistorical response of chaining the variation into ochlospecies. Some time afterwards I put this idea to White. He declined to give a detailed response immediately, but I nevertheless gained the impression that he disagreed with this interpretation of the ochlospecies. Later, he gave me an offprint of an unrelated paper, at the top of which he had written: "Max is partly right about the ochlospecies, but for the wrong reasons". Shortly afterwards he became critically ill and I never was able to discuss this cryptic statement with him. Nevertheless, I think I know what he meant: that there are ochlospecies in temperate regions, but that the strong regionalisation of floristic work in these areas for largely eurocentric and political reasons, has prevented the necessary large scale monographic and floristic work that is now common in tropical plant systematics (for instance, that carried out by White and his colleagues, and under the auspices of *Flora Neotropica* and *Flora Malesiana*). The idea that ochlospecies may live unrecognised in temperate regions because of a psychohistorical artefact comes closer to Walters' central thesis that eurocentric bias is the problem, not the solution. For instance, a possible temperate ochlospecies is *Bupleurum falcatum* (Tutin, 1968) which occurs from central Asia to Western Europe exhibiting a complex variation pattern, but has not been the subject of recent critical study over the whole of its range. In this context it is also interesting to consider *Heracleum sphondylium*, although the constituent variation has been treated (with caveats) as a series of ill-defined subspecies (Brummitt, 1968; 1971). Another candidate is the *Anthyllis vulneraria* complex (Cullen, 1968), although again subspecific treatments have been attempted with varying degrees of success.

Explanations of ochlospecies

The Prance hypothesis

Prance (1982) proposed that the Pleistocene refuge theory could act as an explanation of ochlospecies polymorphism. The refuge theory suggests that drier conditions at times of glacial maxima caused wet forest plants to become discontinuous. Some minor differentiation would then occur in these allopatric populations. On the resumption of humid conditions, ecologically adaptable species with good dispersal would then recolonise and different populations could meet. Thus the variation would no longer be geographically correlated. The variation would be further complicated by infraspecific hybridisation between the formerly geographically isolated populations. This kind of ochlospecies bears a strong relation to the 'syngameon of semispecies' concept of Grant (1971), i.e. the sum of a series of incompletely diverged 'species' linked by natural hybridization'. *Iris* species of the series *Californicae* (Lenz, 1959) are an example, in which 11 'species' are linked in a web of hybridization. Only one species, *I. munzii*, stands apart as a reproductively isolated satellite. This process of isolation and refusion can obviously be driven by climatic change, and there is palaeobotanical evidence for this. For instance, Mason (1949) showed that Pleistocene climatic change was linked to patterns of population isolation and refusion in *Pinus remorata* and *P. muricata* in California. In particular, Prance suggested that the divergence-refusion mechanism might be an appropriate explanation for the variation shown by two ochlospecies, *Licania apetala* and *L. heteromorpha* (Chrysobalanaceae) (Prance, 1972).

The rapid expansion hypothesis

As an alternative to the refuge theory, I suggest that comparatively rapid spread of a taxon could lead to a complex variation pattern such as that seen in ochlospecies, without involving a fragmented, allopatric phase. The expansion of a colonising species with wide ecological tolerance will unleash a number of evolutionary forces within it: genetic drift in founding populations, rapid selection as the species colonises new habitats and climatic zones, and the conservation of mutations in an expanding population. Such recent dispersal will give rise to geographically unstructured variation, which will eventually become ecogeographically sorted with time. Zink & Dittman (1993) showed that despite marked geographic variation in size and plumage colour, mitochondrial DNA variation in the song sparrow (*Melospiza melodia*) was unstructured geographically. They suggested that rapid postglaciation dispersal scattered most DNA haplotypes across the range of the species. Morphological variation appears to have evolved faster and has been ecogeographically sorted, unlike the chaotic distribution of mitochondrial types.

Ecological and chorological transgressors, as many ochlospecies are, will have a tendency to occupy diverse niches at a particular site. Thus they are likely to come under strong disruptive selection. If polymorphism is promoted by niche width (Mather, 1955), then disruptive selection will accentuate the polymorphic divergence (Mather, 1955; Thoday & Boam, 1959). Although sympatric speciation may develop out of the polymorphic divergence (Mather, 1955; Thoday & Gibson, 1962), this will take more time and may not happen in all cases. In the case of a rapidly expanding population, variation will be rapidly exposed to a wide array of selectional forces, over a wide geographical range, and a highly complex polymorphic pattern can arise. However, the rapid expansion hypothesis may not be as plausible for more inert, tropical woody species, as for more vagile herbaceous species (and many animal species). In the contrasting case of a very slow spread of a taxon, this would allow equilibration between the variation and the selectional forces, resulting in sympatric speciation and niche specialization. Evolutionary change would therefore be inevitably incremental and not parallel, leading to highly structured patterns of variation.

A taxon cycle (Wilson, 1959; 1961), if it exists (Liebherr & Hajek, 1990), may be a repeated source of such events. The taxon cycle proposes a primitive and general ecological preference for a group and predicts habitat changes and specialization during diversification with a progression from a few invasive, colonising habitat generalists to many, non-invasive, endemic habitat specialists. The process is cyclical: new colonising generalists can be formed. This 'rapid expansion' or 'colonising generalist' hypothesis, like the Prance hypothesis, is amenable to testing.

Tests of hypotheses

First of all there is the possibility of testing the ochlospecies concept itself: that a particular and characteristic non-hierarchical variation pattern does exist, i.e. is not a sampling artefact, and is caused by processes as yet unknown — not the usual complicating factors of hybridisation between good species or apomixis. Once tested, many putative ochlospecies will perhaps be dismissed. *Schismatoglottis longispatha* (Araceae) appears to be an ochlospecies but nothing is known of its breeding system, and it is possible that it is a clonally-reproducing apomictic complex (A. Hay, pers. comm.)

Next must come, assuming the reality of an ochlospecies, the generation and testing of hypotheses to explain the phenomenon. Two such exemplar hypotheses have been given above, but it is not difficult to imagine others. One of the most powerful ways to test and distinguish between the hypotheses given here would be to examine the mitochondrial genome (mtDNA) or microsatellite regions in the chloroplast genome (cpDNA-SSR). The comparison of haplotype distribution with morphotype distribution would be revealing. If the morphotype mosaic was due to ancient historical processes, such as past allopatry, then deep haplotype clades would be predicted, with probably more geographical structuring of the haplotypes than the morphotypes, since polytopic origin of haplotypes would be less likely. The idea that significant genetic admixture occurred between modern humans and the populations they replaced has been tested by just such methods (Manderscheid & Rogers, 1996).

If on the other hand, the patterns were due to rapid population expansion then a shallow haplotype tree would be expected with haplotypes scattered over the distributional range of the species. Sophisticated models have been developed for predicting population history from haplotype distribution using 'mismatch analysis', which is the frequency distribution of degree of haplotype mismatch in pairwise comparisons of all individuals sampled. Such studies have been used to show that there was a Pleistocene population explosion of *Homo sapiens* (Rogers, 1995).

It is highly probable that a concerted study of an ochlospecies, analysing reproductive biology, morphotype and haplotype across the whole distributional range, would unveil important aspects of plant evolution, as the quotation at the top of this article suggests.

Conclusions and prospects

Taxonomic treatment

While ochlospecies polymorphism may give the impression of nature outwitting the taxonomist, this is only the case if an attempt is made to force the variation into a formal taxonomic hierarchy in the first place. It has been truly said: "Never argue with Nature, her first word is a blow" (Lucretius), and taxonomic treatment should be in agreement with nature. A wide species concept with the main nodes of variation distinguished as *informally named morphotypes*, accommodates the need to acknowledge the existence of locally recognizable "entities" (Vink, 1970). It is difficult to produce workable dichotomous keys to such entities because of the existence of a continuum of intermediates, but multi-access, fuzzy-logic computer keys may be a workable alternative. In *Myrmecodia tuberosa*, Huxley & Jebb (1993) recognise 16 informal taxa, 'bullosa', 'dahlii', 'versteegii' etc. (Figs 2 and 3). Mabberley (1979) similarly recognises ten such informal entities in *Chisocheton.*

However, we must frankly recognise that in a situation of very rapid anthropogenic vegetation change, particularly in the tropics, and the consequent conservation concern, a species carries more political weight than an informal variant. Gentry (1990) went as far as to write: 'variation that is swept under the rug is lost from view, whereas that which is grappled with openly, albeit at great cost of taxonomic time and effort, makes information available to the general public... there is a political imperative as well as a scientific one to accord recognition to taxa that can clearly be demonstrated to behave, either genetically or ecologically, as species'. It is up to the individual taxonomist to find the balance between the political species concept (PoSC) and what is justified from the available evidence. Even if that evidence is scanty because of ignorance, it ill behoves us to invent taxa just because they might, probably do, or should, exist.

The hierarchy of nature

The ochlospecies is at the centre of concerns about the limits to hierarchy in nature: the boundary between reticulating population processes (tokogenesis) and dichotomous lineage formation (cladogenesis). In deer mice (Lansman *et al.*, 1983) a remarkable amount of hierarchical structure has been detected at the population level. In contrast, the ochlospecies appears from morphological evidence alone to be an example where a lack of hierarchical structure has crept up the taxonomic scale to the species and superspecies. A hierarchy of taxa is, of course, more properly a hierarchy of characters, and extending the characters under investigation by including genetic information may reveal an otherwise hidden hierarchy, or not. There is no necessary correlation between DNA variation in specific regions and overall morphology, and thus although the amount of DNA variation cannot be expected to be a guide to the species, the study of genomic regions which are evolving at an appropriate rate may be a valuable help in revealing underlying hierarchy. This can also show up parallel evolution (homoplasy) or phenotypic plasticity, features to which morphological characters are especially prone, if these are the root of observed chaotic variation. If there is to be a synthesis between microevolution and speciation, and between taxonomy and population genetics, then the complex patterns presented by ochlospecies would be an interesting place to start.

Acknowledgements

I thank the many people with whom I have been able to discuss ochlospecies and who gave me copies of papers or other information, particularly: Alistair Hay, Camilla Huxley-Lambrick, Jon Lovett, D.J. Mabberley, Rod Page, Terry Pennington, Toby Pennington, Sir Ghillean Prance and R. Zink. I also thank all the participants at the Frank White Memorial Symposium for much stimulating and constructive comment, particularly C.C. Berg, Dick Brummitt, Helen Fortune-Hopkins, Richard Lester, Peter Linder and Robert Scotland.

References

Beeks, R.M. (1962). Variation and hybridization in southern California populations of *Diplacus* (Scrophulariaceae). *Aliso* **5**: 83–122.

Berg, C.C. and Hijman, M.E.E. (1989). Moraceae. In: R.M. Polhill (editor). Flora of Tropical East Africa. A.A. Balkema, Rotterdam.

Berg, C.C. and Wiebes, J.T.(1992). Africa fig trees and fig wasps. *Verh. Kon. Ned. Akad. Wetensch., Afd. Natuurk.*, Tweede Sect., **89**: 1–298.

Brandbyge, J. (1986). A revision of the genus *Triplaris* (Polygonaceae). *Nordic J. Bot.* **6**: 545–570.

Brummitt, R.K. (1968). *Heracleum.* In: T.G. Tutin *et al.* (editors). Flora Europaea, vol. 2, 364–366. Cambridge University Press.

Brummitt, R.K. (1971). Relationship of *Heracleum lanatum* Michx. of North America to *H. sphondylium* of Europe. *Rhodora* **73**: 578–584.

Cronk, Q.C.B. (1989). Measurement of the historical and biological influences on classifications. *Taxon* **38**: 357–370.

Cronk, Q.C.B. (1990). The name of the pea: a quantitative history of legume classification. *New Phytol.* **116**: 163–175.

Cullen, J. (1968). *Anthyllis.* In: T.G. Tutin *et al.* (editors). Flora Europaea, vol. 2, 177–182. Cambridge University Press.

Gentry, A.H. (1990). Herbarium taxonomy versus field knowledge. Is there an attainable solution? *Fl. Males. Bull., special volume* **1**: 31–35.

Grant, V. (1971). Plant Speciation. Columbia University Press, New York.

Grey-Wilson, C. (1979). *Impatiens* in Papuasia. *Kew Bull.* **34**: 661–668.

Hay, A. (1988). *Cyrtosperma* and its Old World allies. *Blumea* **33**: 427–469.

Hay, A. and Wise, R. (1991). The genus *Alocasia* in Australasia. *Blumea* **35**: 499–545.

Hilliard, O. and Burtt, B.L. (1971). *Streptocarpus* : an African plant study. University of Natal Press.

Huxley, C.R. (1981). Evolution and taxonomy of myrmecophytes with particular reference to *Myrmecodia* and *Hydnophytum* (Rubiaceae). D. Phil. thesis, Oxon.

Huxley, C.R. & Jebb, M.H.P. (1993). The tuberous epiphytes of the Rubiaceae 5 : A revision of *Myrmecodia. Blumea* **37**: 271–334.

Lansman, R.A., Avise, J.C., Aquadro, C.F., Shapira, J.F. & Daniel, S.W. (1983). Extensive genetic variation in mitochondrial DNAs among geographic populations of the deer mouse, *Peromyscus maniculatus. Evolution* **37**: 1–16.

Leenhouts, P.W. (1994). *Allophyllus cobbe* (Sapindaceae). In: F.A.C.B. Adema, P.W. Leenhouts & P.C. van Welzen. Flora Malesiana 11 (3): 459. Rijksherbarium/Hortus Botanicus, Leiden.

Lenz, L.W. (1959). Hybridization and speciation in Pacific coast irises. *Aliso* **4**: 237–309.

Liebherr, J.K. and Hajek, A.E. (1990). A cladistic test of the taxon cycle and the taxon pulse hypothesis. *Cladistics* **6**: 39–59.

Linder, H.P. (1980). An annotated revision of *Schizochilus* Sond. (Orchidaceae). *J. S. African Bot.* **46**: 379–434.

Linder, H.P. and Ellis, R.P. (1990). A revision of *Pentaschistis* (Arundineae: Poaceae). *Contr. Bolus Herb.* **12**: 1–124.

Lourenço, W.R. (1988). Biological diversity and specialisation patterns among Amazon scorpions - *Tityus silvestris* Pocock, a particular case of polymorphism. *Compt. Rend. Acad. Sci. Paris, Sér. 3, Sci. Vie* **306**: 463–466.

Lourenço, W.R. (1994). Biogeographic patterns of tropical South American scorpions. *Stud. Neotrop. Fauna & Environm.* **29**: 219–231.

Mabberley, D.J. (1979). The species of *Chisocheton* (Meliaceae). *Bull. Brit. Mus. (Nat. Hist.), Bot.* **6**: 301–386.

Mabberley, D.J. (1992). Tropical Rain Forest Ecology. Ed. 2. Blackie, Glasgow and London.

Manderscheid, E.J. and Rogers, A.R. (1996). Genetic admixture in the late Pleistocene. *Amer. J. Phys. Anthropol.* **100**: 1–5.

Mason, H.L. (1949). Evidence for the genetic submergence of *Pinus remorata*. In: G.L. Jepsen (editor). Genetics, Palaeontology and Evolution. Princeton University Press, Princeton.

Mather, K. (1955). Polymorphism as an outcome of disruptive selection. *Evolution* **9**: 52–61.

Mayr, E. (1992). A local flora and the biological species concept. *Amer. J. Bot.* **79**: 222–238.

Mello-Silva, R. de (1990). Morphological and anatomical differentiation of *Vellozia hirsuta* populations (Velloziaceae). *Plant Syst. Evol.* **177**: 197–208.

Pennington, T.D. (1969). Materials for a monograph of the Meliaceae 1. A revision of the genus *Vavaea. Blumea* **17**: 351–366.

Pennington, T.D. (1981). A monograph of the Neotropical Meliaceae. *Fl. Neotrop. Monogr.* 28.

Pipoly, J.J. (1983). Contributions toward a monograph of *Cybianthus* (Myrsinaceae): III. A revision of subgenus Laxiflorus. *Brittonia* **35**: 61–80.

Prance, G.T. (1972). Chrysobalanaceae. *Fl. Neotrop. Monogr.* 9.

Prance, G.T. (1982). Forest refuges: evidence from woody Angiosperms. In: G. T. Prance (editor). Biological Diversification in the Tropics, Ch. 11. Pp. 137–158, Columbia University Press, New York.

Rogers, A.R. (1995). Genetic evidence for a Pleistocene population explosion. *Evolution* **49**: 608–615.

Sultan, S.E. (1995). Phenotypic plasticity and plant adaptation. *Acta Bot. Neerl.* **44**: 363–383.

Thoday, J.M. and Boam, T.B. (1959). Effects of disruptive selection. II Polymorphism and divergence without isolation. *Heredity* **13**: 205–218.

Thoday, J.M. & Gibson, J.B. (1962). Isolation by disruptive selection. *Nature* **193**: 1164–1166.

Tutin, T.G. (1968). *Bupleurum.* In: T.G. Tutin *et al.* (editors). Flora Europaea, vol. 2, 345–350. Cambridge University Press.

Vink, W. (1970). The Winteraceae of the Old World. 1. *Pseudowintera* and *Drimys* — morphology and taxonomy. *Blumea* **18**: 225–354.

Walters, S.M. (1963). The shaping of Angiosperm classification. *New Phytol.* **60**: 74–84.

Walters, S.M. (1986). The name of the rose, a review of ideas on the European bias in angiosperm classification. *New Phytol.* **104**: 527–546.

White, F. (1962). Geographic variation and speciation in Africa with particular reference to *Diospyros.* In: Taxonomy and Geography. *Publ. Syst. Assoc.* **4:** 71–103.

White, F. (1988). The taxonomy, ecology and chorology of African Ebenaceae II. The non-Guineo-Congolian species of *Diospyros* (excluding sect. *Royena*). *Bull. Jard. Bot. Nat. Belg.* **58**: 325–448.

White, F. (1993). Twenty–two new and little-known species of *Diospyros* (Ebenaceae) from New Caledonia with comments on section *Maba. Bull. Mus. Natl. Hist. Nat.,* B, *Adansonia* **14 (2)**: 179–222.

White, F. (1998). The vegetative structure of African Ebenaceae and the evolution of rheophytes and ring species. In: H.C. Fortune-Hopkins, C.R. Huxley, C.M. Pannell, G.T. Prance and F. White. The Biological Monograph: The importance of field studies and functional syndromes for taxonomy and evolution of tropical plants. Royal Botanic Gardens, Kew.

Wilson, E.O. (1959). Adaptive shift and dispersal in a tropical ant fauna. *Evolution* **13**: 122–144.

Wilson, E.O. (1961). The nature of the taxon cycle in the Melanesian ant fauna. *Amer. Naturalist* **95**: 169–193.

Zhivotovsky, L.A., Feldman, M.W. and Bergman, A. (1996). On the evolution of phenotypic plasticity in a spatially heterogeneous environment. *Evolution* **50**: 547–558.

Zink, R.M. and Dittman, D.L. (1993). Gene flow, refugia and evolution of geographic variation in the song sparrow (*Melospiza melodia*). *Evolution* **47**: 717–729.

Zink, R.M. and McKitrick, M.C. (1995). The debate over species concepts and its implications for ornithology. *Auk* **112**: 701–719.

Sonké, B. & Lejoly, J. (1998). Biodiversity study in Dja Fauna Reserve (Cameroon): using the transect method. In: C.R. Huxley, J.M. Lock and D.F. Cutler (editors). Chorology, Taxonomy and Ecology of the Floras of Africa and Madagascar. Pp. 171–179. Royal Botanic Gardens, Kew.

12. BIODIVERSITY STUDY IN DJA FAUNA RESERVE (CAMEROON): USING THE TRANSECT METHOD

B. Sonké[1] and J. Lejoly[2]

[1]Département des Sciences Biologiques, Ecole Normale Supérieure de Yaoundé, Université de Yaoundé I, B.P. 047 Yaoundé, Cameroun
[2]Laboratoire de Botanique systématique et de Phytosociologie, Université Libre de Bruxelles, CP 169, Av. F. Roosevelt 50, 1050 Bruxelles, Belgique

Résumé

L'étude de la biodiversité végétale est en cours dans la Réserve de Faune du Dja (Cameroun). Cette reserve est située au sud-est du Cameroun (2°50'N–3°30'N et 12°20'E–13°40'E). Le long d'un transect de 29 km, un inventaire de tous les arbres et lianes à dbh ≥ 10 cm a été realisé. Les données d'une portion de 5 km (24–29 km) de ce transect sont ici analysées. Le choix de cette portion tient au fait qu'elle est représentative de l'hétérogenéité structurale de la forêt du Dja. Cette portion est occupée par 82% de forêts primaires, 14% de forêts marécageuses et 4% de forêts sur rocher. La forêt est caractérisée par une diversité élevée des ligneux qui s'accompagne d'une égale répartition des individus entre les espèces présentes. Les espèces d'arbres signalés sur 2.5 ha sont en majorité des espèces de forêts primaires. La densité moyenne des arbres est de 461.2 troncs/ha pour tous les arbres à dbh ≥ 10 cm. La surface terrière totale est de 30.5 m²/ha et un diamètre moyen de 29.1 cm.

Abstract

A biodiversity study has been made of the vegetation in the Dja fauna reserve located in south-eastern Cameroon (2°50'N–3°30'N and 12°20'E–13°40'E). Along a transect of 29 km, a survey of all trees and lianas of a diameter at breast height of at least 10 cm has been realised. Data from 5 km of this transect are analysed here as they illustrate the heterogeneous structure of this forest. Of this 5 km, 82% is occupied by primary forest, 14% by swamp forest, and 4% by forest on rocky sites. The forest is characterised by a high diversity of trees with dbh ≥ 10 cm, 202 species in 45 families in 2.5 ha (125–138 species/ha), and an even spread of the trees among these species, Pielou's equitability index 0.9. Most of the trees recorded in 2.5 ha belong to primary forest species with the Euphorbiaceae having the highest relative importance. The mean density is 461.2 stems/ha for the trees with dbh ≥ 10 cm. The basal area is 30.5 m²/ha and the mean diameter is 29.1 cm.

Introduction

Extensive areas of tropical rain forest are being cleared every year. The area of forest lost between 1986–1990 is estimated at about 11 million hectares of which 5 million become fallow (FAO in Wolter, 1993). If no action is taken, a considerable number of species are likely to become extinct even before they are known to science. Deforestation is a serious threat to biodiversity. An efficient program of management of the forest ecosystem must to be developed if any project for the conservation of the biological diversity is to be implemented. In this respect, a multidisciplinary monitoring program is underway in the Dja forest reserve, with the aim of assessing the dynamics and condition of this ecosystem.

The Dja reserve (2°50'N–3°30'N and 12°20'E–13°40'E) is located in south-eastern Cameroon within the area known as the southern plateau (Letouzey, 1968). This region belongs to mixed moist semi-evergreen Guineo-Congolian rain forest (White F., 1983). A number of species characteristic of mixed moist semi-evergreen lowland rain forest are represented in the Dja reserve: *Entandrophragma angolense, E. candollei, E. cylindricum, Guarea cedrata, G. thompsonii, Lovoa trichiloides, Maranthes glabra, Parkia bicolor, Petersianthus macrocarpus*, etc. Islands of single-dominant forest no more than a few hectares are present in this reserve. They are usually dominated by *Gilbertiodendron dewevrei*. These islands are called single-dominant moist evergreen and semi-evergreen Guineo-Congolian rain forest (White F., 1983). Most of the reserve is at an altitude of between 600 and 700 m. The monthly average temperature lies between 23.5 and 24.5°C and the annual rainfall between 1180 and 2350 mm (Sonké, 1996).

The botanical survey described in this contribution was initiated in 1993 as a part of this program. The main objective was to obtain, by permanent monitoring, data concerning diversity, growth, mortality, regeneration and forest dynamics.

Study method

For the purpose of the survey, line transects were established in the Dja fauna reserve including a 29 km transect running north–south from the edge to the central part of the reserve. Along this transect, all trees and lianas were measured and labelled on a 5 m band for those with diameter at breast height (dbh, at 1.30 m above ground level) of at least 10 cm and on a 50 m band for those with dbh of at least 70 cm. Only data collected on a 5 km section of transect between PK24 and PK29 are selected here for detailed analysis. This portion of the transect has been considered since it appears to best illustrate the heterogeneous structure of the Dja forest ecosystem. Species richness, which is the number of species recorded in the study area, the diversity index of Shannon-Weaver (Daget, 1976; Frontier & Pichod-Viale, 1993) and the equitability index of Pielou (1966) were used to express the results.

The analyses of abundance, frequency and dominance each illustrate a specific aspect of forest structure. The role of each species in the study area was assessed by its relative importance which included relative abundance, relative dominance and relative frequency. The species distribution among the trees is described using the mathematical models of Motomura and Pareto (in Frontier & Pichod-Viale, 1993). The distribution of trees in size classes was analysed using exponential and power functions.

Results

Vegetation

The transect runs through a diversity of landscape including a series of complex habitats, including swamp forest or forest liable to flooding (14%), primary forest on deep dry soil (82%), and forest on rocky sites (4%); the underlying rocks outcrop in two very localised hilly sites carrying a dense forest of shorter structure and different floristic composition. A total of 19 sections corresponding to different forest types can be distinguished on the transect of 5 km, so that on average there is a change every 263 m.

Taxonomic diversity

Altogether 202 species from 45 families were recorded in the 2.5 ha surveyed along the 5 km of transect analysed here. An average of 60 species were recorded per 0.25 ha, giving 125 to 138 species per hectare on this transect. A total of 70 species with dbh >70 cm belonging to 29 families were recorded on a band 50 m wide which corresponds to

25 ha. The Shannon-Weaver diversity index (ISH) is 5.46 bits and Pielou's equitability index (EQ) is 0.9.

Among the species with dbh ≥10 cm (recorded over 2.5 ha), 40 species (20%) make up about 54% of the relative importance index, the greatest value of relative importance being 3.4% (Table 1). Whereas of the 70 species with dbh ≥ 70 cm (Table 2), 13 make up 50% of the relative importance index of the total, the highest relative importance index being 8.3%.

Table 3 shows that only Euphorbiaceae has more than 10% relative importance, followed by Olacaceae and Mimosaceae with 3.5% for each family. Eleven families among the 40 represented on the 5 m transect make up 57.7% of the relative importance value. Euphorbiaceae are more represented than other families; they contribute 19% of the total number of trees, they occupy 11.7% of the basal area and they are represented by 29 species (Table 3). Olacaceae follow with 7.1% of the trees, 8.1% of basal area, seven species belong to this family. Of the large trees (on the 50 m wide transect), Mimosaceae and Caesalpiniaceae both have a relative importance value of more than 10% (Table 4). Six families among 29 make up 52% of the relative importance value.

Abundance distribution

The distribution curve of species abundance recorded on the 5 m transect is shown in Figure 1. The common species are on the left part of the curve, and the rare species (represented by one individual) are on the right forming the tail. The steps correspond to all species with the same abundance. Adjustment of the equation of abundance distribution to Motomura's model gives $y = -1.5x + 7.96$, with $r^2 = 0.93$; where y is natural logarithm of species abundance and x the logarithm of species rank according to abundance. Adjustment to the Pareto model gives $y = 380.8x^{-1.22}$ as equation function with $r^2 = 0.93$, where y is number of species represented by x or more than x trees in the community.

Vegetation structure

A total of 1154 stems were counted; the mean density of trees on this transect is 461 stems/ha for trees with dbh at least 10 cm and 12.7 stems/ha for trees with dbh at least 70 cm. The size class distribution of trees and lianas is shown in Figure 2. This distribution gives a power function with equation $y = 405.41x^{-2.3}$, with $r^2 = 0.97$, where y is the number of trees belong to size class x, and an exponential function with equation $y = 215.39e^{-0.50x}$, with $r^2 = 0.93$.

Specific structure

Among the species recorded, dbh of some does not exceed 20 cm: *Aulacocalyx jasminiflora*, *Chytranthus atroviolaceus*, *Clausena anisata*, *Cola flavo-velutina*, *Drypetes afzelii*, *D. capillipes*, etc.; some species have an L-shaped distribution: *Calpocalyx dinklagei*, *Corynanthe pachyceras*, *Santiria trimera*, *Uapaca guineensis*, etc.; some have an S-shaped distribution: *Pentaclethra macrophylla*, *Petersianthus macrocarpus*; finally some like *Desbordesia glaucescens* have an erratic distribution.

Basal area

The value of the mean basal area was calculated directly using dbh measurements of all stems with dbh ≥ 10 cm, this gave 31 m²/ha. The mean diameter of stems is 29 cm for all trees recorded on 5 m width of the transect, while all trees with dbh ≥ 70 cm have 9.0 m²/ha as the basal area. The total basal area is equal to 30.5 m²/ha. The distribution of basal area by stratum shows that all trees with dbh < 40 cm is 13.1 m²/ha while all trees with 40 cm >dbh < 70 cm includes about 8.4 m²/ha, and all trees with dbh ≥ 70 cm includes 8.99 m²/ha.

TABLE 1. List of 40 species with dbh $\geq$ 10 cm recorded in 2.5 ha at Dja forest with the greatest values of relative importance. a, number of stems/ha; b, basal area (m²/ha); c, relative density (abundance) = percentage ratio of number of individuals of a species over total number of individuals; d, relative dominance = percentage of ratio of basal area of a species over total basal area; e, specific contribution of a species = ratio of frequency of a species over the sum of all frequencies; f, relative importance = (c + d + f)/3.

SPECIES	FAMILY	a	b	c	d	e	f
Pentaclethra macrophylla	Mimosaceae	8.8	2.02	1.9	6.5	1.7	3.4
Desbordesia glaucescens	Irvingiaceae	16.8	1.49	3.6	4.8	1.5	3.3
Tabernaemontana crassa	Apocynaceae	26.4	0.70	5.7	2.3	1.7	3.2
Strombosiopsis tetrandra	Olacaceae	14.0	1.02	3.0	3.3	1.7	2.7
Dichostemma glaucescens	Euphorbiaceae	11.6	0.77	2.5	2.5	1.7	2.2
Santiria trimera	Burseraceae	14.4	0.47	3.1	1.5	1.3	2.0
Pancovia pedicellaris	Sapindaceae	12.4	0.52	2.7	1.7	1.0	1.8
Polyalthia suaveolens	Annonaceae	11.2	0.37	2.4	1.2	1.7	1.8
Erythrophleum suaveolens	Caesalpiniaceae	2.4	1.27	0.5	4.1	0.5	1.7
Strombosia pustulata	Olacaceae	10.8	0.42	2.3	1.4	1.5	1.7
Antidesma laciniatum var. *membranaceum*	Euphorbiaceae	12.0	0.25	2.6	0.8	1.5	1.7
Duboscia macrocarpa	Tiliaceae	4.0	0.77	0.9	2.5	1.0	1.5
Corynanthe pachyceras	Rubiaceae	9.2	0.28	2.0	0.9	1.2	1.4
Uapaca guineensis	Euphorbiaceae	8.8	0.36	1.9	1.2	1.2	1.4
Petersianthus macrocarpus	Lecythidaceae	3.2	0.64	0.7	2.1	1.2	1.3
Calpocalyx dinklagei	Mimosaceae	8.0	0.26	1.7	0.8	1.2	1.2
Cavacoa quintasii	Euphorbiaceae	6.8	0.53	1.5	1.7	0.5	1.2
Dialium zenkeri	Caesalpiniaceae	6.4	0.28	1.4	0.9	1.3	1.2
Entandrophragma candollei	Meliaceae	0.4	1.05	0.1	3.4	0.2	1.2
Allanblackia floribunda	Clusiaceae	3.2	0.63	0.7	2.0	0.5	1.1
Mareyopsis longifolia	Euphorbiaceae	8.0	0.09	1.7	0.3	1.3	1.1
Panda oleosa	Pandaceae	4.4	0.38	1.0	1.2	1.2	1.1
Pteleopsis hylodendron	Combretaceae	1.6	0.76	0.3	2.5	0.5	1.1
Trichoscypha acuminata	Anacardiaceae	6.8	0.16	1.5	0.5	1.3	1.1
Tridesmostemon omphalocarpoides	Sapotaceae	1.6	0.71	0.3	2.3	0.5	1.0
Alstonia boonei	Apocynaceae	1.6	0.63	0.3	2.0	0.5	0.9
Anthonotha macrophylla	Caesalpiniaceae	5.6	0.11	1.2	0.4	1.2	0.9
Blighia welwitschii	Sapindaceae	4.8	0.30	1.0	1.0	0.8	0.9
Heisteria trillesiana	Olacaceae	2.8	0.41	0.6	1.3	0.7	0.9
Sorindeia grandifolia	Anacardiaceae	5.2	0.11	1.1	0.4	1.3	0.9
Homalium discophylum	Flacourtiaceae	3.6	0.27	0.8	0.9	0.8	0.8
Lasiodiscus mannii	Rhamnaceae	4.4	0.09	1.0	0.3	1.0	0.8
Pycnanthus angolensis	Myristicaceae	1.2	0.54	0.3	1.7	0.3	0.8
Strombosia scheffleri	Olacaceae	3.6	0.28	0.8	0.9	0.8	0.8
Trichoscypha arborea	Anacardiaceae	1.2	0.52	0.3	1.7	0.3	0.8
Caloncoba glauca	Flacourtiaceae	3.2	0.17	0.7	0.5	1.0	0.7
Celtis tessmannii	Ulmaceae	2.0	0.31	0.4	1.0	0.8	0.7
Irvingia gabonensis	Irvingiaceae	3.2	0.24	0.7	0.8	0.7	0.7
Pterocarpus soyauxii	Fabaceae	2.0	0.32	0.4	1.0	0.8	0.7
Trichilia rubescens	Meliaceae	3.2	0.21	0.7	0.7	0.8	0.7
Sub Total (40 species)		260.8	20.7	56.3	67.0	40.7	54.4
Others (162 species)			10.26	43.7	23.0	59.3	45.6
TOTAL: 202 species		461.6	30.96	100.0	100.0	100.0	100.0

TABLE 2. List of 13 species dbh ≥ 70 cm recorded in 25 ha at Dja forest with the greatest values of relative importance. a, number of stems/ha; b, basal area (m²/ha); c, relative density (abundance) = percentage ratio of number of individuals of a species over total number of individuals; d, relative dominance = percentage of ratio of basal area of a species over total basal area; e, specific contribution of a species = ratio of frequency of a species over the sum of all frequencies; f, relative importance = (c + d + f)/3.

SPECIES	FAMILY	a	b	c	d	e	f
Pentaclethra macrophylla	Mimosaceae	1.40	0.83	10.7	9.2	5.0	8.3
Alstonia boonei	Apocynaceae	0.70	0.58	5.7	6.5	4.5	5.6
Petersianthus macrocarpus	Lecythidaceae	0.80	0.49	6.6	5.5	4.5	5.5
Desbordesia glaucescens	Irvingiaceae	0.60	0.36	5.0	4.0	3.5	4.2
Duboscia macrocarpa	Tiliaceae	0.50	0.39	3.8	4.3	3.5	3.9
Erythrophleum suaveolens	Caesalpiniaceae	0.40	0.42	3.5	4.7	2.5	3.6
Piptadeniastrum africanum	Mimosaceae	0.40	0.39	3.1	4.3	3.0	3.5
Distemonanthus benthamianus	Caesalpiniaceae	0.40	0.26	3.5	2.9	3.5	3.3
Cylicodiscus gabunensis	Mimosaceae	0.30	0.34	2.2	3.8	2.5	2.8
Klainedoxa gabonensis	Irvingiaceae	0.30	0.22	2.5	2.4	3.0	2.6
Pteleopsis hylodendron	Combretaceae	0.30	0.28	2.5	3.1	2.0	2.5
Terminalia superba	Combretaceae	0.40	0.22	3.1	2.4	2.0	2.5
Uapaca paludosa	Euphorbiaceae	0.30	0.21	2.2	2.3	3.0	2.5
Sub Total (13 species)		6.80	4.99	54.40	55.40	42.50	50.80
Others (57 species)		5.90	4.00	45.6	44.6	57.5	49.2
TOTAL: 70 species		12.70	8.99	100.0	100.0	100.0	100.0

TABLE 3. List of 11 families with stems of dbh ≥ 10 cm (recorded in 2.5 ha) with the greatest values of relative importance. a, basal area (m²/ha); b, relative density (abundance) (%); c, relative dominance (%); d, family contribution (%); e, relative importance (%); f, number of species belonging to the family; g, relative diversity = the number of species of that family as a percentage of the total number of species.

FAMILY	a	b	c	d	e	f	g
Euphorbiaceae	3.61	19.0	11.7	3.6	11.4	29	14.4
Olacaceae	2.50	7.1	8.1	3.6	6.3	7	3.5
Mimosaceae	2.90	4.2	9.4	3.6	5.7	7	3.5
Apocynaceae	1.42	6.8	4.6	3.6	5.0	6	3.0
Caesalpiniaceae	2.22	4.2	7.2	3.6	5.0	8	4.0
Irvingiaceae	1.88	5.3	6.1	3.2	4.9	4	2.0
Annonaceae	1.11	6.2	3.6	3.6	4.5	13	6.4
Sapindaceae	1.26	5.7	4.1	3.2	4.3	8	4.0
Meliaceae	1.78	2.0	5.8	3.2	3.7	10	5.0
Rubiaceae	0.70	5.3	2.3	3.2	3.6	17	8.4
Anacardiaceae	0.91	3.6	2.9	3.6	3.4	7	3.5
Sub Total (11 families)	20.29	69.40	65.80	38.00	57.80	116	57.70
Others (34 families)	10.66	30.6	34.2	62.0	42.2	86	42.3
TOTAL : 45 families	30.95	100.0	100.0	100.0	100.0	202	100.0

TABLE 4. List of six families with stems of dbh ≥ 70 cm recorded in 25 ha at Dja with the greatest values of relative importance. a, basal area (m²/ha); b, relative density (abundance) (%); c, relative dominance (%); d, family contribution (%); e, relative importance (%); f, number of species belonging to the family; g, relative diversity of the family = percentage of ratio of number of the species for the families over total number of the species.

FAMILY	a	b	c	d	e	f	g
Mimosaceae	1.66	17.0	18.5	7.1	14.2	6	8.7
Caesalpiniaceae	1.05	12.3	11.7	7.1	10.4	7	10.1
Irvingiaceae	0.81	10.4	9.0	7.1	8.8	5	7.2
Meliaceae	0.78	6.0	8.7	4.3	6.3	8	11.6
Apocynaceae	0.58	5.7	6.5	6.4	6.2	1	1.4
Lecythidaceae	0.49	6.6	5.5	6.4	6.2	1	1.4
Sub Total (6 families)	5.37	58.0	59.9	38.4	52.1	28	40.4
Others (23 families)	3.59	42.0	39.1	61.6	47.9	52	59.6
TOTAL : 29 families	8.96	100.0	100.0	100.0	100.0	80	100.0

Discussion and Conclusion

According to Barbault (1992), a species diversity index which takes into account the total number of the species and their relative abundance offers a good description of communities and also allows comparison between them. This definition is similar to that of Frontier & Pichod-Viale (1993), which takes into account the number of taxa and how the individuals are distributed among the taxa. The equitability index of Pielou (EQ), varies between 0 and 1. It is close to 0 when only one species has been recorded or for a community dominated by one or few species and close to 1 when many species are represented by a similar number of individuals. The number of species (202 in 2.5 ha) recorded on this transect is greater than in other parts of the reserve. This value is indicative of a relatively young forest; however, it is important to point out that in young forest there is an unequal distribution of abundance among species. In the present case, the equitability of Pielou is 0.9, suggesting that there is an even representation of abundance among the species. The diversity index (ISH) is close to the values calculated in other parts of the reserve, between 5.24 and 5.26 bits (Sonké, 1996); it is greater than the values obtained by White, J.T.L. (1992) in Gabon. The analysis of Shannon-Weaver's index and the equitability suggest that the forest ecosystem is at a pre-climax stage where species diversity is maximal. This arises from the occurrence of both shade-tolerant and light demanding species typical of early successional stages of forest (Ramade, 1994).

The results of floristic analysis shows that no one species is dominant here. It is important to point out that in the forests which show the phenomenon of specific dominance, dominant species can comprise up to 75% of basal area and 58% of the total individuals (Wolter, 1993), but that is not the case here. Floristic analysis tends to confirm the impressive heterogeneity of the Dja forest. It also underlines the importance of Euphorbiaceae among trees with dbh ≥ 10 cm and also Mimosaceae and Caesalpiniaceae in the trees size classes with dbh ≥ 70 cm.

Statistical analysis of the species abundance within a given community is of great importance as it allows interpretation of community structure and interaction between species, and suggests factors which may influence their relative abundance. It will be suitable to consider a more detailed description which is given by the shape of the abundance distribution of species within the study area. Species abundance curves

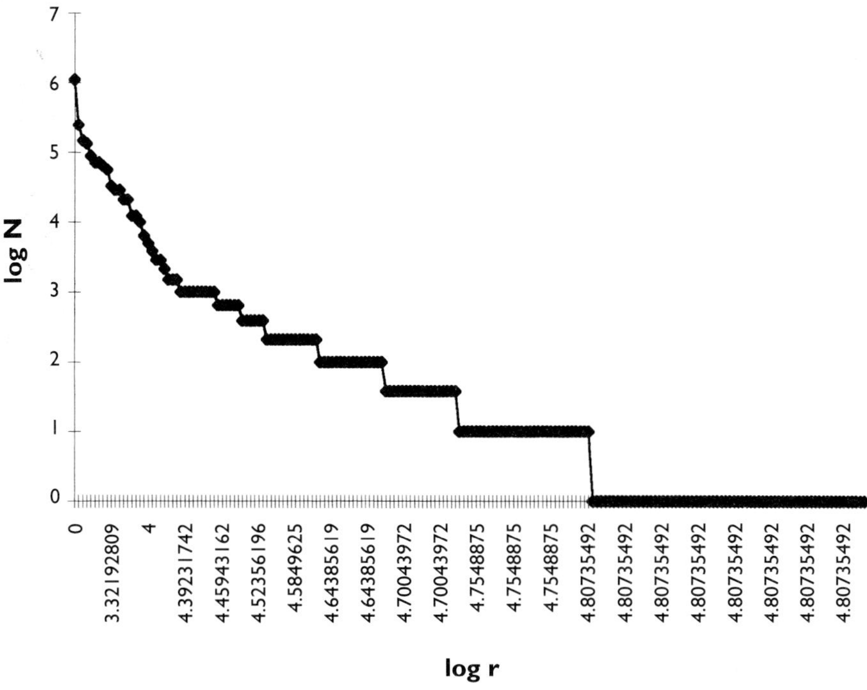

FIG. 1. Species abundance curve for all trees and lianas with dbh ≥ 10 cm on 2.5 ha at Dja forest, Cameroon. Species rank (r) according to abundance (N).

(species ranked according to diversity) corresponding to the models of distribution allow us to understand the species richness of the vegetation community and then calculate numerical indices.

The two fundamental components of diversity (number of species and regularity of the distribution of individuals among species) are directly legible on the curve. The number of species is shown by the extension of the curve to the right and the regularity, by the convex or concave shape. The construction of these curves shows that the abundance distributions obtained for most communities studied correspond in general to three major categories described respectively by MacArthur, Motomura and Preston (Ramade, 1994). We have tried to fit our distributions of species abundance to the Motomura and Pareto models. These models have been used by other authors to study tropical forests (Rollet, 1974; Devineau, 1984). The values of the correlation coefficient are 0.93 for the two models. These high values indicate a high degree of fit. According to Devineau (1984), a weak value of slope of Pareto corresponds to a relatively great importance of well-represented species, showing the phenomenon of dominance and low diversity. A high value of the slope corresponds to elongated distributions, with many scarce species and a greater diversity; this is the case for the forest studied here where the value of slope is -1.5 for Motomura model and -1.22 for Pareto model. These values allow us to conclude that the diversity is great here.

In any vegetation community, the size class distribution of trees takes the shape of a decreasing curve; many authors are interested in the study of this distribution curve which is important in management. De Liocourt (1898) pointed out that the stems of a "normally constituted" fir plantation should be distributed between size class following a

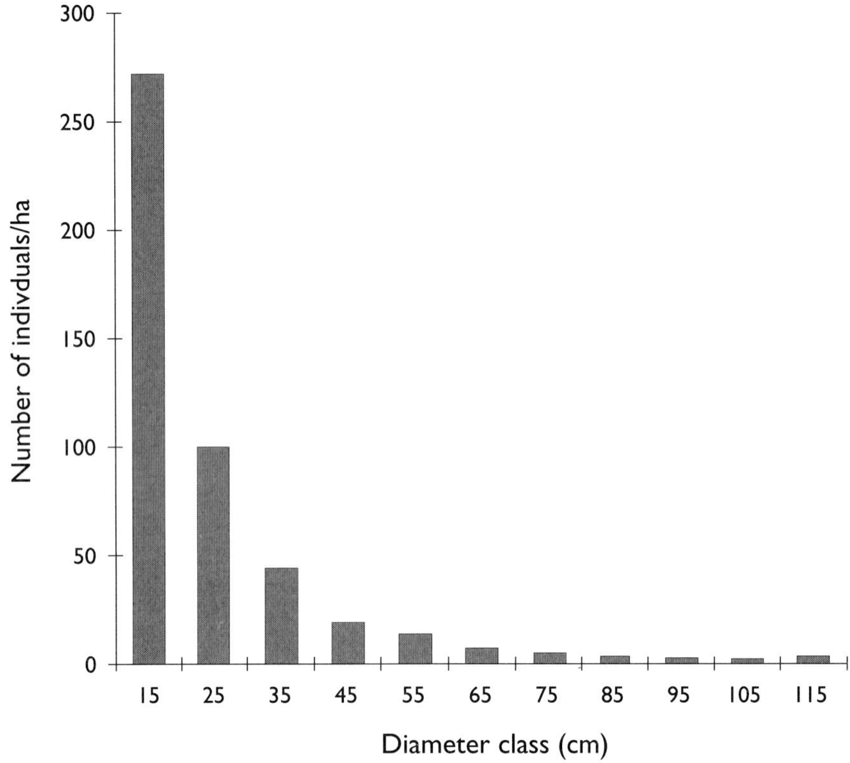

F‌IG. 2. Size class distribution of trees and lianas at Dja forest.

geometric series with a ratio less than 1. Meyer (1933–1952) and Lenger (1954) proposed an exponential expression to describe the structure of vegetation communities. Lecacheux (1955) has suggested a log-normal law for frequency distribution of diameter (D); which means the logarithm of D follows a normal distribution (or Laplace-Gauss). Pierlot (1966) observed that the adjustment to an exponential distribution does not agree for small diameters. He suggested that the adjustment to an hyperbola for better understanding of the size class distribution of trees. According to Rollet (1974) all forests in equilibrium show a tendency to an exponential distribution. We have fitted our structure to power and exponential functions; these fit very well, with values of differential determination coefficient, $r^2 = 0.97$ for the power function and $r^2 = 0.93$ for the exponential function. It is possible to make a correlation between distribution and ecological preference of species; the shade trees have an L-shaped distribution, whereas the sun species show an erratic shape or that of a flat bell (Rollet, 1974).

The total basal area taking into account all trees of dbh at least 10 cm was 31 m^2/ha and 9 m^2/ha for trees with dbh at least 70 cm. The difference between the values for the different widths of transect is due to inadequate sampling of big trees (dbh $\geq$ 70 cm) on the 5 m transect.

This great diversity of the trees in central part of the Dja reserve (as studied by transect) and in the Dja region in general, can be related to the variable nature of the topography. The transect method takes into account this characteristic and quantifies the juxtaposition of dry, swampy, and rocky forests. Of the 202 species of trees recorded in 2.5 ha of this forest, most are from primary forest; in addition to these species there are successional species and species specialised to different types of habitat. The origin of the strongly disturbed areas along this transect is still uncertain. Possible causes of forest gaps include wind, large mammals and human activity such as clearing for farming. In conclusion the transect method gives an idea of the global diversity of an area integrating different types of forest.

Acknowledgements

This study was sponsored by the ECOFAC (EEC) program. We would also like to thank the Cameroonians and Europeans who helped us and provided facilities. We are especially grateful to Dr. R. Fotso and Dr. G. Achoundong.

References

Barbault, R. (1992). Ecologie des peuplements. Structure, dynamique et évolution. Masson, Paris.

Daget, J. (1976). Les méthodes mathématiques en écologie. Masson, Collection d'Ecologie, Paris.

De Liocourt, F. (1898). De l'aménagement des sapinières. *Bull. Soc. Forest. Franche-Comté & Belfort* **4 (6)**: 396–409.

Devineau, J.L. (1984). Structure et dynamique de quelques forêts tropophiles de l'ouest africain (Côte d'Ivoire), Programme MAB savane. Université d'Abidjan 5: 249.

Frontier, S. and Pichod-Viale, D. (1993). Ecosystèmes: structure, fonctionnement, évolution. Masson, Collection d'Ecologie 21, Paris.

Lecacheux, P. (1955). Analyse statistique de la structure de la forêt tropicale en vue de son utilisation pour la production de cellulose. *J. Agric. Trop. Bot. Appl.* **2**: 1–17.

Lenger, A. (1954). A propos d'une loi mathématique simple concernant la structure équilibrée des peuplements forestiers. *Bull. Inst. Agron. État. Gembloux* **22**: 241–254.

Letouzey, R. (1968). Etude phytogéographique du Cameroun. Editions P. Lechevalier, Paris.

Meyer, H.A. (1952). Structure, growth and drain in balanced uneven-aged forest. *J. Forest. (Washington)* **50**: 85–95.

Pielou, E.C. (1966). Species diversity and patten diversity in study of ecological succession. *J. Theor. Biol.* **10**: 370–383.

Pierlot, R. (1966). Structure et composition des forêts denses d'Afrique centrale, spécialement celles du Kivu. *Mém. Acad. Roy. Sci. Outre-Mer, Cl. Sci. Nat. Méd., Collect. 8vo.* **16**: 1–137.

Ramade, F. (1994). Eléments d'écologie; écologie fondamentale. 2e édition, Ediscience Internationale.

Rollet, B. (1974). L'architecture des forêts denses humides sempervirentes de plaines. C.T.F.T., Paris.

Sonké, B. (1996). Synthèse des données des inventaires floristiques dans la Réserve de Faune du Dja (Cameroun). Projet Ecofac, Agreco-CTFT.

White, F. (1983). The Vegetation of Africa: a descriptive memoir to accompany the UNESCO/AETFAT/UNSO vegetation map of Africa. *Natural Resources Research* **20**. UNESCO, Paris.

White, J.T.L. (1992). Vegetation history and logging disturbance: Effects on rain forest mammals in the Lopé reserve, Gabon (with special emphasis on elephants and apes). Ph.D. thesis. University of Edinburgh.

Wolter, F. (1993). Etude des possibilités techniques, économiques et financières d'un aménagement des forêts tropicales denses humides de la cuvette centrale du Zaïre, basé sur ses capacités naturelles. Thèse de doctorat. Université de Louvain.

Lock, J.M. (1998). Aspects of fire in tropical African vegetation. In: C.R. Huxley, J.M. Lock and D.F. Cutler (editors). Chorology, Taxonomy and Ecology of the Floras of Africa and Madagascar. Pp. 181–189. Royal Botanic Gardens, Kew.

13. ASPECTS OF FIRE IN TROPICAL AFRICAN VEGETATION

J.M. Lock

Royal Botanic Gardens, Kew, Richmond, Surrey, TW9 3AB, UK

Abstract

Frank White mentioned the fire factor in African vegetation in various papers. This contribution covers those aspects which are particularly relevant to his work, and those on which his work has substantial bearing. His *Vegetation Map of Africa* can be used to estimate quantities of material burned and carbon dioxide released each year. Research relevant to the balance between grassland and woodland is summarised. Some adaptations to fire, promoting regeneration of both woody and herbaceous species, are described. White's ideas on geoxylic suffrutices are discussed and it is concluded that he was mistaken; this growth-form is much more likely to be an adaptation to fire than, as he suggested, to seasonally waterlogged, nutrient-poor soils.

Introduction

Fire in vegetation in Africa is both a real problem and an emotive subject. Few countries do not have some kind of legislation, usually hopelessly impractical and totally unenforceable, which is supposed to limit when fires occur or, more often, when they should not. There is also an extensive literature but much of this is based on personal opinion; the number of actual long-term experiments that have been carried out in Africa is small. For Africa, the earlier literature was reviewed by West (1965). The whole topic of fire ecology has been treated recently by Whelan (1995), although this work has some Australian bias, and by Bond & van Wilgen (1996); this latter is somewhat slanted by the experience of the authors in the Mediterranean-type ecosystems of the Cape flora. The views in the present paper are based on experience in the seasonally dry regions of tropical Africa.

Fire is important at a range of scales. On a global scale, the burning of grassland contributes a very substantial amount of carbon dioxide and other gases to the atmosphere each year, although, of course, all or most of the carbon dioxide is fixed again in the next growing season. At the ecosystem scale, fire has undoubtedly had great effects on the distribution of vegetation types in Africa. At even smaller scales, many plant species show features which allow them to survive and reproduce successfully under a regime of regular fires. Frank White made contributions to our understanding of fire at all these scales. This paper tries to evaluate these contributions, rather than attempting a full review.

The global scale

Crutzen & Andreae (1990) and Helas (1995), among others, have drawn attention to the substantial contribution to the global CO_2 equation made by the burning of tropical vegetation. Helas (1995) also gave figures for other gases released by biomass burning; for example, of the total annual inputs to the atmosphere, 26% of carbon

monoxide (CO), 24% of nitrogen oxides (NO_x), and 39% of ozone (O_3) derives from biomass burning. Helas (1995) estimated that vegetation burning releases more than 4000 Tg (g × 10^{12}) of carbon per annum world-wide. It is possible, thanks to Frank White's classification and mapping of African vegetation (1983), to make estimates for Africa alone. To estimate the quantity of carbon released, the following data are needed:
– the standing crop of dry matter at the end of the growing season
– the proportion of the standing crop that is burned
– the potential area burned
– the proportion of that area that is burned in each year.

Of these four variables, by far the largest is the area burned. White's (1983) estimates of the area of each of his phytochoria, together with inspection of his accompanying maps, allow an estimate to be made for this. Table 1 shows the estimates that can be derived in this way. The potential area burned can be taken from the figures for whole phytochoria in White (1983), corrected to take account of the proportion of that phytochorion which is likely to burn. Thus 90% of the Zambezian and Sudanian regions are covered by vegetation that potentially burns every year, as against 10% for the Guineo-Congolian region. As a working assumption, 75% of the possible area is probably burned each year. The percentage of the area burned is at least partly dependent on the numbers of people using the area, but the effects of people are of two contrasting kinds. More people means more hunters, honey-gatherers and pastoralists, all of whom start fires, but in many areas it also means more

TABLE 1. Areas, and areas liable to burning, in White's (1983) African phytochoria.

Phytochorion	Area (km²)	Grassland (%)	Area of grassland (km²)
Guineo-Congolian	2 800 000	10	280 000
Zambezian	3 770 000	90	3 393 000
Sudanian	3 731 000	90	3 357 900
Somalia-Masai	1 873 000	30	561 900
Cape	71 000	–	–
Karroo-Namib	661 000	–	–
Mediterranean	330 000	–	–
Afromontane	715 000	10	71 500
Guineo-Congolian / Zambezian	705 000	80	564 000
Guineo-Congolian / Sudanian	1 165 000	90	1 048 500
Lake Victoria	224 000	50	112 000
Zanzibar-Inhambane	336 000	40	134 400
Kalahari-Highveld	1 223 000	50	612 500
Togaland-Pondoland	148 000	40	59 200
Sahel	2 482 000	10	248 200
Sahara		–	–
TOTAL			**10 443 100**

Assume: 75% of potential area is burned each year.
　　　　All burned material is carbohydrate (CH_2O)
　　　　Standing crop (dry matter) of 500 g m⁻²
　　　　80% combustion of dry matter
(For discussion, see text).
Then: 4700 Tg CO_2 would be released into the atmosphere annually by biomass burning in Africa.

cultivation which destroys the natural vegetation and creates potential fire-breaks. The standing crop of dry matter at the end of the growing season can be taken as averaging 500 t km^{-2}, based on figures in César & Menaut (1974), Strugnell & Pigott (1977), Penning de Vries & Djiteye (1982), Howell *et al.* (1988) and Long *et al.* (1992). Of this, about 80% is burned in a fire (based on figures in Howell *et al.* (1988)). Using these figures (see Table 1), the total annual combustible savanna biomass of Africa can be estimated as 5222 Tg. Of this, not all will be burned; if 80% of the material is combusted and 75% of the potential area is burned, then 3134 Tg would be burned each year. This is well within the estimates (from different authors and methodologies) listed by Helas (1995): 3822, 2428 and 2520 Tg. Without the detailed mapping and description of the vegetation of the continent given by White these estimates could not be made.

The ecosystem scale

Fire has considerable effects on the balance between vegetation types in Africa, particularly on the balance between forest and savanna. This balance has been the subject of a number of long-term studies (e.g., Charter & Keay, 1960 (Nigeria); Trapnell, 1959, and Lawton, this volume (Zambia); Swaine *et al.*, 1992 (Ghana); Louppe *et al.*, 1995 (Côte d'Ivoire)). The study sites for these experiments were close to the forest-savanna boundary, and in all of them exclusion of fire allowed forest trees and shrubs to colonise the grassland. At other sites, further from the forest boundary (e.g. Brookman-Amissah *et al.*, 1980 (Ghana); Chidumayo, 1988 (Zambia)), colonisation is slower or does not occur at all, although there tends to be an increase in the proportion of woody vegetation. All of these studies compared not only the effect of complete exclusion of fire, but also the effects of burning at the beginning of the dry season ('early burning') and at the end ('late burning'). Early fires are relatively cool and incomplete while late fires are hot and leave few residues. In the sites near the forest-savanna boundary, the effects of early burning were generally similar to those of fire exclusion although slower in taking effect.

The individual plant scale

The plants of regularly burned sites have developed characteristics or strategies which allow them to both survive and reproduce in spite of the occurrence of regular fires. Most of these depend on the insulating properties of either bark or soil. It has often (e.g. Hopkins, 1965) been pointed out that savanna trees have thicker bark than those of forest — but I am not aware of any sets of comparative measurements, although Gill & Ashton (1968) measured the rate of heat conduction in eucalypts with different bark types.

The regeneration of plants, particularly woody ones, in regularly burned sites, is difficult. The first stage at which the effects of fire may be avoided are in the dispersal and placement of the seeds. Soil is a poor conductor of heat (Priestley, 1959) and a covering of only a few centimetres of soil will protect a seed from the heat of a fire. Indeed, in tropical grasslands the increased heating of the soil by the sun after the removal of the grass cover and the blackening of the soil surface by fire, may well produce greater soil heating than the fire itself (Tothill, 1969; Lock & Milburn, 1971). It has been shown that the awned dispersal units of many savanna grasses can bury themselves in the ground by hygroscopic movements of the awn (Darwin, 1876; Lock & Milburn, 1971). Do any trees possess similar mechanisms — or does the burying of tree seeds by scatter-hoarding rodents perform the same function?

Increased germination of many species after fire has also often been noticed (e.g. Tothill, 1969; Lock & Milburn, 1971). The precise factor that stimulates germination remains, however, elusive. The above authors believed that temperature was responsible — either the increased maxima, or the increased daily range that follow removal of the grass cover. Cresswell & Nelson (1971) found that boron broke dormancy and stimulated germination of the grass *Themeda triandra* but unless significant quantities of boron are released into ash after burning, it is hard to see how this could be significant in the field. Recent work has tended to emphasise the role of substances in smoke as a stimulant to germination (e.g., Baxter *et al.*, 1994).

Study of the germination of several savanna trees in West Africa has shown that their young stages are well adapted to resist fire. Jackson (1968) has described the germination of *Vitellaria paradoxa* (*Butyrospermum paradoxum)* (Sapotaceae). This tree has a large seed which germinates on the surface of the ground. The cotyledons remain within the seed coat while what appears to be a radicle emerges and grows downwards into the soil. From this, below soil level, arises the plumule, which breaks out from the apparent radicle and grows upwards, bearing several scale leaves before emerging from the ground and producing green leaves. As a result of this pattern of growth, there are buds below ground which can develop if the first above-ground shoot is destroyed by fire. Jackson (1974) described similar behaviour in several species of *Combretum* (Combretaceae), although in this genus it is the fused stalks of the cotyledons which elongate and carry the plumular bud below the surface. He found similar, if less marked, seedling morphology in other woody savanna species in Combretaceae (*Guiera, Quisqualis*), Ochnaceae (*Lophira*), Rubiaceae (*Gardenia*) and Leguminosae (*Pterocarpus, Piliostigma*). Various South American palms also have seedlings in which the plumule first grows downwards into the soil before turning upwards (Corner, 1966: 90–91), so that the growing point ends up buried beneath the soil.

Subsequent development of the seedling may also produce structures which allow survival after fire. In *Cussonia barteri* (Araliaceae), a West African savanna tree, the hypocotyl of the seeding swells to form a subterranean tuber (César & Menaut, 1974). When the aerial shoot is killed by fire, new shoots are produced from the apex of the tuber, which thus resembles the lignotubers of many species of *Eucalyptus* (Myrtaceae), which can be extremely large (Whelan, 1995). Similar structures are found in many woody plants of African savanna (see, e.g., Lebrun, 1947: 595) although their development has only been studied in *Cussonia*. Plants with substantial woody underground parts which produce annual above-ground shoots which flower and fruit were referred to as geoxylic suffrutices and discussed at length by White (1977) (Fig. 1).

White restricted his account to those species which occur in genera which consist otherwise of large woody plants, excluding the many suffrutices which belong to groups of shrubby or herbaceous habit, and those species whose aerial shoots are only facultatively annual. He pointed out that there appear to be more plants of this habit in Africa than anywhere else in the tropics, although they are frequent in the dry savanna vegetation of tropical South America and recent work on this region may have shown that they are commoner than he believed. He listed many more species from the Zambezian Region than from the Sudanian; while this is probably a true reflection of relative numbers of species, he omits Sudanian taxa such as *Cochlospermum planchonii* (Lawson *et al.*, 1968) and *Annona glauca.*

White (1977) considered three possible ecological factors as favouring the geoxylic suffrutex — frost, fire, and oligotrophic seasonally waterlogged soils. Burtt Davy (1922) had suggested that frost was an important factor but, as White (1977) pointed out, frost may well be a factor in the highveld region where Burtt Davy worked, but no more than 10% of southern African suffrutices occur in the region and another explanation must be sought for the rest. He rejected fire as a causal factor, mainly because of the very large differences between the numbers of suffrutices in the floras of the Sudanian and Zambezian Regions, which have similar fire regimes. White (1977) favoured the combination of extremely nutrient-poor soils and seasonal waterlogging that characterises the regions of southern Africa covered by a mantle of Kalahari Sand.

FIG. 1. *Euclea crispa*, a geoxylic suffrutex, from White (1977). Reproduced from Gardens Bulletin, Singapore 29: 60 (1977), with permission.

Two points need addressing here: why are there more suffrutices in the Zambesian Region; and to what factor is the geoxylic suffructicose habit an adaptation? White accepted that Pleistocene climatic fluctuations would probably have caused greater extinction in the Sudanian Region, but he did not feel that this was enough to explain the very large differences in numbers of suffrutices between two regions with similar fire regimes. In the Sudanian Region the opportunities for north-south migration are limited, and the region lacks extensive mountains which could have provided refugia in very dry periods. In the Zambezian Region north-south migration would have been much less restricted during periods of climatic fluctuation. A parallel to this is the comparison between the impoverished tree flora of Europe where southward migration was blocked by the Mediterranean, and the rich forests of eastern Asia and eastern North America, where unimpeded north-south migration was possible. At a smaller scale, the Kalahari Sands of the Zambezian region provide dry, nutrient-poor habitats which display complex interdigitating patterns of forest, woodland and grassland. Such conditions could well favour fragmentation of populations, and thus, speciation. During the climatic fluctuations of the Pleistocene, the areas of grassland would have expanded during dry periods and contracted during wet ones, but the low nutrient status and seasonal waterlogging of the grassland soils could well have led to patches remaining free of trees in even the wettest periods, thus providing refugia for grassland species. Further refugia seem likely to have developed on sites rich in heavy metals, which often bear few or no trees (Wild, 1978). The importance of forest-free refugia in allowing the survival of open-country species at times of forest expansion has been highlighted for the British Isles by Pigott & Walters (1954). In their surveys of Ghanaian forests, Hall & Swaine (1981) recorded savanna tree species on rocky hills within forest far from the forest-savanna boundary, and the wider role of inselbergs as refugia for species of open grasslands was also emphasised by Barthlott & Porembski (1996). The Kalahari Sands may well have functioned in a similar way, and the abundance of geoxylic suffrutices in this region may be best explained by the presence of extensive refugia during periods of forest expansion. Vollesen (1981) came to similar conclusions.

White (1977) claimed that suffrutices in the Zambezian Region occurred almost exclusively on seasonally waterlogged oligotrophic soils. Vollesen (1981) pointed out that the nutrient-poor sandy soils of SE Tanzania, far from being treeless, in fact carry the best-developed *Brachystegia-Julbernardia* woodlands of the Zanzibar-Inhambane Regional Mosaic. There are also many examples of forests in different parts of the world growing on extremely nutrient-poor sandy soils. Seasonal water logging excludes all but the most specialised trees and shrubs. In the Sudd Region of the Sudan, trees are confined to higher, better-drained sites, and only *Balanites aegyptiaca* (Balanitaceae) and *Acacia seyal* (Leguminosae-Mimosoideae) can survive in seasonally flooded areas. Increased flood duration leads to the death of even these species (Howell *et al.*, 1988). Adaptations to waterlogged soils are both structural (presence of aerenchyma in the roots allowing oxygen diffusion from the aerial parts) and physiological (increased resistance to anaerobic conditions). It must be doubtful whether the massive woody underground bases of geoxylic suffrutices would be resistant to prolonged water logging. Vollesen (1981) has noted that in the Zanzibar-Inhambane Regional Mosaic geoxylic suffrutices are rare or absent from the seasonally waterlogged 'mbuga' grasslands, and are mostly confined to the well-drained sandy soils of the *Brachystegia-Julbernardia* woodlands. He also noted that many species are facultatively suffruticose, according to the frequency and severity of fires — as was pointed out by White (1977). Clearly more work is needed to clarify the soil conditions under which geoxylic suffrutices are abundant, but for the present it is hard to avoid the conclusion that White was wrong here, and that regular burning is much more likely to be the factor favouring their development, as postulated by Exell & Stace (1972) and Vollesen (1981).

Conclusions

Fire is certainly an extremely important ecological factor in African vegetation at the present time. Natural fires, started by lightning and volcanic activity, have a long history, but it seems a reasonable hypothesis that the frequency of fire increased when man began to use fire (probably 1.0–1.5 Myr BP — Brain & Sillen (1988)), again when fire could be made, and yet again with the introduction of the safety match. How has this long history of burning affected vegetation in Africa, and are the effects of fire more marked in Africa than in other continents? White (1977) noted that geoxylic suffructices are more abundant in Africa than in other continents. Awned grass 'seeds' like those of *Themeda triandra*, capable of burying themselves in the soil (Lock & Milburn, 1971), are a distinctive feature of African grasslands. Comparison of the grass flora of Tanzania (Clayton, 1970; Clayton *et al.*, 1974; Clayton & Renvoize, 1982) with that of the Brazilian state of Bahia (Renvoize, 1984), shows that awned species make up a higher proportion of the flora in Africa (Table 2). The complex taxonomy of some awned African genera, such as *Hyparrhenia* (Clayton, 1969), is suggestive of recent expansion and spread into new habitats, as is the morphological and cytological variability of other genera such as *Themeda* (Chippindall, 1955). It is tempting to see the effects of increased fire frequency in these trends but more work is still needed.

TABLE 2. Awned grasses in Tanzania and Bahia (Brazil).

	Area (km^2 × 10^3)	No. of Grass species	No. with awns	%
TANZANIA	945	452	139	30.7
BAHIA	560	208	24	11.5

For Tanzania, only those species occurring below 1500 m are included.
Only species recorded as occurring in 'grassland' or other habitats likely to burn are included.
Data from Renvoize (1984) (Bahia), and Clayton (1970), Clayton *et al.* (1974), and Clayton & Renvoize (1982) (Tanzania).

White's work can also help us to understand the dynamics of the world's atmosphere — something he probably never envisaged. His ideas on suffrutices, although I believe them to be wrong, have provided a stimulus to further thought about this unusual growth form — what is now needed is careful field observation, extending throughout the year, combined with some simple measurements and experiments. The change from flowering as an adult to flowering as a one-year-old shoot is a process of neoteny, and may well be based on a single biochemical switch — surely amenable to experiment. White, in his ecological thinking, seems generally to have kept to the broad view; he did not carry out any detailed autecological work on individual species and, indeed, he did not have the opportunity. It is here that additional work can easily be done by residents of Africa. Studies of the behaviour of species in relation to fire — their seed dispersal, seed germination, establishment, and interactions with fire as adult plants — are all best done by those who can examine plants regularly. Much work remains to be done.

References

Barthlott, W. and Porembski, S. (1996). Biodiversity of arid islands in tropical Africa: the succulents of inselbergs. In: The Biodiversity of African Plants (Proc. XIVth Congr. AETFAT). Pp. 49–53. Kluwer, Dordrecht.

Baxter, B.J.M., van Staden, J., Granger, J.E. and Brown, A.J.C. (1994). Plant-derived smoke and smoke extracts stimulate seed germination of the fire-climax grass *Themeda triandra. Environm. Exp. Bot.* **34**: 217–223.

Bond, W.J. and van Wilgen, B.W. (1996). Fire and plants. Chapman and Hall, London.

Brain, C.K. and Sillen, A. (1988). Evidence from the Swartkrans cave for the earliest use of fire. *Nature* **336**: 464–466

Brookman-Amissah, J., Hall, J.B., Swaine, M.D. and Attakorah, J.Y. (1980). A reassessment of a fire protection experiment in northeastern Ghana savanna. *J. Appl. Ecol.* **17**: 85–89.

Burtt Davy, J. (1922). The suffrutescent habit as an adaptation to environment. *J. Ecol.* **10**: 211–219.

César, J. and Menaut, J.C. (1974). Le peuplement végétal des savanes de Lamto (Côte d'Ivoire). *Bull. Liais. Cherch. de Lamto*, Numéro Spécial 1974, Fasc. II.

Charter, J.R. and Keay, R.W.J. (1960). Assessment of the Olokemeji fire-control experiment (investigation 254) 28 years after establishment. *Nigerian Forest Inform. Bull.* **3**: 1–32.

Chidumayo, E.N. (1988). A reassessment of effects of fire on miombo vegetation in the Zambian copperbelt. *J. Trop. Ecol.* **4**: 361–372.

Chippindall, L.K.A. (editor) (1955). A Guide to the Identification of Grasses in South Africa. The Grasses and Pastures of South Africa. Central News Agency, Johannesburg.

Clayton, W.D. (1969). A revision of the genus *Hyparrhenia. Kew Bull., Addit. Ser.* **2**: 1–196.

Clayton, W.D. (1970). Gramineae, Part 1. In: E. Milne-Redhead and R.M. Polhill (editors). Flora of Tropical East Africa. Crown Agents, London.

Clayton, W.D., Phillips, S.M. and Renvoize, S.A. (1974). Gramineae, Part 2. In: R.M. Polhill (editor). Flora of Tropical East Africa. Crown Agents, London.

Clayton, W.D. and Renvoize, S.A. (1982). Gramineae, Part 3. In: R.M. Polhill (editor). Flora of Tropical East Africa. A.A. Balkema, Rotterdam.

Corner, E.J.H. (1966) The Natural History of Palms. Weidenfeld and Nicolson, London.

Cresswell, C.F. and Nelson, H. (1971). The effect of boron on the breaking of dormancy, and germination of rooigras, *Themeda triandra* Forsk. *S. African J. Sci.* **67**: 471–474.

Crutzen, P.J. and Andreae, M.O. (1990). Biomass burning in the tropics: impact on atmospheric chemistry and biogeochemical cycles. *Science* **250**: 1669–1678.

Darwin, F. (1876). On the hygroscopic mechanism by which certain seeds are enabled to bury themselves in the ground. *Trans. Linn. Soc. London, Bot.* **1**: 149–167.

Exell, A.W. and Stace, C.A. (1972). Patterns of distribution in the Combretaceae. In: D.H. Valentine (editor). Taxonomy, phytogeography and evolution. Pp. 307–323. Academic Press, London.

Gill, A.M. and Ashton, D.H. (1968). The role of bark type in relative tolerance to fire of three central Victorian eucalypts. *Austral. J. Bot.* **16**: 491–498.

Hall, J.B. and Swaine, M.D. (1981). Distribution and ecology of vascular plants in a tropical rain forest. Forest vegetation in Ghana. W. Junk, The Hague.

Helas, G. (1995). Emissions of atmospheric trace gases from vegetation burning. *Philos. Trans., Ser.* A **351**: 297–312.

Hopkins, B. (1965). Forest and savanna. Heinemann, London.

Howell, P.P, Lock, J.M. and Cobb, S. (1988). The Jonglei Canal. Impact and Opportunity. Cambridge University Press.

Jackson, G. (1968). Notes on West African Vegetation — III. The seedling morphology of *Butyrospermum paradoxum* (Gaertn. f.) Hepper. *J. W. African Sci. Assoc.* **13**: 215–222.

Jackson, G. (1974). Cryptogeal germination and other seedling adaptations to the burning of vegetation in savanna regions: the origin of the pyrophytic habit. *New Phytol.* **73**: 771–780.

Lawson, G.W., Jeník, J. and Armstrong-Mensah, K.O. (1968). A study of a vegetation catena in Guinea savanna at Mole Game Reserve (Ghana). *J. Ecol.* **56**: 505–522.

Lebrun, J. (1947). La végétation de la plaine alluviale au sud du Lac Edouard. Exploration du Parc National Albert, Mission J. Lebrun (1937–1938). 1: 1–800. Institut des Parcs Nationaux du Congo Belge.

Lock, J.M. and Milburn, T.R. (1971). The seed biology of *Themeda triandra* in relation to fire. In: E. Duffey and A.S. Watt (editors). The Scientific Management of Animal and Plant Communities for Conservation. Pp. 337–349. Blackwell Scientific Publications, Oxford.

Long, S.P., Jones, M.B. and Roberts, M.J. (editors)(1992). Primary Productivity of Grassland Ecosystems of the tropics and sub-tropics. Chapman and Hall.

Louppe, D., Oattara, N. and Coulibaly, A. (1995). The effects of brush fires on vegetation: the Aubréville fire plots after 60 years. *Commonw. Forest. Rev.* **74**: 288–292.

Penning de Vries, F.W.T. and Djiteye, M.A. (1982). La productivité des pâturages sahéliennes: une étude des sols, des végétations et de l'exploitation de cette resource naturelle. PUDOC, Wageningen.

Pigott, C.D. and Walters, S.M. (1954). On the interpretation of the discontinuous distribution shown by certain British species of open habitats. *J. Ecol.* **42**: 95–116.

Priestley, C.H.B. (1959). Heat conduction and temperature profiles in air and soil. *J. Austral. Inst. Agric. Sci.* **25**: 94–107.

Renvoize, S.A. (1984). The Grasses of Bahia. Royal Botanic Gardens, Kew.

Strugnell, R.G. and Pigott, C.D. (1977). Biomass, shoot-production and grazing of two grasslands in the Rwenzori National Park, Uganda. *J. Ecol.* **66**: 73–96

Swaine, M.D., Hawthorne W.D. and Orgle, T.K. (1992). The effects of fire exclusion on savanna vegetation at Kpong, Ghana. *Biotropica* **24**: 166–172.

Tothill, J.C. (1969). Soil temperatures and seed burial in relation to the performance of *Heteropogon contortus* and *Themeda australis* in burned native woodland pastures in Eastern Queensland. *Austral. J. Bot.* **17**: 269–275

Trapnell, C.G. (1959). Ecological results of woodland burning experiments in Northern Rhodesia. *J. Ecol.* **47**: 129–168.

Vollesen, K. (1981). *Catunaregam pygmaea* (Rubiaceae) — a new geoxylic suffrutex from the woodlands of SE Tanzania. *Nordic J. Bot.* **1**: 735–740.

West, O. (1965). Fire in vegetation and its use in pasture management with special reference to tropical and subtropical Africa. Mimeo. Publ. No. 1/1965, Commonwealth Bureau of Pastures and Field Crops: 1–53. Commonwealth Agricultural Bureaux, Farnham Royal, England.

Whelan, R.J. (1995). The ecology of fire. Cambridge University Press.

White, F. (1977). The underground forests of Africa: a preliminary review. *Gard. Bull. Singapore* **29**: 57–71.

White, F. (1983). The Vegetation of Africa: a descriptive memoir to accompany the UNESCO/AETFAT/UNSO vegetation map of Africa. *Natural Resources Research* **20**. UNESCO, Paris.

Wild, H. (1978). The vegetation of heavy metal and other toxic soils. In: M.J.A. Werger (editor). Biogeography and Ecology of Southern Africa. Pp. 1301–32. W. Junk, The Hague.

Lawton, R.M. (1998). The ecology of the wetter miombo woodlands of north-eastern Zambia and a review of the Ndola demonstration plots, Zambia. In: C.R. Huxley, J.M. Lock and D.F. Cutler (editors). Chorology, Taxonomy and Ecology of the Floras of African and Madagascar. Pp. 191–205. Royal Botanic Gardens, Kew.

14. THE ECOLOGY OF THE WETTER MIOMBO WOODLANDS OF NORTH-EASTERN ZAMBIA AND A REVIEW OF THE NDOLA DEMONSTRATION PLOTS, ZAMBIA

R.M. Lawton

Mulberry House, Stanville Road, Cumnor Hill, Oxford, OX2 9JF, UK

Abstract

White's phytochoria are a source of background information for biodiversity studies and a framework for detailed ecological work. Sometimes they are divided into drier or wetter zones, and this may be of ecological significance. In the wetter woodlands of the Zambezian regional centre of endemism, species of the genus *Uapaca* form colonies, or clones, of coppice, derived from root suckers, that shade out the 2 m tall woodland grasses and create a fire-free habitat in which the canopy species are able to regenerate. The pattern of root systems is of ecological importance. The Ndola demonstration plots, the oldest in Africa, show that miombo woodland can be managed under an early burning regime, but annual late burning will destroy it. The presence of *Maesopsis eminii*, distributed by hornbills from avenues in Ndola, is an interesting development. The plots have recently been damaged by fire, but they could still yield something of ecological value if they were fenced, maintained and patrolled. It is recommended that they should be designated a Biosphere Reserve under the UNESCO Man and the Biosphere Programme.

Introduction

The vegetation map of Africa and the accompanying memoir by Dr Frank White (1983), are based on chorology. The classification of the phytochoria is "based on richness (or otherwise) of their endemic floras at the <u>species</u> level." The map and memoir provide a framework for detailed ecological studies. They are a source of initial background information for studies of biodiversity, where a knowledge of endemism in plant and animal communities is considered essential (Gentry, 1986; Sayer *et al.*, 1995).

The vegetation is mapped and described in terms of Regional Centres of Endemism (hereafter referred to as Regions), separated by transition zones and mosaics. The Regions are divided into wetter and drier zones and this division is often of ecological significance. For example, in the Sudanian Region the characteristic species of the drier northern and wetter southern zones are listed, and those that are endemic to the Sudanian Region are marked (S) (White, 1983: 105). The other characteristic species, with wider distributions, may occur in the Zambezian Region, or in the drier Somalia-Masai Region. The ecology of the miombo woodlands in the Zambezian Region will be described to illustrate the use of this map and memoir.

The ecology of the miombo woodlands in the Zambezian region

The miombo woodland has been divided into wetter miombo, where the annual rainfall is 1,000 mm or more, and the drier miombo, where the annual rainfall is less than 1,000 mm. The rainy season is from November to March; the rest of the year is dry. The first half of the dry season from April to mid August is cool, when the daytime temperatures range from 18–24°C. In the second half of the dry season the daytime temperature gradually rises to over 30°C and humidity increases before the onset of the rains in November.

Dry season fires are of ecological importance, particularly in the wetter miombo, where the woodland grasses, bracken (*Pteridium aquilinum*) and *Aframomum alboviolaceum* (Zingiberaceae), may grow to a height of 2 m. The time of burning is important; if the grasses are burnt during the cool season, when most of the woody plants are dormant, the fire is less destructive than a fire during the latter half of the dry season, when the trees and shrubs have come into new leaf and probably flowered. The effects of time of burning will be discussed later.

The woodland canopy is dominated by species of the genera *Brachystegia* and *Julbernardia* (Caesalpiniaceae) (Plate 1). In comparison with the drier miombo, the wetter is species rich; for example, there are more species of *Brachystegia* in the wetter zone, and an evergreen element, including *Marquesia macroura* (Dipterocarpaceae), a feature of the wetter, is absent from drier miombo. The ecology of the wetter miombo is different to that of the drier zone, with particular reference to species of the genus *Uapaca* (Euphorbiaceae).

The ecology of *Uapaca* species in wetter miombo woodland

The genus *Uapaca* is found throughout the miombo; it plays an important role in the ecology of the wetter zone. *Uapaca kirkiana*, and the other species also, form coppice colonies up to 20 m or more in diameter that suppress, or shade out, the grasses, and so create a patch of fire-free ground cover (Plate 2). The species of *Uapaca* have shallow lateral root systems at a depth of 10–20 cm, with sinker roots (Timberlake & Calvert, 1993) (Fig. 1 and Plate 3). Excavation of the root systems indicates that the *Uapaca* colonies do not originate from individual seedlings, but they are in fact root suckers that have grown from the shallow lateral roots (Fig. 2) (Trapnell, 1996). They are clones, like the clones of elm in temperate woodlands (Rackham, 1980).

The lateral roots connecting the suckers form an open network — a favourable micro-site for seed of the canopy species, which will germinate within two weeks of sowing, and by then the tap-root will be about 15 cm long (Munyamziza, 1994) (Fig. 2). During the dry season the seedlings may die back, for up to 8 to 10 years, until the tap-root is able to sustain the shoot. The canopy species then grow up under the protection of *Uapaca* and eventually form a higher canopy that will suppress the *Uapaca* (Plates 4–6). Established canopy trees have large storage tap-roots, that provide sufficient water and nutrients to support the flush of new growth in August when the temperature rises, although this is at least two months before the rains (Fig. 3) (Timberlake & Calvert, 1993).

The roots of the miombo canopy species and *Uapaca* spp. form ectomycorrhizal associations (Högberg & Piearce, 1986). In Southern Malawi, where most of the miombo has been cleared for cultivation, it was decided to try and grow shelter-belts of *U. kirkiana* along the contours, to check the serious soil erosion. The seed germinated in the nursery, but the seedlings failed to grow more than a few centimetres, until the soil was inoculated with soil from a remnant patch of miombo; then they grew about 2 m in one season (Lawton, 1993a). The period of cultivation had apparently destroyed the mycorrhizal fungus.

PLATE 1. A typical stand of wetter miombo woodland. Luwingu District, North-eastern Zambia.

PLATE 2. A coppice colony of *Uapaca* spp. in wetter miombo. North-eastern Zambia.

PLATE 3. The open network of the lateral roots of *Uapaca sansibarica*. Miombo woodland, North-eastern Zambia.

In the wetter miombo, *Uapaca* spp. create conditions that enable the canopy species to regenerate. Similarly in the wetter southern Sudanian Region, *U. togoensis* protects the regeneration of *Isoberlinia doka* (Caesalpiniaceae) from grass fires. *Uapaca togoensis* grows as a woodland fringe around forest outliers in Central Nigeria (Jones, 1963; Lawton, 1976). In the wetter zones, species of *Uapaca* would appear to be precursors of woodlands and high forest. Where *Uapaca* spp. are absent, other species, including some geoxylic suffrutices (see Lock, this volume) provide a favourable environment for the regeneration of the canopy. A knowledge of the distribution, pattern and function of root systems should form an essential part of any ecological study (Lawton, 1996).

The management of miombo woodland

In the past, miombo has been the fallow crop for traditional agricultural systems. It has been cleared for other forms of land-use; for the cultivation of cash crops, for example, tobacco. Before electricity was available, large amounts of cordwood were cut to power the copper mines. This was done under a form of controlled exploitation, by coupes. But until recently no effort has been made to utilize miombo on a sustained yield basis.

Attempts are now being made to introduce a form of coppice-with-standards management (Werren *et al.*, 1995; Lawton, 1993a). It is important to introduce some form of control, and the results so far are encouraging. Two years after coppicing, the re-growth was between 1 and 2 m high. A knowledge of the dynamic ecology of the woodland has some practical value; where there are clones of *Uapaca* ground cover, seed of the canopy species could be sown to regenerate and enrich the canopy.

PLATE 4. *Brachystegia floribunda* regenerating in *Uapaca kirkiana* coppice. North-eastern Zambia.

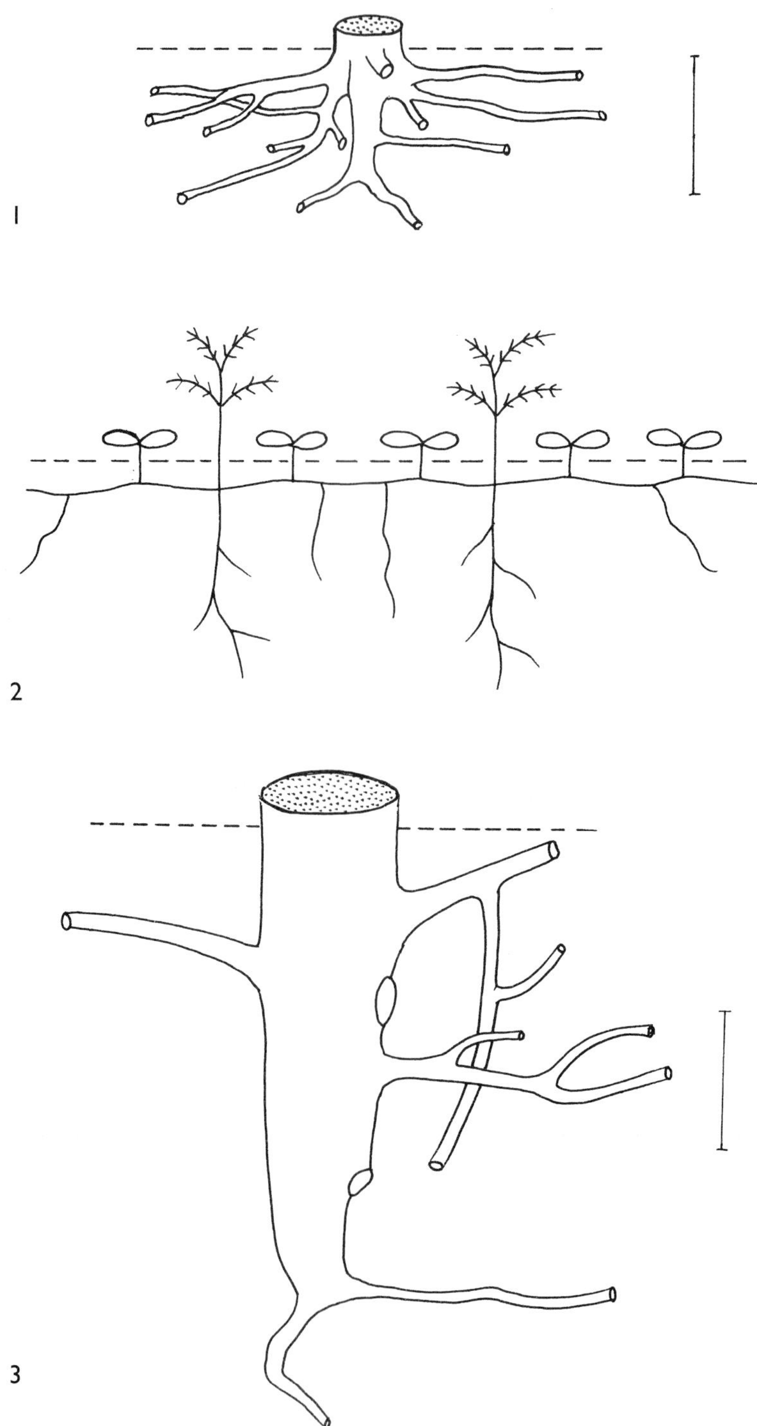

PLATE 5. *Brachystegia glaberrima* growing up with *Uapaca sansibarica.* North-eastern Zambia.

FIG. 1. The root system of *Uapaca kirkiana.* (From Timberlake & Calvert, 1993). The laterals have been truncated. Scale bar 100 cm. Redrawn by Rosemary Wise.

FIG. 2. A schematic drawing of a clone of *Uapaca* spp. root suckers with seedlings of the canopy species in the gaps. Redrawn by Rosemary Wise.

FIG. 3. The root system of *Brachystegia spiciformis.* (From Timberlake & Calvert, 1993). Scale bar 100 cm. Redrawn by Rosemary Wise.

PLATE 6. *Brachystegia floribunda* above the *Uapaca* canopy. North-eastern Zambia.

A review of the Ndola burning demonstration plots

The demonstration plots, near Ndola in the copper belt of Zambia, are the oldest in Africa. They were established by Mr C.E. Duff CBE in 1933–34, two years before the Aubréville plots in the Ivory Coast (Louppe *et al.*, 1995). The main objective of the experiment was to determine the effects of burning the woodland grasses at different times, on the growth and girth increment of the woodland canopy trees. Three treatments were prescribed:–

1. *Early burning*: that is, burning the grasses as soon as possible during the dry season, usually in the period May to July.

2. *Late burning*: burning the grasses just before the beginning of the rains, usually in October.
3. *Complete protection*: complete protection from fire.

Four 0.4 ha (40 m × 100 m) plots were demarcated in mature miombo. These are known as the 'Woodland Plots'; the prescribed treatment is:–
Woodland Plot 1:   early burning
Woodland Plot 2:   late burning
Woodland Plots 3 and 4:   complete protection.

Another series of eight plots of the same size were laid out, and the woodland was coppiced. These are known as the 'Coppice Plots'. Plots 1 to 4 were cut at breast height (127 cm) and the remainder were coppiced at ground level. The treatments were similar to those for the woodland plots.
A further series of four larger plots of c. 4.4 ha (50 m × 740 m) were demarcated in a mixture of natural miombo woodland and dry evergreen forest. These are the Demonstration Plots and were given the same treatment as the woodland plots.
The initial enumerations of 1933, followed by a layer count in 1944, a re-measurement of the plots in 1946 and a simple visual appraisal in 1956, have been recorded and discussed by Trapnell (1959). He noted that "only *Uapaca nitida* effected any natural increase under late burning", and there was a "striking increase in *U. kirkiana*" in the early burnt plots. In one of the early burning coppice plots it was observed that the canopy species *Julbernardia paniculata*, had begun to overtop *Uapaca* spp. to form a higher canopy. These observations have been confirmed by later studies (Lawton, 1978).

Trapnell's conclusions can be briefly summarized:–

1. Persistent late burning will destroy the woodland canopy;
2. Miombo woodland can be maintained and will regenerate under an early burning regime;
3. Complete protection from fire will lead to the entry of a fire-sensitive shade-tolerant evergreen understorey below the miombo canopy, and eventually to a mixture of evergreen canopy species;
4. It is possible to group species according to their degree of fire tolerance.

In 1969 some random 10 m × 10 m samples were enumerated in the woodland and coppice plots (Lawton, 1972). These samples confirmed Trapnell's conclusions. One interesting development was observed; a number of saplings of *Maesopsis eminii* (Rhamnaceae) were recorded in the complete protection Demonstration Plots (Plate 7). The saplings were mainly found under the crowns of *Syzygium guineense* subsp. *afromontanum* (Myrtaceae). *Maesopsis eminii* grows as an avenue tree in Ndola, about 2 km away, and it is thought that hornbills have distributed the fruit. It is known that

PLATE 7. A sapling of *Maesopsis eminii* under the crown of *Syzygium guineense* subsp. *afromontanum* in complete protection demonstration plot, Ndola, 1969.

PLATE 8. A *Maesopsis eminii* has grown up in a gap. It is damaged at the base by fire. Demonstration plot, Ndola, October 1992.

hornbills feed off the fruits of *M. eminii* and *S. guineense* subsp. *afromontanum* (Eggeling & Dale, 1951). *Maesopsis eminii* is a colonizer of the fringe communities on the edge of the high forest in Uganda (Eggeling, 1947), where it has also been used successfully as a shade tree for coffee (Eggeling & Harris, 1939). Only one indigenous specimen of *M. eminii* has been recorded in a riverine forest in north-eastern Zambia (Lawton, 1969).

In 1973 some of the treatments in the coppice plots and Demonstration plots were changed. A number of the coppice plots were sampled in 1982 (Chidumayo, 1988), and it was concluded that early burning is a practical method of management for miombo woodland.

In the late dry season of 1987, a disastrous fire swept through the complete protection plots, destroying some of the fire-sensitive evergreen understorey and canopy species. A large *Entandrophragma delevoyi* (Meliaceae) was killed, and the fire burnt underground along the root systems. Although the damage was severe, remnants of the dry evergreen forest and mature miombo have survived, and the Forestry Department has continued to carry out the prescribed treatments, so it was possible to evaluate the state of the plots in 1992 (Lawton, 1993b).

Dense thickets of the noxious weed, *Lantana camara* (Verbenaceae), have invaded the complete protection and early burning plots; these create a fire hazard. In the late burnt coppice plots it is of interest to note that after 59 years of mainly late burning, the root systems of the canopy species are still alive, and send up coppice shoots each year. Although the parent tree has now been killed by the fire, there is a good scatter of saplings, poles and even small timber trees of *Entandrophragma delevoyi* in the complete protection plots. Some of the pole-sized *E. delevoyi* have been cut, but they sprout again, sometimes from below ground level. Many of the *M. eminii* saplings

PLATE 9. The crown of *Maesopsis eminii* in Plate 10 reaching through a gap to emerge above the canopy. *Albizia adianthifolia* is to the left. Complete protection demonstration plot, Ndola, October 1992.

PLATE 10. A *Maesopsis eminii* (53 cm DBH) protected from fire by the understorey. Complete protection demonstration plot, Ndola, October 1992.

growing in 1969 (Plate 7) have either been shaded out or destroyed by fire, but some have survived to grow through gaps and emerge above the miombo canopy (Plates 8, 9). Some have been protected from fire by the understorey (Plate 10).

Although the plots are patrolled by Forest Rangers and Forest Guards, they are not fenced. Firewood, poles and bamboo are cut illegally. Even the remains of a charcoal clamp were seen in one plot. Despite the fire damage, it is possible to salvage something. With complete protection from fire, a dry evergreen forest of *E. delevoyi* and *M. eminii* as emergents, would become established.

It is recommended that these plots should be made a Biosphere Reserve, under the UNESCO Man and the Biosphere Programme (MAB). This would give them the international status they deserve, and attract funds to pay for fencing, maintenance and patrolling. The creation of a Biosphere Reserve does not alter the sovereignty, or ownership, of the experiment.

Conclusions

The chorological map and description of the vegetation of Africa by Dr Frank White (1983) could be used as a model for the classification of other continental floras. The map, based on the richness, or otherwise, of the endemic floras at the *species* level, provides essential background information for biodiversity studies. It is also a framework for detailed ecological work, that may be of practical value in the management of natural forests and woodland, as in the case of the wetter miombo woodland.

The demonstration plots, near Ndola, have provided important insights into the effect of fire on miombo woodland. Although they have been badly damaged by fire, if they were fenced and patrolled, they could still yield further results of great ecological value. It is recommended that they should be made a Biosphere Reserve under the UNESCO MAB Programme.

References

Chidumayo, E.N. (1988). A re-assessment of effects of fire on miombo regeneration in the Zambian Copperbelt. *J. Trop. Ecol.* **4**: 361–372.

Eggeling, W.J. (1947). Observations on the ecology of the Budongo rain forest, Uganda. *J. Ecol.* **34**: 20–87.

Eggeling, W.J. and Dale, I.R. (1951). The indigenous trees of the Uganda Protectorate, 2nd edn. University Press, Glasgow.

Eggeling, W.J. and Harris, C.M. (1939). Fifteen Ugandan timbers. In: L. Chalk, J. Burtt-Davy and A. C. Hoyle (editors). No. 4 of Forest trees and timbers of the British Empire. Imperial Forestry Institute, Clarendon Press, Oxford.

Gentry, A.H. (1986). Endemism in tropical versus temperate plant communities. In: M.E. Soulé (editor). Conservation biology. The science of scarcity and diversity. Pp. 286–303. Sinauer Associates, Sunderland, Massachusetts.

Högberg, P. and Piearce, G.D. (1986). Mycorrhiza in Zambian trees in relation to host taxonomy, vegetation type and successional patterns. *J. Ecol.* **74**: 775–785.

Jones, E.W. (1963). The forest outliers in the Guinea zone of Northern Nigeria. *J. Ecol.* **51**: 415–434.

Lawton, R.M. (1969). A new record, *Maesopsis eminii* Engl. for Zambia. *Kirkia* **7** **(1)**: 145–146.

Lawton, R.M. (1972). An ecological study of miombo and chipya woodland with particular reference to Zambia. Unpublished D. Phil. thesis, University of Oxford.

Lawton, R.M. (1976). The floristic composition and ecology of the forests and woodlands of the Kaduna Plains. Central Nigeria Project, Overseas Development Administration.

Lawton, R.M. (1978). A study of the dynamic ecology of Zambian vegetation. *J. Ecol.* **66**: 175–198.

Lawton, R.M. (1993a). Management of miombo woodlands. In: Management of miombo as community woodlots, ODA/OFI project no. R4599.

Lawton, R.M. (1993b). An ecological evaluation of the Ndola woodland burning experiments. In: Management of miombo as community woodlots, ODA/OFI project no. R4599.

Lawton, R.M. (1996). The ecological importance of the root systems in the woodlands of Central Africa and in the desert plants of Oman. International Society of Root Research Fourth Symposium on 'Root systems and natural vegetation', Almaty, Kazakhstan, 5–11 September 1994. *Acta Phytogeogr. Suec.* **81**: 24–28.

Louppe, D., Oattara, N. and Coulibaly, A. (1995). The effects of brush fires on vegetation: the Aubréville fire plots after 60 years. *Commonw. Forest Rev.* **74 (4)**: 288–292.

Munyamziza, E. (1994). (pers. comm.) on growth of *Julbernardia globiflora* seedlings.

Rackham, O. (1980). Ancient woodland: its history, vegetation and uses in England. Edward Arnold.

Sayer, J.A., Zuidema, P.A. and Rijks, M.H. (1995). Managing for biodiversity in humid tropical forests. *Commonw. Forest Rev.* **74 (4)**: 282–287.

Timberlake, J.R. and Calvert, E.M. (1993). Preliminary Root Atlas for Zimbabwe and Zambia. *Zimbabwe Bull. Forest. Res.* **10**.

Trapnell, C.G. (1959). Ecological results of woodland burning experiments in Northern Rhodesia. *J. Ecol.* **47**: 129–168.

Trapnell, C.G. (1996). (pers. comm.) The root systems of *Uapaca* spp.

Werren, M., Lowore, J., Abbot, P. and Siddle, B. (1995). Management of miombo by local communities. Forest Research Project R4599. Forest Research Institute of Malawi and Department of Forestry, University of Aberdeen. Funded by ODA/OFI.

White, F. (1983). The Vegetation of Africa: a descriptive memoir to accompany the UNESCO/AETFAT/UNSO vegetation map of Africa. *Natural Resources Research* **20**. UNESCO, Paris.

Grimshaw, J.M. (1998). Disturbance, pioneers and the Afromontane Archipelago. In: C.R. Huxley, J.M. Lock and D.F. Cutler (editors). Chorology, Taxonomy and Ecology of the Floras of Africa and Madagascar. Pp. 207–220. Royal Botanic Gardens, Kew.

15. DISTURBANCE, PIONEERS AND THE AFROMANTANE ARCHIPELAGO

JOHN M. GRIMSHAW

Animal Ecology Research Group, Department of Zoology, University of Oxford, South Parks Rd, Oxford, OX1 3PS, UK*

Abstract

Fieldwork carried out on the northern slope of Mt. Kilimanjaro, Tanzania, during 1992–1994 found a clear pattern of altitudinal zonation, reflected in both physiognomy and floristics, that agrees well with the concept of an Afromontane phytochorion. Superimposed upon this pattern are the effects of burning, and case studies of characteristic Afromontane species (*Erica excelsa*, *Hagenia abyssinica*, *Juniperus procera*, *Ocotea usambarensis* and *Olea europaea* subsp. *cuspidata*) are used to demonstrate that large areas of Afromontane forest are the result of burning and clearance during the 'Little Ice Age' and before. Ecological studies show that many Afromontane forest trees can be considered to be pioneers or aggressive colonisers characterised by easily dispersed seeds, although it is suggested that amazing feats of long-distance dispersal are not needed for an explanation of distribution within the Afromontane Archipelago.

This extension of the same [temperate] forms . . . across the whole continent of Africa . . . is one of the most astonishing facts ever recorded in the distribution of plants.

C. Darwin, *On the Origin of Species*, 5ᵗʰ Edn., 1869.

Introduction

The vegetation of African mountains has excited biogeographers since the early collections of Mann, New, Schimper and others revealed similarities in the floras of widely isolated mountains (e.g. Hooker, 1862, 1864, 1874). This combination of similarity and disjunction, posing so many questions about the origins of the flora and its dispersal, seems to have held a particular fascination for Frank White, whose publications on the subject extend over nearly 30 years (White, 1965, 1970 (in Chapman & White), 1978, 1981, 1983a, 1983b, 1993a, 1993b). He first proposed an Afromontane Region of endemism in 1965, subsequently refining and clarifying the concept and its position in the grand phytochorological division of Africa (White, 1978, 1983a), summing-up that *"The Afromontane flora is more complex in its origin than that of any other African phytochorion, but it resembles them sufficiently, in its internal cohesiveness and degree of difference from adjacent phytochoria, to fully justify the recognition of an archipelago-like Afromontane Region."* (White, 1978: 507–508).

A fundamental tenet of montane biogeography is that there is a distinction between montane and non-montane, instituting the concept of vegetation zonation with altitude. On many African mountains the montane/non-montane disjunction is easily

*Present address: 35 Wessex Way, Cox Green, Maidenhead, Berks., SL6 3BP, U.K.

visible, as are further primary distinctions, defined by Hedberg (1951) as vegetation belts of global validity. However, as Moreau (1933) early pointed out, distinctions between montane and non-montane in Africa are exacerbated by human disturbance, which may have created an artificial distinction of biota, reinforced by subjective perception of the topography. In more recent years the existence of a distinct, altitudinally zoned Afromontane flora has often been questioned (Lebrun, 1947; Monod, 1957; Hamilton, 1975; Lovett, 1993b, 1996), although others have found evidence to support the validity of White's Afromontane archipelago-like centre of endemism and the existence of clear floristic zonation within it (Dowsett-Lemaire, 1985, 1988a, 1989a, 1990; Friis, 1992).

Fieldwork carried out between 1992 and 1994 on the northern slope of Mt. Kilimanjaro, Tanzania, found a clear pattern of zonation, and data gathered were subsequently used to investigate the nature of the Afromontane Region and to determine the validity of the observed zonation (Grimshaw, 1996).

Methods

The vegetation of the northern slope of Kilimanjaro was surveyed using a stratified sampling method adapted from Kigomo *et al.* (1990), locating 30 × 50 m quadrats at 200 m intervals on baseline contours 100 m apart (1,700, 1,800 m etc). Within these quadrats the dbh of all trees (>3 cm dbh) was measured, with shrub and sapling stems being counted in five 5 × 10 m sub-quadrats, and all herbs and seedling trees being counted in five 1 × 2 m sub-quadrats. A total of 96 30 × 50 m quadrats was surveyed.

Data were analysed by TWINSPAN (Hill, 1979). Separate iterations of the program used relative dominance for trees, and occurrence in sub-quadrats (on the scale 1–5, as a crude estimate of abundance) for herbs and shrubs, as indications of abundance in each quadrat surveyed.

The phytochorological position of plants found in the quadrat survey, with the exception of ferns, was assessed from the literature, following the framework of Friis (1992).

Vegetation Zonation on Kilimanjaro and its causes

In the field four major vegetation types and two transition zones are noticeable. A dry forest type occurs at 1,700–1,800 m, dominated by *Olea europaea* ssp. *cuspidata* at 1,800 m, but by *Diospyros abyssinica* at 1,700 m. A transition zone at 1,900 m leads to forest dominated by *Cassipourea malosana* between 2,000–2,300 m. Physiognomically this vegetation resembles a true rain forest, with occasional huge emergents of *Entandrophragma excelsum.* There is a sharp shift in floristic composition in the interval 2,300–2,400 m, and by 2,400 m *Cassipourea* is completely absent and very open forest (strictly 'woodland', if White's 1983a definitions are to be rigorously applied), naturally dominated by *Podocarpus latifolius,* occurs. Logging has removed much of this species and places of intense logging disturbance may be occupied by *Hagenia abyssinica.* At about 2,700 m there is a very sharp transition to *Erica excelsa* forest, which occurs in a narrow band below altimontane scrubland.

TWINSPAN analysis (for both the trees and herbs/shrubs iterations) indicates that there are marked floristic discontinuities at 1,900 m and 2,700 m (second division of the data), with the primary division falling between 2,300 and 2,400 m, thus confirming the four principal vegetation types recognised in the field as distinct entities. Despite this it is necessary to assess the validity of the divisions, two of which represent post-fire successional states.

Demographic studies of the trees demonstrated that the *Olea-* and *Erica-*dominated vegetation types are stages in a fire-induced succession, whose later stages could also be traced (Fig. 1). In neither case was the dominant species regenerating. In *Olea* forest, the monodominant canopy of mature *Olea* is replaced by a mixture of species, especially *Diospyros abyssinica*, sometimes with *Calodendrum capense* and *Croton megalocarpus* as emergents. These species are non-pioneer light-demanders (NPLD) *sensu* Hawthorne (1995); others such as *Teclea simplicifolia* and *Turraea floribunda* are obligate shade species. The succession in *Erica* forest is almost parallel. Following a fire, mass germination results in a dense even-aged stand, in which are a few *Hagenia* seedlings that grow up with the *Erica* canopy. *Podocarpus latifolius* seedlings establish in the light shade beneath closed *Erica* canopy, and are joined by other NPLD species such as *Ilex mitis* and *Prunus africana* as the succession continues. The *Erica* senesces and a mixed forest of *Hagenia* and *Podocarpus* with *Ilex* and *Prunus* results. Eventually, in the absence of further disturbance, *Hagenia* dies out and a *Podocarpus*-dominated forest with balanced regeneration is left.

Bussmann's (1994) work on Mt. Kenya has permitted the age of some species of Afromontane trees to be determined, and from his data it was possible to age relict *Olea* trees at 1,700 m to c. 320 years (1650–1700 AD). Younger stands at 1,800 m (c. 165 years) and 1,900 m (c. 250 years), as well as recent regeneration at 1,700 m, illustrate a mosaic of fires occurring at different times and long intervals. A similar pattern may be detected in the upper parts of the forest, where dating from *Podocarpus* and *Hagenia* is possible; this succession is more rapid however, and pure *Podocarpus* forest may develop in less than 250 years. *Erica excelsa* seems to be relatively short-lived, becoming senescent within 100 years.

Cassipourea-dominated forest does not, at first, seem likely to burn, but relict individuals of *Entandrophragma excelsum*, *Juniperus procera*, *Nuxia congesta* and *Olea capensis*, all species requiring open conditions for establishment, provide indications of extensive fires at mid-altitudes on the northern and western slopes of Kilimanjaro. The ecology of *Juniperus* is particularly well known, and seedlings can only establish in open conditions on mineral soils (Hall, 1984), thus providing proof of former forest clearance. As with the *Olea* at 1,700 m, the relict *Juniperus* trees are datable to the late 17th Century. This falls in the middle of the period 1500–1800, known as the 'Little Ice Age', when global temperatures were depressed. In East Africa mean temperatures were reduced by 0.5°C (Osmaston, 1989), and as with earlier glacial periods, rainfall also diminished (Hamilton, 1982; Vincent *et al.*, 1989). Dale (1954) has described widespread human migrations in East Africa at this time; agriculturists moved upslope to avoid increasing aridity on the plains, while pastoralists occupied the vacated land. Forest descriptions from across eastern Africa (either explicitly, or implicitly from the description and known ecology of the species concerned) indicate that large areas of forest were cleared at this time, and have since regenerated. Clearance, probably by fire, has also affected the wet southern slope of Kilimanjaro, and the even wetter Usambara mountains, as demonstrated by the presence of *Ocotea usambarensis*, another species requiring open conditions for establishment (Abraham, 1958; Bussmann, 1994; Hamilton, 1989; Lovett, 1996; Mugasha, 1978; Willan 1965), although the Kilimanjaro stands probably represent a clearance event some centuries before those of the 'Little Ice Age' (Grimshaw, 1996).

Many individuals of the assumed late seventeenth century cohort of *Juniperus* on Kilimanjaro are now found only as unrotten trunks on the forest floor, beneath a mature *Cassipourea* canopy. Since *Cassipourea* requires shade for establishment and development it must be assumed that this apparently 'climax' Undifferentiated Afromontane forest has developed in less than 300 years. *Juniperus* is now too rare to demonstrate the stages of this succession conclusively, but at 1,900 m the succession from *Olea*- to *Cassipourea*-dominated forest was

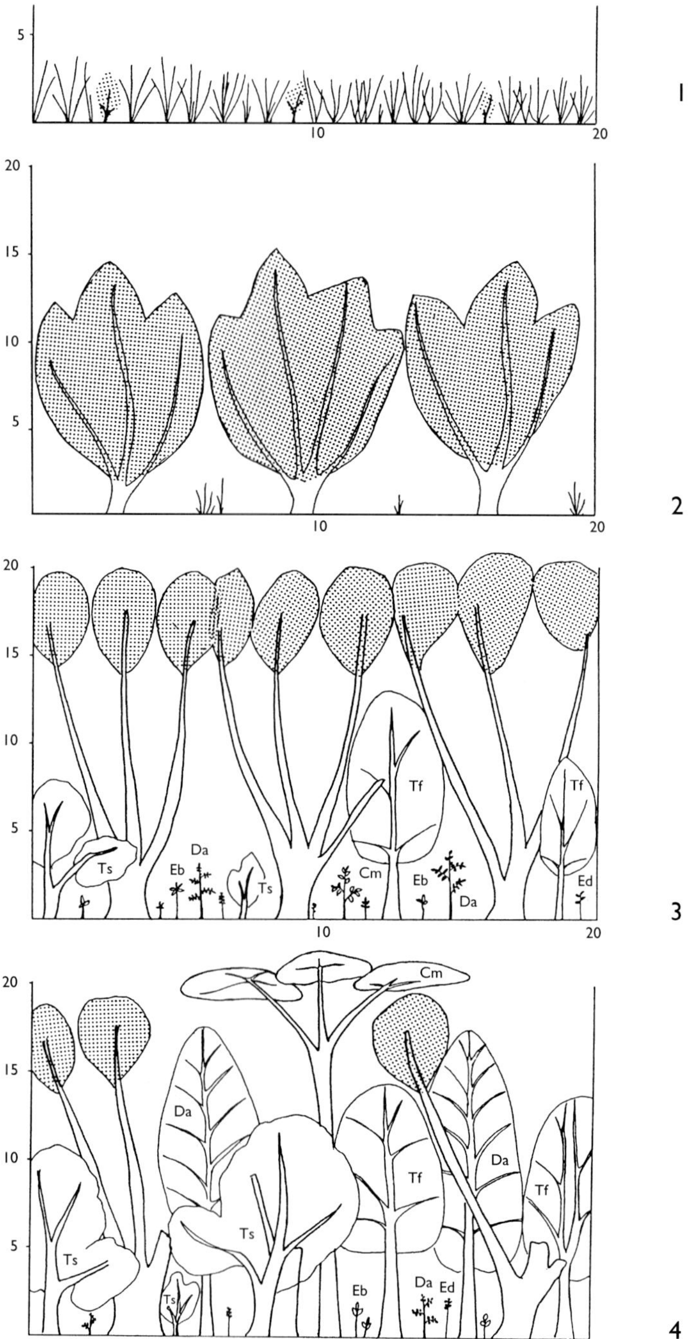
5
10
20
1
20
15
10
5
10
20
2
20
15
10
5
Tf
Tf
Da
Ts
Eb
Cm
Ts
Eb
Ed
Da
10
20
3
20
Cm
15
Da
10
Da
Tf
Tf
Ts
Da
5
Ts
Ts
Eb
Da Ed
Ts
4

visible. Here relict *Olea* trees were dated at 250 years, while the oldest of the replacing *Cassipourea* was c. 100 years. At 150 years old the *Olea* canopy is mature and provides the dense shade needed for *Cassipourea* establishment (demonstrated by the few *Cassipourea* saplings at 1,800 m below 165 year old *Olea*): I believe that the rapidly growing *Cassipourea* soon out-competes the *Olea* and hastens its senescence. *Cassipourea* does best when rainfall is in excess of 875 mm/yr (Gilchrist, 1952), reached on the northern slope of Kilimanjaro at about 1,900 m (mean annual rainfall calculated to be 850–950 mm (Grimshaw, 1996)). Rainfall at 1,800 mm is 800–900 mm/yr, probably marginal for this species.

The transition from *Olea-* to *Cassipourea*-dominated forest at 1,900 m could be seen as a successional process of no floristic significance, leading to (in the absence of further fires) a climax continuum from *Diospyros-* to *Cassipourea*-dominated forest along the rainfall gradient. This is undoubtedly the case: *Diospyros abyssinica* and many other species are absent, or present in minute numbers at 2,000 m (and not at all thereafter), and are presumed to be excluded from their shade-obligate or NPLD niches by the extremely competitive *Cassipourea* under conditions of higher rainfall (950–1000 mm/yr at 2,000 m).

However, a phytochorological analysis of the flora demonstrated that the TWINSPAN-recognised intervals 1,700–1,900 m and 2,000–2,300 m, were floristically distinct. In the lower interval only 33% of the total flora was composed of Afromontane endemics or near-endemics (*sensu* White 1983a; Friis 1992), while widely distributed species and aliens accounted for 48%. The remainder of the flora (19%) are plants confined to the Somalia-Masai Region in East Africa, often occurring in upland but not strictly montane areas. In the 2,000–2,300 m interval 64% were Afromontane endemics, 31% widespread species, and only 5% were submontane East African species.

Using White's criterion of 50% endemics to define a phytochorion it is clear that the lower forest cannot be regarded as Afromontane. Similar vegetation was described by White (1983a: 166) as 'dry transitional montane forest'; Friis (1992) has described a similar vegetation type in Northeast Tropical Africa as a local phytochorion with the name 'Transitional Afromontane-Somalia-Masai area'. Similar forest is widespread, although now often only as threatened relics, throughout eastern Africa; the forests around Nairobi are the best known examples (White, 1983a; Kigomo *et al.*, 1990).

Thus at 1,900 m there is a meeting of two phytochoria, whose relative position is a function of the rainfall gradient, and competition upon it, although the actual composition of the vegetation is determined by the processes of succession.

The floristic disjunction between 2,300–2,400 m is very pronounced, although phytochorological analysis indicates that the flora of both areas belongs to the Afromontane Region, despite the discontinuity in specific composition. The 2,000–2,300 m interval (with 64.3% Afromontane endemics) is, from the definitions

FIG. 1. Post-fire succession in dry forest at 1,700–1,800 m on Mt. Kilimanjaro.

1) 5–10 years after the fire. Young *Olea* seedlings growing amongst *Lippia*-dominated scrub. This phase continues for 40–50 years, as the *Olea* trees increase in size and gradually suppress other shrubs.

2) 60–90 years. Young *Olea* trees form a monospecific forest, with few understorey shrubs.

3) 150–200 years. *Olea* trees now mature, forming a high monolayer canopy. *Teclea simplicifolia* (Ts) and *Turraea floribunda* (Tf) are established in the shade, forming a loose subcanopy and reproducing there. In the understorey there are numerous saplings and seedlings of *Calodendrum capense*, *Croton megalocarpus* (Cm), *Diospyros abyssinica* (Da), *Euclea divinorum* (Ed), and *Elaeodendron buchananii* (Eb), awaiting release to the canopy when a gap occurs.

4) 250–300 years. *Olea* trees now senescent, while the other species have grown up in the gaps formed by the dieback of the *Olea*. A relatively open canopy permits the development of stable populations of most species.

and descriptions of White (1983a) and Friis (1992), best termed Undifferentiated Afromontane forest, containing 5 out of 11 characteristic Afromontane species (White, 1983a: 165). The upper interval (2,400–2,700 m), with 88.5% Afromontane endemics, could also be included as Undifferentiated Afromontane forest, but in view of the floristic distinction, I feel it is better provisionally to apply the term Upper (undifferentiated) Afromontane forest to this vegetation type.

The cause of the floristic discontinuity at the 2,300–2,400 m boundary is almost certainly the effect of the daily formation of a cloud-layer with its lower surface at this altitude. The development of this cloud-layer, a characteristic feature of most tropical mountains, has been described by Troll & Wien (1949) and Coetzee (1967) in terms of the daily convection of warmed air upwards from the plains; condensation occurs at between 2,400–2,700 (–3,000) m as a function of the lapse rate, while the temperature inversion caused by the protrusion of the mountain into the lower atmosphere limits the upwards development of the cloud and creates a cloud stratum (Coutts, 1969). The upper surface of this stratum can be shown to correspond with the tree line, dividing the humid forest from the dry altimontane belt with its marked diurnal climate (Troll, 1958; Coetzee, 1967; Coetzee & van Zinderen Bakker, 1970).

It is no coincidence that the lower surface of the cloud stratum coincides with the primary floristic discontinuity. Mist has a profound influence on many aspects of plant physiology, affecting photosynthesis, respiration, transpiration and morphogenesis (Pendry & Proctor, 1996), while the lower temperatures and lower evaporation potential under misty conditions contribute to a tendency towards waterlogging and peatiness in the soil. This inhibits the mineralisation process and available nitrogen and phosphorus have been widely shown to diminish with altitude (Edwards, 1977; Grubb, 1971, 1977; Lieberman *et al.*, 1996; Tanner *et al.*, 1990; Vitousek, 1984), the resulting infertility often imposing a nutrient cap on species' distributions, even if temperature is not limiting (Grubb, 1971). Upper Afromontane forest species must be 'mist-tolerant'; species, such as *Cassipourea*, that apparently cannot cope with the effects of mist — for whatever reason — are totally excluded from the vegetation.

The fourth horizon, recognised as distinct by TWINSPAN, is the *Erica*-dominated forest at 2,700 m. Since this remarkable vegetation can be shown to be merely an early stage of the post-fire regeneration of *Podocarpus*-dominated forest on the same site, I regard it as a temporary facies of Upper Afromontane forest with a transiently distinct flora; a temporal, rather than spatial transition, that does not warrant phytochorological distinction. The transition at about 1,900 m is, however, truly spatial, even if occasionally perturbed by fire.

Implications for the study of the Afromontane Archipelago

These observations from Kilimanjaro, a classic island mountain, have several implications for the study of the vegetation of the Afromontane archipelago. This unique phytochorion is distinguished by two overwhelming factors: disturbance and distance. Disturbance is used in the broadest sense of any perturbation causing an alteration in vegetation type, from the scale of a local fire to the devastation of a volcanic eruption and the consequences of climatic change during glacial periods. All of these have effects that necessitate recolonisation from what is almost inevitably a distant source. The flora can be shown to be well adapted to these dual challenges, being composed of tolerant species and pioneers adapted to long-distance dispersal and rapid establishment in unspecialized conditions.

Detailed demographic and ecological study of the trees demonstrates that these forests are exceptionally dynamic, with many species being dependent on fire or disturbance for successful mass regeneration. (In the absence of fire they are usually forest-edge species, or occupy unstable slopes). This is not surprising in a flora largely occupying volcanic mountains, and studies in other volcanic regions have shown very similar patterns of pioneer trees forming monodominant stands before giving way to later successional mixed forest (Mueller-Dombois *et al.*, 1983; Veblen *et al.*, 1981). Hedberg (1970) noted the role of volcanic disturbance in providing empty 'agar plates' for potential colonisation, in competition-free conditions, by incoming Afroalpines. I suggest that this effect extends beyond the Afroalpine situation to much of the Afromontane forest flora, especially as the development of the Afromontane and Afroalpine floras during the Tertiary (Friis, 1992; Hedberg,1961) coincided with the onset of African volcanism (Griffiths, 1993).

The other perturbations experienced by the Afromontane archipelago are the climatic consequences of global glaciations. There has been much controversy over these, and White (1981, 1993b) persistently criticised the models of Livingstone (1975) and Hamilton (1982, 1988) which proposed widespread aridity during the last main glaciation, with the montane flora being confined to climatically favourable refugia, or being lost altogether from some mountains. It was this, and the necessary corollary of extremely rapid forest spread following deglacierization (*c.* 14,000 BP), that White objected to. He rightly pointed out that observed distribution patterns are far too complex to be explicable in terms of only a few thousand years, as an extreme reading of Livingstone and Hamilton might suggest. Despite White's objections, the theory of glacial aridity in inland tropical Africa is supported by a vast literature (reviewed by Lovett, 1993a) and now widely accepted.

It is easy to demonstrate the effects of a diminution in temperature and rainfall on the vegetation of the northern slope of Kilimanjaro. At 2,000 m, now occupied by *Cassipourea*-dominated forest, recent mean annual rainfall is *c.* 975 mm/yr (Grimshaw, 1996). A 20% diminution in rainfall would render conditions marginal for *Cassipourea*; a 30% reduction (to *c.* 700 mm) would exclude it. Bonnefille *et al.* (1990) have calculated that East African rainfall diminished, at the last glacial maximum, by 30%. Likewise, with a lapse rate of -0.59°C/100 m in eastern Tanzania (Lovett, 1996), a diminution in mean temperature by 2.36°C would create conditions at 2,000 m equivalent to those now experienced at 2,400 m, where the effects of mist exclude *Cassipourea* and most of its associated species. Bonnefille *et al.* (1990) estimated a temperature decrease of 4 ± 2°C at the last glacial maximum; Osmaston (1989), using equilibrium line altitudes, calculated a reduction in temperature for Kilimanjaro of approximately 6°C.

Thus a reduction in mean rainfall of 30%, and in mean temperature of less than 3°C, can be shown to eliminate at least one of the main vegetation types of Kilimanjaro. Hamilton (1982, 1987, 1988) believes that forest disappeared almost entirely from some mountains, especially those like Mt. Elgon and Cherangani that receive rainfall from the Lake Victoria local climate, which failed as the lake dried out. Mountains closer to the coast and experiencing monsoon rainfall, such as Mt. Kenya and Kilimanjaro, are believed to have retained forest cover in some areas at least; that this is the case is demonstrated by the Sacred Lake (Mt. Kenya) pollen diagram (Coetzee, 1967), which extends over nearly 35,000 years and reveals the presence of the same assemblage of species before and after the last glaciation.

There can be no doubt that during glacial periods there have been widespread changes in vegetation type, or that contraction of the forest to refugia occurred. This is dictated by the plants' innate physiological responses to physical phenomena. I believe that to explain the effects of glacial periods on the Afromontane flora, it is necessary to postulate the persistence of most species on most mountains, even if they were confined to very small, climatically favourable areas, or former scrubland communities of individuals that failed to reach full stature under harsh conditions, easily observed today in the case of trees from lower altitudes growing away from the forest.

Recolonisation by forest types was rapid, once again demonstrating the tolerance of unusual physical conditions and the facility for rapid dispersal characteristic of

Afromontane trees. Hamilton (1982, 1988) found that although Mt. Elgon was apparently deforested in the last glaciation, forest cover was extensive by 11,000 BP, even though expansion may not have commenced until 12,500 BP. The demographic study of the relationships between pioneer and shade-dependent species in the forest of the northern slope of Kilimanjaro, demonstrated how Undifferentiated Afromontane forest could re-establish itself after a fire within three hundred years. In this case the regenerating area was surrounded by similar forest acting as a seed source, but a similar succession following climatic amelioration can be invoked if robust pioneers were present at the margins of the refuge and spread rapidly outwards.

Rapid spread following a disaster requires effective seed dispersal and tolerance of a wide range of conditions for germination and establishment. Hedberg (1961; 1969) concluded that Afroalpine distributions can be explained only as the product of independent long distance dispersal, and this seems to be also true of Afromontane species. The isolated nature of most mountains in the Afromontane archipelago, even when distances between them are short, is sufficient to prevent species of poor dispersal capability reaching them, even over extended periods of time. Dowsett-Lemaire (1988b; 1989b) has demonstrated the importance of birds as seed-dispersal agents in Afromontane forests in Malawi, and from her work it can be shown that there are sufficient migratory frugivores capable of making inter-mountain journeys to collectively achieve the dispersal of Afromontane trees throughout the archipelago (Grimshaw, 1996). On Kilimanjaro 64% of the woody flora found in the survey quadrats has fleshy fruit capable of being swallowed by this guild of migratory species; a further 13% has dry, dehiscent seeds that lack any obvious adaptation for dispersal, but many of these are recorded to be consumed by the migratory dove *Aplopelia larvata*. 22% (26 species) are adapted for wind-dispersal; these are mostly shrubby composites such as *Vernonia* occupying disturbed sites, but six species of wind-dispersed tree (including *Entandrophragma, Erica, Hagenia, Nuxia*) are also pioneers on disturbed ground and demonstrate the rapid germination characteristic of wind-dispersed species exploiting temporarily available niches (Hladik & Miquel, 1990). Only one species, the shrub *Pavonia urens*, has bristles indicating adaptation to exozooic dispersal. Large fruited, large seeded species, such as the 30–44% of trees in Guineo-Congolian forests recorded to be dispersed by elephants (Alexandre, 1978; Gautier-Hion *et al.*, 1985; Feer, 1995), are not found on Kilimanjaro or on other isolated mountains, indicating that they are poorly adapted for long distance transport, or require specialised conditions for germination and establishment.

Inter-mountain dispersal journeys are not necessarily long. The isolation of many African mountains is largely mythological, and may be an artefact of the European preoccupation with the Afroalpine flora; attention has been focused on the 'botanical big game' (Hedberg, 1969) of the few large massifs, while I believe that the many smaller peaks hold the key to species distribution in the Afromontane archipelago. The widest gap in the mountain chain between the Imatong Mountains (Sudan) and Mt. Mulanje (Malawi) is only 140 km, which is within easy reach for some birds (Wickens, 1976). At times of greater rainfall than at present, such as the wet period 9,000–8,000 BP (Kutzbach & Street-Perrott, 1985), even low hills must have been forest-covered, forming a series of potential stepping stones for the migration of forest organisms (Lovett, 1993a).

White (1981; 1990; 1993b) postulated the existence of a southern migratory track, by which the Afromontane flora of eastern Africa reached the very isolated West African mountains, showing how Afromontane species can be found in scattered small pockets of (often locally unique) favourable habitat. Many of these are nutrient-deficient, either on oligotrophic substrates or in swamps (Hamilton, 1974; Lind & Morrison, 1974; White, 1978; 1981; 1983a). These observations support the hypothesis that Afromontane trees, especially those from the higher areas, are adapted to nutrient-poor conditions. Swamps, as well as mountains, can thus be invoked as stepping stones for the migration of Afromontane species, and may be the means by which they have crossed the wide lowland gaps between the eastern and western rift systems, or the Congo basin. During pluvials the area of such swamps may have been much greater, further facilitating dispersal by reducing the distances involved.

During the past 2.3 million years there have been 21 glacial or near-glacial periods (Van Donk, 1976), or 20 cycles of forest spread and retreat (Hamilton, 1988), long enough for a flora evolving during the Miocene to achieve, by individual migration, the complicated distribution patterns observed today, even if a species may have recolonised the same massif on several occasions. Although we can show how dispersal can be achieved, current conditions do not provide us with all the information needed for untangling distributional conundra; these can only be ascribed to the chances of age and climate creating the necessary stepping stones for an ecologically versatile, yet still specialised flora to colonise the Afromontane archipelago across the continent.

The versatility of Afromontane trees is demonstrated by a comparison with the flora of the Eastern Arc mountains of Tanzania, which are exceptional in the Afromontane archipelago for their stability. The Eastern Arc mountains are ancient fault blocks (Griffiths, 1993), located close to the Indian Ocean, which has given them meteorological conditions largely unaffected by the glacial aridity experienced further inland (Lovett, 1993a). The biota has long been recognized as remarkable, having large numbers of endemic plants and animals, with many taxa being shared with West Africa (e.g. Hamilton & Bensted-Smith, 1989; Lovett & Wasser, 1993; Rodgers & Homewood, 1982), including many of the elephant-dispersed Guineo-Congolian tree genera not found on other East African mountains (Grimshaw, 1996). This pattern is explained by the existence in the mid-Tertiary of a forest connection across Africa from west to east, later disrupted by the onset of Miocene tectonic activity that warped and uplifted the ancient peneplain, creating an arid corridor between tracts of moist forest (Lovett, 1993a).

Lovett (1992; 1993b) has shown that a majority of tree species in the Eastern Arc mountains are of Guineo-Congolian affinity, and argues that these mountains should be regarded as an outlier of the Guineo-Congolian Region, rather than as a part of the Afromontane Region, into which they would be placed if White's methodology of determining a phytochorion by reference to the current distribution of individual species was applied. White (1978; 1981) acknowledged that Afromontane trees are closely related to lowland species, but emphasized that they now occupy a distinctly different ecological niche. Although Lovett (1993b; 1996) found no evidence for sharply defined critical altitudes in the Eastern Arc forests, it is noticeable from the species lists (Lovett, 1993b) that the higher parts of these mountains contain a much higher proportion of widespread Afromontane endemics than Guineo-Congolian survivors. Where disturbance has occurred in the Eastern Arc, the successional forest is composed of such widespread Afromontane species (Hamilton, 1989; Lovett, 1993b; 1996), exploiting the niche available to them as tolerant pioneers but not to the specialized, poorly dispersed species of Guineo-Congolian origin that have survived in the stable climatic enclave of the Eastern Arc.

It is clear that the vegetation of the Eastern Arc is a special case in the Afromontane archipelago, but it should not be allowed to mask the validity of the continent-wide Afromontane Region whose flora is distinguished by its tolerance of the effects of altitude, the regime of repeated disturbance and the need for easy recolonisation. It is not necessary slavishly to attempt to bend every patch of vegetation to one of White's continental phytochoria; local phytochoria can be defined with equal validity, as Friis (1992) did with the forest flora of Northeast Tropical Africa, or as Dowsett-Lemaire (1985; 1988a; 1989a) found in Malawi, when the local patterns were found to complement, not conflict with, White's continental view.

Conclusion

The recurrent theme of this paper is the dynamism of Afromontane vegetation in response to physical influences, whether they be local and short-term, such as a fire, or recurrently devastating, as in periodic glaciations, or fundamental in the uplift and creation of the mountains themselves. Hedberg (1957) has described the Afroalpine vegetation as a great 'natural experiment', but it is only an extreme product of the influences that have shaped the vegetation of the Afromontane Archipelago as a whole: dispersal, adaptation and speciation. As a continent Africa is remarkable for its geological stability, which has not permitted the development of diverse landscapes facilitating speciation (Pannell & White, 1988). The exception has been the 'islands' of the Afromontane Archipelago, mostly uplifted since the Miocene, and here a remarkable flora has evolved in response to the physical challenges these mountains present.

In its tolerance to harsher montane climates the Afromontane flora is specialist, having evolved as a montane vicariant of lowland Guineo-Congolian vegetation, tolerant of the lower temperatures and reduced soil fertility of a montane environment. The evolutionary pressures have also favoured tolerance of disturbance (at all levels) and 'easy' dispersal to new habitats, together permitting rapid recolonisation of deforested areas, a strategy that might be described as generalist. In view of these adaptations, which characterize an immediately recognisable vegetation type found across Africa, I believe that to consider Afromontane vegetation to be merely a depauperate version of the Guineo-Congolian region (Lovett, 1993b, 1993c) is wide of the mark and I uphold its claim to distinction.

Acknowledgements

I should like to thank the Tanzania Commission for Science and Technology for permission to work in Tanzania, and Tanzania National Parks and Officers of Kilimanjaro Region for their assistance. My research on Kilimanjaro was funded by several organisations, whom I thank for their support: Magdalen College, Oxford; Care for the Wild; Wildlife Conservation Society; National Geographic Society (grant #4965-93).

References

Abraham, M.F.H. (1958). The East African Camphor Forests of Mt. Kenya. *E. African Agric. J. Kenya* **24**: 139–141.

Alexandre, D.-Y. (1978). Le rôle disséminateur des éléphants en forêt de Tai, Côte d'Ivoire. *Terre & Vie* **32**: 47–72.

Bonnefille, R., Roeland, J.C. and Guiot, J. (1990). Temperature and rainfall estimates for the past 40,000 years in equatorial Africa. *Nature* **346**: 347–349.

Bussmann, R.W. (1994). The forests of Mt. Kenya, Kenya. Vegetation, Ecology, Destruction and management of a Tropical Mountain Forest Ecosystem. Published doctoral thesis, Bayreuth University.

Chapman, J.D. and White, F. (1970). The Evergreen Forests of Malawi. Commonwealth Forestry Institute, University of Oxford, Oxford.

Coetzee, J.A. (1967). Pollen analytical studies in East and Southern Africa. Palaeoecology of Africa 3. A.A Balkema, Cape Town.

Coetzee, J.A. and Bakker, E.M. van Zinderen (1970). Palaeoecological problems of the quaternary of Africa. *S. African J. Sci.* **4**: 78–84.

Coutts, H.H. (1969). Rainfall of the Kilimanjaro area. *Weather* **24**: 66–69.

Dale, I.R. (1954). Forest spread and climatic change in Uganda during the Christian era. *Empire Forest. Rev.* **33**: 23–29.

Dowsett-Lemaire, F. (1985). The forest vegetation of the Nyika Plateau (Malawi-Zambia): ecological and phenological studies. *Bull. Jard. Bot. Belg.* **55**: 301–392.

Dowsett-Lemaire, F. (1988a). The forest vegetation of Mt. Mulanje (Malawi): a floristic and chorological study along an altitudinal gradient (650–1950 m). *Bull. Jard. Bot. Belg.* **58**: 77–107.

Dowsett-Lemaire, F. (1988b). Fruit choice and seed dissemination by birds and mammals in the evergreen forests of upland Malawi. *Rev. Écol.* **43**: 251–285.

Dowsett-Lemaire, F. (1989a). The flora and phytogeography of the evergreen forests of Malawi I. Afromontane and mid-altitude forests. *Bull. Jard. Bot. Belg.* **59**: 3–131.

Dowsett-Lemaire, F. (1989b). Ecological and biogeographical aspects of forest bird communities in Malawi. *Scopus* **13**: 1–80.

Dowsett-Lemaire, F. (1990). The flora and phytogeography of the evergreen forests of Malawi II. Lowland forests. *Bull. Jard. Bot. Belg.* **60**: 9–71.

Edwards, P.J. (1977). Studies of mineral cycling in a montane rain forest in New Guinea. Part 2. The production and disappearance of litter. *J. Ecol.* **65**: 971–992.

Feer, F. (1995). Morphology of fruits dispersed by African forest elephants. *African J. Ecol.* **33**: 279–284.

Friis, I. (1992). Forests and Forest Trees of Northeast Tropical Africa. *Kew Bull., Addit. Ser.* **15**: 1–396.

Gautier-Hion, A., Duplantier, J.-M., Quris, R., Feer, F., Sourd, C., Decoux, J.-P., Dubost, G., Emmons, L., Erard, C., Hecketsweiler, P., Moungazi, A., Roussilhon, C. and Thiollay, J.-M. (1985). Fruit characters as a basis of fruit choice and seed dispersal in a tropical forest vertebrate community. *Oecologia* **65**: 324–337.

Gilchrist, B. (1952). In: Report of Central African rail link — Development Survey 1-2. Overseas Consultants Inc. and Sir Alexander Gibb and Partners. Colonial Office, London.

Griffiths, C.J. (1993). The geological evolution of East Africa. In: J.C. Lovett and S.K.Wasser (editors). Biogeography and Ecology of the Rain Forests of Eastern Africa. Pp. 9–21. Cambridge University Press, Cambridge.

Grimshaw, J.M. (1996). Aspects of the Ecology and Biogeography of the Forest of the Northern Slope of Mt. Kilimanjaro, Tanzania. Unpubl. D.Phil. thesis, University of Oxford.

Grubb, P.J. (1971). Interpretation of the 'Massenerhebung' effect on tropical mountains. *Nature* **229**: 44–45.

Grubb, P.J. (1977). Control of forest growth and distribution on wet tropical mountains. *Annual Rev. Ecol. Syst.* **8**: 83–107.

Hall, J.B. (1984). *Juniperus excelsa* in Africa: a biogeographical study of an Afromontane tree. *J. Biogeogr.* **11**: 47–161.

Hamilton, A.C. (1974). Distribution patterns of forest trees in Uganda and their historical significance. *Vegetatio* **29**: 21–35.

Hamilton, A.C. (1975). A quantitative analysis of altitudinal vegetation in Uganda forests. *Vegetatio* **30**: 99–106.

Hamilton, A.C. (1982). Environmental history of East Africa: a study of the Quaternary. Academic Press, London.

Hamilton, A.C. (1987). Vegetation and climate of Mt. Elgon during the late Pleistocene and Holocene. *Palaeoecol. Africa* **18**: 283–304.

Hamilton, A.C. (1988). Guenon evolution and forest history. In: A. Gautier-Hion, F. Bourlière, J.P. Gautier and J. Kingdon (editors). A Primate Radiation: Evolutionary Biology of the African Guenons. Pp. 13–34. Cambridge University Press, Cambridge.

Hamilton, A.C. (1989). History of resource utilisation and management. The pre-colonial period. In: A.C. Hamilton and R. Bensted-Smith (editors). Forest Conservation in the East Usambara Mountains Tanzania. Pp. 35–38. IUCN, Gland, Switzerland and Cambridge, UK.

Hamilton, A.C. and Bensted-Smith, R. (editors) (1989). Forest Conservation in the East Usambara Mountains Tanzania. IUCN, Gland, Switzerland and Cambridge, UK.

Hawthorne, W.J. (1995). Ecological Profiles of Ghanaian Forest Trees. Tropical Forestry Papers 29. Oxford Forestry Institute, Oxford.

Hedberg, O. (1951). Vegetation Belts of the East African Mountains. *Svensk Bot. Tidskr.* **45 (1)**: 140–204.

Hedberg, O. (1957). Afroalpine Vascular Plants. *Symb. Bot. Upsal.* **15 (1)**: 1–411.

Hedberg, O. (1961). The phytogeographical position of the Afroalpine flora. *Recent Advances Bot.* **1**: 914–919.

Hedberg, O. (1969). Evolution and speciation in a tropical high mountain flora. *Biol. J. Linn. Soc.* **1**: 135–148.

Hedberg, O. (1970). Evolution of the Afroalpine flora. *Biotropica* **2**: 16–23.

Hill, M.O. (1979). TWINSPAN — a FORTRAN Program for arranging Multivariate Data in an Ordered Two Way Table by Classification of the Individuals and the Attributes. Cornell University, Department of Ecology and Systematics, Ithaca, NY.

Hladik, A. and Miquel, S. (1990). Seedling types and plant establishment in an African rain forest. In: K.S. Bawa and M. Hadley (editors). Reproductive Ecology of Tropical Forest Plants. Pp. 261–282. Unesco, Paris.

Hooker, J.D. (1862). On the vegetation of Clarence Peak, Fernando Po, with descriptions of the plants collected by Mr. Gustav Mann on the higher peaks of that mountain. *J. Proc. Linn. Soc., Bot.* **6**: 1–23.

Hooker, J.D. (1864). On the plants of the temperate regions of the Cameroons Mountains and islands in the Bight of Benin, collected by Mr. Gustav Mann, Government Botanist. *J. Proc. Linn. Soc., Bot.* **7**: 171–240.

Hooker, J.D. (1874). On the subalpine vegetation of Kilima-Njaro, East Africa. *J. Linn. Soc. Bot.* **14**: 141–146.

Kigomo, B.N., Savill, P.S. and Woodell, S.R.J. (1990). Forest composition and its regeneration dynamics; a case study of semi-deciduous tropical forests in Kenya. *African J. Ecol.* **28**: 174–188.

Kutzbach, J.E. and Street-Perrott, F.A. (1985). Milankovitch forcing of fluctuations in the level of tropical lakes from 18 to 0 kyr BP. *Nature* **317**: 130–134.

Lebrun, J. (1947). La végétation de la plaine alluviale au sud du Lac Édouard. Exploration du Parc National Albert. Mission J. Lebrun (1937–1938). Pt 1 (2 vols). Inst. Parcs Nationaux Congo belge: Brussels.

Lieberman, D., Lieberman, M., Peralta, R. and Hartshorn, G.S. (1996). Tropical forest structure on a large-scale altitudinal gradient in Costa Rica. *J. Ecol.* **84**: 137–152.

Lind, E.M. and Morrison, M.E.S. (1974). East African Vegetation. Longman Group, London.

Livingstone, D.A. (1975). Late Quaternary climatic change in Africa. *Annual Rev. Ecol. Syst.* **6**: 249–280.

Lovett, J.C. (1992). Classification and Affinities of the Eastern Arc Moist Forests of Tanzania. Unpubl. Ph.D. thesis, University College of North Wales, Bangor.

Lovett, J.C. (1993a). Climatic history and forest distribution in eastern Africa. In: J.C. Lovett and S.K. Wasser (editors). Biogeography and ecology of the rain forests of eastern Africa. Pp. 23–29. Cambridge University Press, Cambridge.

Lovett J.C. (1993b). Eastern Arc moist forest flora. In: J.C. Lovett and S.K. Wasser (editors). Biogeography and ecology of the rain forests of eastern Africa. Pp. 33–55. Cambridge University Press, Cambridge.

Lovett, J.C. (1993c). Temperate and tropical floras in Tanzania. *Opera Bot.* **121**: 217–227.

Lovett, J.C. (1996). Elevational and latitudinal changes in tree associations and diversity in the Eastern Arc mountains of Tanzania. *J. Trop. Ecol.* **12**: 629–650.

Lovett, J.C. and Wasser, S.K (editors) (1993). Biogeography and ecology of the rain forests of eastern Africa. Cambridge University Press, Cambridge.

Monod, T. (1957). Les grandes divisions chorologiques de l'Afrique. CCTA/CSA Publ. No. 24: 1–147, London.

Moreau, R.E. (1933). Pleistocene climatic changes and the distribution of life in East Africa. *J. Ecol.* **21**: 415–435.

Mueller-Dombois, D., Canfield, J.E., Holt, R.A. and Bigelow, G.P. (1983). Tree-group death in North American and Hawaiian forests: a pathological problem or a new problem for vegetation ecology. *Phytocoenologia* : 117–137.

Mugasha, A.G. (1978). Tanzania Natural Forests Silvicultural Research — Review Report. *Tanzania Silvic. Techn. Note* **39**: 1–41.

Osmaston, H. (1989). Glaciers, glaciations and equilibrium line altitudes on Kilimanjaro. In: W.C. Mahaney (editor). Quaternary and Environmental Research on East African Mountains. Pp. 7–30. Balkema, Rotterdam and Brookfield.

Pannell, C.M. and White, F. (1988). Patterns of speciation in Africa, Madagascar, and the Tropical Far East: Regional faunas and cryptic evolution in vertebrate-dispersed plants. *Monogr. Syst. Bot. Missouri Bot. Gard.* **25**: 639–659.

Pendry, C.A. and Proctor, J. (1996). The causes of altitudinal zonation of rain forests on Bukit Belalong, Brunei. *J. Ecol.* **84**: 407–418.

Rodgers, W.A. and Homewood, K.M. (1982). Species richness and endemism in the Usambara Mountain Forests, Tanzania. *Biol. J. Linn. Soc.* **18**: 197–242.

Tanner, E.V.J., Kapos, K., Freskos, S., Healey, J.R. and Theobald, A.M. (1990). Nitrogen and phosphorus fertilisation of Jamaican montane forest trees. *J. Trop. Ecol.* **6**: 231–238.

Troll, C. (1958). Tropical Mountain Vegetation. In: Proceedings of the 9th Pacific Scientific Congress 20.

Troll, C. and Wien, K. (1949). Der Lewisgletscher am Mount Kenya. *Geogr. Ann. Svenska Sällsk. Antropol.* **31**: 257–274.

Van Donk, J. (1976). An ^{18}O record of the Atlantic Ocean for the entire Pleistocene. *Mem. Geol. Soc. Amer.* **145**: 147–164.

Veblen, T.T., Donoso Z., C., Schlegel, F.M. and Escobar R., B. (1981). Forest dynamics in south-central Chile. *J. Biogeogr.* **8**: 211–247.

Vincent, C.E., Davies, T.D., Brimblecombe, P. and Beresford, A.K.C. (1989). Lake levels and glaciers: Indicators of the changing rainfall in the mountains of East Africa. In: W.C. Mahaney (editor). Quaternary and Environmental Research on East African Mountains. Pp. 199–216. A.A. Balkema, Rotterdam and Brookfield.

Vitousek, P.M. (1984). Litterfall, nutrient cycling and nutrient limitation in tropical forests. *Ecology* **65**: 285–298.

White, F. (1965). The savanna woodlands of the Zambesian and Sudanian domains. An ecological and phytogeographical comparison. *Webbia* **19**: 651–681.

White, F. (1978). The Afromontane Region. In: M.J.A. Werger (editor). Biogeography and ecology of Southern Africa. Pp. 463–513. W. Junk, The Hague.

White, F. (1981). The history of the Afromontane archipelago and the scientific need for its conservation. *African J. Ecol.* **19**: 33–54.

White, F. (1983a). The Vegetation of Africa: a descriptive memoir to accompany the UNESCO/AETFAT/UNSO vegetation map of Africa. *Natural Resources Research* **20**. UNESCO, Paris.

White, F. (1983b). Long-distance dispersal and the origins of the Afromontane flora. *Sonderb. Naturwiss. Vereins Hamburg* **7**: 87–116.

White, F. (1990) *Ptaeroxylon obliquum* (Ptaeroxylaceae), some other disjuncts, and the Quaternary history of African vegetation. *Bull. Mus. Natl. Hist. Nat., B, Adansonia* **12(2)**: 139–185.

White, F. (1993a). African Myricaceae and the history of the Afromontane flora. *Opera Bot.* **121**: 173–188.

White, F. (1993b). Refuge theory, ice-age aridity and the history of tropical biotas; an essay in plant geography. *Fragm. Florist. Geobot. Suppl.* **2(2):** 385–409.
Wickens, G.E. (1976). Speculations on long distance dispersal and the flora of Jebel Marra, Sudan Republic. *Kew Bull.* **31**: 105–150.
Willan, R.L. (1965). Natural regeneration of high forest in Tanganyika. *E. African Agric. Forest. J.* **31**: 43–53.

Cutler, D.F. (1998). Dry woodlands, their utilisation and conservation. In: C.R. Huxley, J.M. Lock and D.F. Cutler (editors). Chorology, Taxonomy and Ecology of the Floras of Africa and Madagascar. Pp. 221–226. Royal Botanic Gardens, Kew.

16. DRY WOODLANDS, THEIR UTILISATION AND CONSERVATION

D.F. CUTLER

Jodrell Laboratory, Royal Botanic Gardens, Kew, Richmond, TW9 3AB, UK

Abstract

Natural woodlands in dryland regions of southern Africa, where rainfall is typically ≤ 650 mm, are being depleted at the rate of about 1 million hectares a year. Population increase in these regions is around 3.5% a year. Sustainable management of the key fuelwood species, which include *Acacia karroo, A. tortilis, Colophospermum mopane, Combretum apiculatum, Combretum imberbe* and *Terminalia sericea,* is of prime importance. The economic significance of these indigenous woodlands has been recognised for some time, but only recently have those responsible for state forestry in southern Africa decided that they merited study. This paper discusses two approaches that aim to address the problems of overuse. One concerns the selection of drought-resistant genotypes of four of the species listed above for future enrichment planting or the re-establishment of woodland. The other considers cropping experiments designed to discover which practice gives the maximum sustainable production of fuelwood.

Introduction

In the face of population pressure, the future of many of the vegetation types in Africa recognised by Frank White depends on effective management of the ecosystems concerned. In many regions it is very difficult or even impossible to maintain wilderness reserves. For most of Africa the plants and the people must coexist. Natural vegetation must be seen to contribute to an acceptable standard of living or it will be under threat. This is particularly evident in the fragile ecosystems of the dry woodlands.

For many years exotics, particularly *Eucalyptus* species, have been grown by Forestry Departments in arid and semiarid savanna regions in southern Africa, either in plantations or in woodlots. This has been very effective in some areas. It has produced timber with a range of uses at faster rates than those normally associated with the indigenous species. However, in these managed areas, there is often restricted access for villagers and their ubiquitous cows and goats. Traditionally, villagers regard fuelwood and fodder as a free resource. Furthermore, in countries where oral traditions still prevail, well-known multipurpose trees from the natural, local woodlands are most popular. However, a more disciplined approach to harvesting local woodlands may prove to be necessary for sustainability. There is little satisfactory information about the most effective cropping techniques for many of the more popular native fuelwood species. When optimal techniques for cropping are understood, they will probably be accepted and fit readily into the traditions of village life.

The native species from low rainfall areas are more drought resistant than *Eucalyptus* species, and do not dry the soil as extensively. This often permits adjacent planting of crops which is not possible in the presence of *Eucalyptus.* There are ecological benefits in using indigenous tree species in a sustainable way. The susceptibility of *Eucalyptus,* even *Eucalyptus camaldulensis*, to prolonged, severe drought and to indigenous pests has been evident in recent years. The rapid diminution of the natural savanna woodlands, through severe over-utilisation by a rapidly increasing population, has emphasised the urgent need for their preservation. This is seen as a major problem throughout the SADC (Southern Africa Development Community) area, where wood usage is calculated to be 1.5 cubic metres per person per annum.

At an international symposium on the Ecology and Management of Indigenous Forests in Southern Africa, held in Zimbabwe in 1992, the urgent need for increased research in indigenous species was reiterated by most delegates. This need had already been identified by the World Bank, who set up Phase II of their Rural Afforestation Programme in Zimbabwe, specifically to deal with the performance of selected indigenous species in management trials. Similar work is also being funded by the Swedish Agency for Research cooperation with developing Countries (SAREC) and the Ford Foundation in Zimbabwe. Gesellschaft für Technische Zusammenarbeit (GTZ), the German Agency is funding woodland inventories in Zimbabwe. The UK ODA (Overseas Development Agency, now the Department for International Development) is funding an evaluation of the field and genetic performance of six African acacias. Scandinavian aid agencies and the Ford Foundation are funding small projects in Namibia and Mozambique respectively. There is now an urgent need for a well co-ordinated regional project that would look at the possibilities for sustainable biomass supplies.

Traditional use of communal land woodlands varies from country to country in Africa. It is usual for women and children to have prior claim on trees nearest to villages, for domestic use. Generally the men have to work further out to gather wood for sale. However, conflict of interest can arise as the trees near to villages become over-used, and distances travelled for domestic users increase. Overuse near to the villages leads to soil erosion, increased dust levels and a general reduction in amenity. Serious problems arise more quickly when wood is cut for local industrial processes, such as brick firing. In Malawi for example, there is considerable cutting of mopane (*Colophospermum mopane*) for charcoal production for use in the towns and cities.

Major research reports have been prepared on fuelwood supplies; see, for example, the 1985 United Nations Food and Agriculture Organisation Tropical Forestry Action Plan. Clearly there is a need for large-scale research, but there are benefits in small, local projects. These can be taken up by the developing countries themselves, and continued after the external agencies and funding have gone away. Success depends to a considerable extent on local involvement and 'ownership'. Another ingredient for success is the availability of popular local means of getting information to the villages in a way that is not threatening.

Sustainable use of fuelwood

Two approaches leading to sustainable use of natural woodlands in semi-arid parts of Africa are considered in this paper. Both methods have potential use in other parts of the world with similar conditions. First, a project that sought out genotypes of indigenous trees with stress-induced adaptations will be described. If the natural woodland can be managed, the chances of retaining what is left of the biodiversity are strengthened. Replanting, or enrichment planting with drought-resistant genotypes of the indigenous species in areas where there has been extensive degradation could lead to improved fuelwood supply and encourage the re-establishment of the natural ecosystem. Where the use of exotic trees in these fragile ecosystems is possible, that is in the less dry woodlands, it may improve the fuelwood supply, but generally leads to a loss of natural biodiversity.

This account will be followed by a consideration of small-scale, local trials on sustainable cropping of selected native trees. This has a real chance of succeeding in the Communal Lands of Zimbabwe, where initial plans are being made. Similar projects could be undertaken in most of the SADC countries, with local management variations.

Trees with stress-induced adaptations

From 1989–91 a pilot study was conducted in Zimbabwe with the objective of finding drought-resistant genotypes of four indigenous tree species used preferentially by local people for fuelwood and browse. This project, entitled 'The improved productivity of African fuelwoods by the use of trees with stress-induced adaptations', was funded by the EU. It involved a European partnership with the Forestry Commission of Zimbabwe and the University of Zimbabwe, Harare (Prior *et al.*, 1991).

The selected species were *Acacia karroo* and *A. tortilis* (Leguminosae: Mimosoideae) which are deep-rooted, nodulating pioneers, and *Colophospermum mopane* (Leguminosae: Caesalpinoideae), non-nodulating, and *Combretum apiculatum*, from the Combretaceae, both generally shallow-rooting, secondary colonisers. All are capable of tolerating ≤ 650 mm of rain p.a., and in extreme circumstances some of them will survive a few years of < 250 mm. The wood of all these species has burning properties that are particularly desirable: slow and hot, with little smoke. This is thought to be due in part to their density, and also to the presence of calcium oxalate crystals (Prior & Cutler, 1992).

Four sites were selected that were sufficiently distant from one another for there to be little or no chance of gene flow between them. Table 1 gives details of the sites. A minimum of three out of the four species were present at each site, and these were represented by trees of different ages forming pure or mixed stands. The soils were of the types in which the species would normally occur. There were considerable differences in rainfall and/or temperature between the sites. Three sites were temperature or water stressed, and one was more moderate.

Seeds were collected from each species growing at the four sites. These were stored at the Forestry Commission seed bank in Harare. Germination tests at different temperatures and degrees of water stress (regulated by the use of polyethylene glycol) showed differences in response that were related to the temperature or water regimes of the provenances. For example, mopane seeds collected at Chiredzi continued to germinate at water potentials lower than -1.03 Mpa, whereas those from the Zambezi Valley ceased to germinate at potentials below -0.29 Mpa. Conversely, the Zambezi Valley seeds were more tolerant to heat (Tuohy, reported in Prior *et al.*, 1991). These results indicated that there was clear evidence of genotypic differences in stress adaptations between populations.

TABLE 1. Location, altitude and climatic details of the four sites used for the drought resistance studies in Zimbabwe.

SITE	Altitude (m a.s.l.)	Annual rainfall (mm)	Annual temperature °C		Latitude and Longitude
			Mean max.	Mean min.	
Zambezi Valley	550	750	34.1	19.6	16°08'S 29°24'E
Kadoma	1140	780	28.1	14.4	18°33'S 29°56'E
Matopos	1360	250–1400	25.9	10.8	20°25'S 28°29'E
Chiredzi	600	564	29.8	15.6	21°03'S 31°57'E

Prof. M. Popp, a biochemist on the EU team, has demonstrated higher levels of the stress protectant pinotol or quaternary ammonium compounds in the seeds, leaves and bark of trees from the genotypes from stressed than those from more moderate sites. She will be continuing studies on tannin content in the leaves and twigs used as fodder from the specimens. High tannins can make the material unpalatable and may lead to digestive problems in livestock. The trees from the field trial will also be assessed in this respect. An account of the range of these and other experimental researches pursued during the EU project has been given by Prior *et al.*, 1991.

Anatomical studies showed a reduced stomatal count on the abaxial surface of leaves of *Combretum apiculatum* from the drier sites when compared with those from less stressed conditions (Prior *et al.*, 1991). Biomass measurements were made at the end of the experiment, and the results published in Mushove *et al.* (1995), together with a preliminary analysis of growth increments. However, although it was found that growth rings that probably equate with annual rings were measurable in mopane and *C. apiculatum*, the same was not true for the acacias (Prior & Cutler, 1996). The main observation made in the latter paper was that no correlation could be shown between rainfall and growth ring width in mopane or *C. apiculatum*. The models that operate for north temperate species did not seem to apply. It is possible that recorded rainfall does not relate directly to rain water available to the trees. It was clear that even in stressed sites the two last species perform better than general opinion would suggest. Average radial increments for mopane, even in hot, dry soils in the Zambezi Valley were found to be 0.91 mm (average annual rainfall 750 mm). They reached 1.05 mm at Chiredzi, where it is drier but somewhat cooler (average annual rainfall 564 mm). *C. apiculatum* performed better at both sites, with 0.95 and 1.26 mm average radial increments respectively (Prior & Cutler, 1996, Table 1).

About 600 seedlings of the four species from the different provenances were raised under normal nursery conditions in polythene sleeves, and planted on into a trial field at the Chesa Research Station, Forestry Commission, near Matopos. This work is continuing under the direction of the author, and Dr Juliet Prior (Director of the EU project), generously funded by Glaxo (now Glaxo-Wellcome). The seedlings were planted at the beginning of the rains in 1993, and watered-in. No other maintenance has been provided apart from keeping the general vegetation cut to about 50 cm and the soil 1 m around each tree regularly weeded. No additional watering was done. Following the initial two years of drought, over 95% of the seedlings survived. Rainfall has been above average in the past two years. There have been some additional failures, due mainly to some of the young trees having been eaten, and others (acacias) having been attacked by bark beetles. Even so it seems probable that the survival rate may be about 80–85% when the trial is concluded in 1999. The trees will then be harvested, and their productivity measured. Initial observations on their progress, particularly of the acacias, suggest that there is not a trade-off between productivity and stress resistance. The results to date indicate that it may be preferable to use seeds from trees in stressed sites when replanting or doing enrichment planting. The seedlings might be expected to survive neglect and the mature trees the apparent climatic changes that could be leading to lower annual rainfall. At the time of harvesting, leaves and bark will be collected by Prof. M. Popp. They will be analysed for stress compound levels. In addition, tannin levels will be measured in relation to palatability and the subsequent value of the leafy twigs as fodder.

Controlled harvesting

During the last few years, several research projects have concentrated on miombo woodland. Yet the majority of rural people in much of southern Africa are confined to drier areas, where mopane and acacias are often the dominant species. The little research that has been done on savanna species has been small scale and has not provided statistically sound results with wider application (Mushove & Muchichawa, 1996).

It is important to use appropriate, statistical plot design to maximise useful information from compact cropping trials. Analysis of results from such plots would lead to increased ability to manage the vital natural dry woodland resources and bring added value. It could also enable high quality future sources of fuelwood and browse to be identified rapidly and cheaply through the use of mathematical modelling and indicator or marker characters.

There is a pressing need to achieve sustainable production of high quality fuelwood and browse through simple, inexpensive management systems. In order to be sustainable in impoverished parts of the world, after its initiation, a programme should not depend for its continuation on expensive methods. If it is necessary to use complex equipment, its use should be confined to the early stages where it can be allowed for in the primary funding.

There is work in progress in Zimbabwe, funded by, for example, the World Bank Rural Afforestation Programme Phase II, and SAREC. This has produced some interesting results. Soil suffered rain compaction in fenced plots in southern Zimbabwe, leading to lack of aeration and slower than expected regrowth. The persistence of the surface panning appears to be one of the consequences of excluding grazing animals. In open, unfenced woodland their hooves break such crusts, improving water penetration. Free movement of grazing or browsing animals in woodland is normal in villages, so any research with the object of improving natural woodland needs to take this into account. It is very expensive to fence off areas, so trees around villages are likely to be browsed. Fencing out animals from trial plots gives a false impression of what might be achieved under normal circumstances.

Delegates at an international, EU funded workshop organised by D. Cutler and J. Prior held in Zimbabwe in July 1994 were much in favour of setting up cropping trials on a wider basis. Following the meeting, and subsequent discussion with the delegates, it was recommended that an effective approach would be to apply and evaluate different cropping methods using common fuelwood tree species. These might be, for example *Acacia karroo, Combretum apiculatum, Colophospermum mopane* and *Terminalia sericea* found in savanna woodlands of Southern Africa. The objective would be to achieve optimal, sustainable production of high quality fuelwood and palatable, digestible browse. An effective method that also protected the environment and reduced loss of biodiversity would be sought. The most successful cropping method would be recommended to rural communities where continuing droughts and rapid population increases (up to 3.5% per annum) are leading to soil erosion and a severe decrease in the availability of vital, local, multipurpose trees. The potential for wider application of the results in other dry woodland areas of the tropics would be of considerable interest. Fine tuning to local requirements may be necessary.

Such a scheme would involve setting up experimental plots of about 2 hectares in area in secure, natural woodlands with similar species composition. The plots would need to be sited across southern Africa in areas where the maximum rainfall is 650 mm per annum, from Namibia to Mozambique. Cropping methods could include: a) coppicing, b) pollarding, and c) thinning to a selected number of stems. It has not yet been established when it would be best to cut the trees to obtain maximum regrowth. Cutting some trees at the end of the dormant period and others in the middle of the growing season could provide some valuable information on this aspect.

Wood and leaves of cropped and uncropped, control material could be compared using the following techniques: 1. Ecophysiology (photosynthetic and transpiration rates, growth analyses and biomass allocation studies). 2. Anatomy (amount of thick walled fibrous material in the wood; internal cell/air surface area of the leaves and stomatal distribution and frequency). 3. Biochemistry (analysis of stress compounds such as cyclitols and tannins and an evaluation of the relative importance of these compounds on browse palatability and digestibility). 4. Biomass estimation of standing trees and a comparison of the amount of biomass produced by variously cropped and control trees. 5. A comparison of the burning quality of cropped and control fuelwood.

Different stocking densities of cows and goats, simulating the conditions found in villages, could be used at additional plots while using the same tree species, cropped using the same range of techniques. The Matopos Research Station Zimbabwe has knowledge relating to work of this type that is unique within Africa. The effect of animal density could be investigated on a) the types of browse eaten; b) the quantity of new biomass over a 3 year period and c) soil structure and possible degradation.

If properly organised, these approaches to the management problems of regeneration could make a significant impact. They could provide training schemes in experiment design, fieldwork techniques, data collection and analysis, so that the low cost research could be continued by local personnel within each country. Innovative methodology required for future predictions of woodland productivity would be developed. The use of the best cropping methods would give added value to the natural woodlands.

There are no adverse environmental impacts visualised from the tree species used, since all selected would be indigenous, and worked on in established woodlands. The work involved would not increase the fire risk in the sites concerned.

Conclusions

The outcome of research in enrichment or replacement planting using drought-resistant genotypes, and different management methods of existing woodland, would be to indicate management systems for dry woodland that could minimise environmental degradation whilst providing sustainable supplies of high quality fuelwood and browse. Their application would benefit both the environment and the local people who depend on the resources of the woodland for their survival.

References

Mushove, P.T. & Muchichawa, J. (1996). Appropriate silvicultural intervention for improved growth in *Colophospermum mopane*. In: Proceedings of the International Conference on Sustainable Management of Indigenous Forests in the Dry Tropics, Kadoma, Zimbabwe, 28 May to 1 June, 1996.

Mushove, P.T., Prior J.A.B., Gumbie, C. & Cutler, D.F. (1995). The effects of different environments on diameter growth increments of *Colophospermum mopane* and *Combretum apiculatum*. *Forest Ecol. Managem.* **72**: 287–292.

Prior, J. & Cutler, D.F. (1992). Trees to fuel Africa's fires. *New Sci.* **135**: 35–39.

Prior, J.A.B. & Cutler, D.F. (1996). Radial increments in four tropical, drought tolerant firewood species. *Commonw. Forest. Rev.* **75(3)**: 227–233.

Prior, J.A.B., Cutler, D.F., Mushove, P.T., Pearson, J., Popp, M., Stewart, G.R. & Thouy, J.M. (1991). The improved productivity of African Fuelwoods by the use of trees with stress induced adaptations. Final Report of Southern African wood studies project. (EEC TS2 0211) Imperial College, London.

United Nations Food and Agriculture Organisation, 1985. Tropical Forestry Action Plan.

Tutin, C.E.G. (1998). Gorillas and their food plants in the Lopé Reserve, Gabon. In: C.R. Huxley, J.M. Lock and D.F. Cutler (editors). Chorology, Taxonomy and Ecology of the Floras of Africa and Madagascar. Pp. 227–243. Royal Botanic Gardens, Kew.

17. GORILLAS AND THEIR FOOD PLANTS IN THE LOPÉ RESERVE, GABON

CAROLINE E.G. TUTIN

Centre International de Recherche Médicales de Franceville, Gabon
and
Department of Biological and Molecular Sciences, University of Stirling, Scotland
Correspondence Address: S.E.G.C., B.P. 7847, Libreville, Gabon

Abstract

Western lowland gorillas (*Gorilla gorilla gorilla*) are the most numerous of the three gorilla sub-species and occur, at low population densities, in the tropical rain forests of west central Africa. In these botanically diverse forests, gorilla diet is dominated by fruit and vegetative plant parts (pith and leaves) but smaller quantities of seeds, bark and insects are also eaten. Gorillas are not typical frugivores and their large size and 'sloppy' feeding habits confer both benefits and costs to trees whose fruit they eat. Gorillas are high quality dispersers for the seeds of a range of tree species: not only are large numbers of intact seeds carried from the parent tree, but also many of these seeds are deposited in favourable microhabitats for germination and early seedling growth, i.e. in light-gaps where gorillas prefer to nest at night. On the other hand, the weight and strength of gorillas can inflict substantial damage on the trees in which they feed (and sometimes nest) with breaking of branches resulting in up to 30% canopy death. Predation on immature seeds by gorillas, which occurs when succulent fruit (their preferred food) is scarce, can deplete entire crops before fruit ripens. Interactions between gorillas and their food plants, observed at Lopé over 13 years, provide examples of the complexity of plant-animal interactions and give clues about selective pressures that may have shaped the evolution both of morphological characters of fruit and seeds and phenological rhythms of gorilla food plants. Understanding plant-animal interactions requires a multi-disciplinary approach and can contribute to understanding plant and animal biogeography and to conserving tropical rain forests.

Introduction

Western lowland gorillas (*Gorilla gorilla gorilla*) are the most numerous of the three gorilla sub-species and occur, generally at low population densities, in the tropical rain forests of west central Africa. In these botanically diverse forests, gorilla diets are dominated by fruit flesh and vegetative plant parts (pith and leaves) but smaller quantities of seeds, bark and insects are also eaten (Nishihara, 1992; Remis, 1994; Tutin & Fernandez, 1993a; Williamson *et al.*, 1990). The well-known mountain gorillas (*G. g. beringei*) of the Virunga Volcanoes are folivores but few plants with succulent fruit occur in the montane habitat (Vedder, 1984; Watts,1984). Field studies of gorillas in a range of habitats in equatorial Africa have shown diet to the determined by habitat: when succulent fruit are available they are the preferred food but when they are absent, gorillas are able to subsist on an entirely vegetative diet (Watts, 1990; 1996).

In the Lopé Reserve, in central Gabon, we have studied the ecology and behaviour of sympatric gorillas and chimpanzees since 1984. The apes have remained shy and we have not been able to habituate them to our presence in the way that has been possible for mountain gorillas in Rwanda and for chimpanzees in Tanzania (Tutin & Fernandez, 1991). Thus, the emphasis of our research at Lopé has been ecological and over the past decade we have documented gorillas' interactions with their tropical rain forest habitat. Gorillas are one of eight species of diurnal primate that occur at Lopé. The primate community makes up only about 10% of total mammalian biomass (White, L.J.T., 1994a) but, as is the case in many tropical rain forests in SE Asia and in South America, primates are the most important arboreal consumers in the forests of central Africa (Emmons *et al.*, 1983; Galat & Galat-Luong, 1985; Struhsaker & Leland, 1979). The diets of all primates at Lopé are dominated by fruit (for seven of the species) or seeds (one species) (Tutin *et al.*, 1997a) and, as almost all fruit are produced seasonally, although not necessarily annually, patterns of fruit availability and abundance have a major influence on primate behaviour. We have collected data on the phenology of 63 species of tree that produce fruit eaten by gorillas (and other primates) since October 1986 (Tutin & Fernandez, 1993b; Tutin & White, L.J.T., in press; White, L.J.T., 1994b). These data quantify variations in food availability within and between years, and allow an analysis of gorilla diet in terms of choice of foods. Lopé gorillas show marked preferences for certain foods which are eaten whenever available, while other foods are eaten only when the preferred foods are scarce or absent. Of 22 'important', or preferred, foods of Lopé gorillas, 15 are fruit (all produced seasonally) compared to fewer species of leaves (five) or pith (two) (Tutin & Fernandez, 1993a). From the plant's point of view, feeding on leaves, bark and pith, as well as the destruction of seeds, are impositions unlikely to be of benefit, but succulent fruit are designed to be eaten — the flesh being the bait to encourage seed dispersal.

Recent research at Lopé has shown that gorillas are high quality seed dispersers for a number of tree species as, in addition to swallowing large quantities of seeds, they deposit many of them in dung at nest-sites which are favourable micro-habitats for germination and early seedling survival and growth (Rogers *et al.*, in press; Voysey, 1995). It is clear that gorillas depend entirely on plants for their survival and the reverse appears to be true for one species of tree, *Cola lizae*, Sterculiaceae, an endemic with a limited geographical distribution which is the commonest tree within our study area (Hallé, 1987; Tutin *et al.*, 1994). Gorillas are the only consumer to swallow the large seeds of *Cola lizae* and they disperse large numbers during the annual two month fruiting season (Tutin *et al.*, 1991a).

However, dependence on a single disperser is rare and most fruit attract a range of frugivores many of which disperse at least some seeds intact. It is interesting to consider these plant-animal interactions from both sides, i.e., from the plant's and the animal's point of view, in order to look at function and outcomes. Gorillas are not typical frugivores, on the contrary, they are an extreme case: the largest bodied arboreal frugivore in Africa. The weight and strength of gorillas means that they can inflict substantial damage on the trees in which they feed, by deliberate or accidental breakage of branches. Seed predation by gorillas can be extreme and in some circumstances leads to the destruction of entire crops (Tutin *et al.*, 1996). Thus gorillas can be both good and bad news for the plants.

In this paper I review gorilla diet at Lopé and then focus on the interactions between gorillas and their food plants. Seeing things from the point of view of both the gorillas and the plants is an approach fostered by interactions with Frank White who firmly believed that an understanding of plant taxonomy, evolution and biogeography, was dependent on a knowledge of fauna as well as flora (e.g. Pannell & White, F., 1988). As he pointed out (White, F., 1993a), primatologists, unlike botanists, enjoy the privilege of spending long periods of time in the same small area of forest. Frank's encouragement and enthusiasm were precious but sadly, our direct collaboration began only in 1991 and produced a single fruit before his death (Tutin *et al.*, 1996). The present paper bears witness to the lasting inspiration that Frank White's input gave to our research at Lopé.

Study area and methods

The study area of the S.E.G.C. (Station d'Etudes des Gorilles et Chimpanzés) covers approximately 50 km² of tropical rain forest in the Lopé Reserve in central Gabon (0°10'S 11°35'E) (Fig. 1). Within the study area, the forest is heterogeneous in both floristic composition and structure reflecting a dynamic history of change linked to both climate and human activity (White, L.J.T., in press; White, L.J.T. & Abernethy, 1996; White, L.J.T. *et al.*, in press). Areas with a dense understorey of herbs, dominated

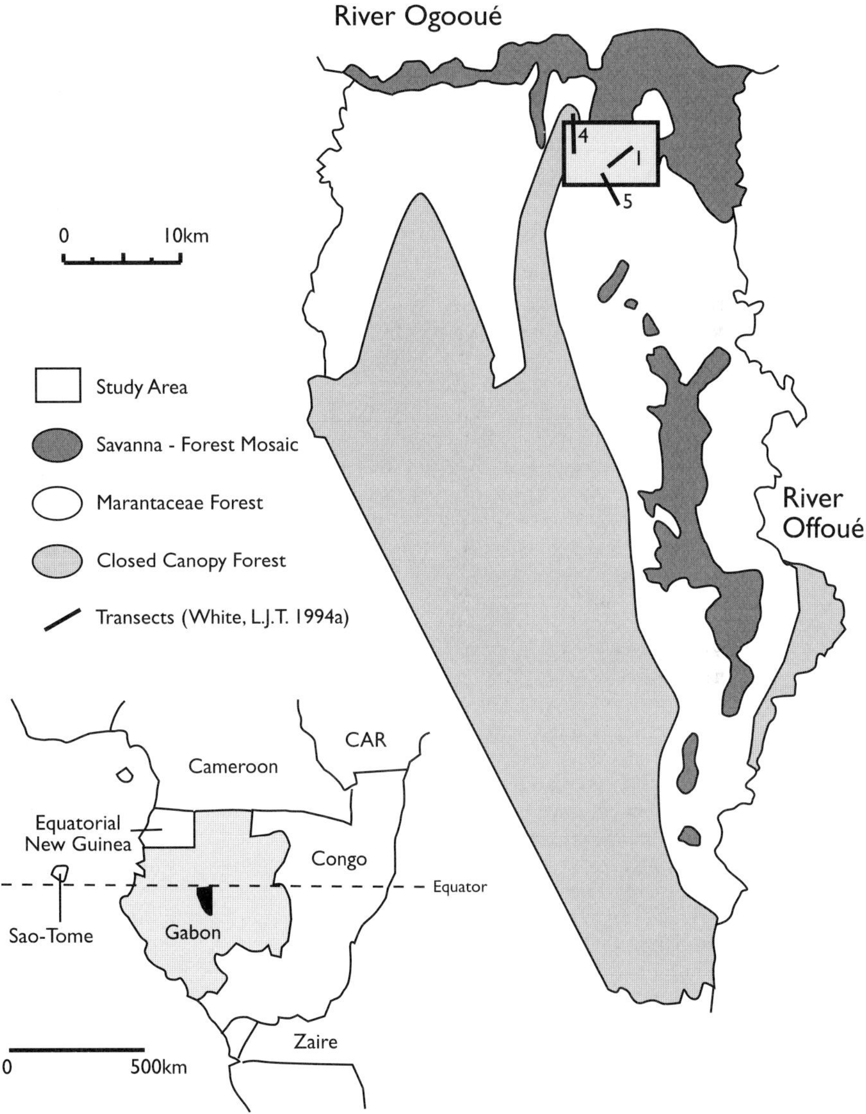

FIG. 1. Map of the Lopé Reserve showing the SEGC study area and major vegetation types.

by species of Marantaceae and Zingiberaceae, are common and are favoured by gorillas (Rogers & Williamson, 1987; White, L.J.T. *et al.*, 1995). To date, over 1,300 plant species have been identified: trees of the families Caesalpiniaceae, Burseraceae, Euphorbiaceae, Mimosaceae, Ochnaceae and Sterculiaceae dominate the upper canopy and the commonest smaller trees are members of the Annonaceae, Ebenaceae and Rubiaceae (Tutin *et al.*, 1994; L.J.T. White, unpublished data).

Average annual rainfall is 1548 mm (1984–1995) and the climate is characterised by a three month dry season lasting from about mid-June to mid-September. Mean monthly minimum and maximum air temperatures vary from 20.1 to 23.2°C and 27.0 to 32.8°C, respectively. During the dry season constant low cloud cover results in consistently cool and humid weather which explains the predominantly evergreen forest of the area despite the low rainfall.

Research began at Lopé in 1984. Gorillas occur at a population density of about one individual per km^2 and share the habitat with chimpanzees (*Pan t. troglodytes*) and six diurnal species of monkey (*Mandrillus sphinx, Colobus satanas, Cercocebus albigena, Cercopithecus nictitans, C. pogonias* and *C. cephus*). Total mammalian biomass varies between 1000 and 6000 kg/km^2 in different vegetation types (Tutin *et al.*, 1997b; White, L.J.T., 1994a). Field procedure involves searching through the study area for gorillas, or for indirect signs of their presence and activities. In addition to observation, many data on diet come from analysis of faeces (Tutin & Fernandez, 1994a), which allows dietary monitoring. Intact seeds in faeces show gorillas have fed on fruit, while fragments of seed cases reflect predation on seeds. Seeds have distinct features and those found in faeces can be determined to at least generic, and usually to species level. A reference collection of seeds has been established at the field station.

Samples of all plant foods are collected and specimens identified at the Missouri Botanical Garden, St. Louis, the Royal Botanic Gardens, Kew, the Museum of Natural History, Paris, and the National Herbarium of Gabon, Libreville.

Data are collected each month on leafing, flowering and fruiting patterns of 63 tree species that provide important food for primates. The size of the fruit crops of individual trees is quantified on a 10-point scale and fruit are classed as immature or ripe (see Tutin & Fernandez, 1993b for details of phenology methods).

Results

Gorilla diet at Lopé

Table 1 summarises the diet of Lopé gorillas by the number of foods recorded in each food class and by the percentage of time spent feeding on the different classes of foods. Gorilla diet is diverse with 220 different foods recorded to date. Overall, fruit is the most important class of food for Lopé gorillas in terms both of the number of different species eaten and the proportion of time spent feeding. On average, the remains of three different species of fruit were found per faecal sample (range 0–10, N = 4301 faecal samples) and remains of at least one species of fruit were found in 96% of analysed faeces. Leaves are the second most important food class, followed by seeds then piths.

Diet varies both within and between years in ways clearly related to the availability of fruit (Tutin *et al.*, 1991b). Succulent fruit is scarce each year during the dry season and fruit failures of some species in some years lead to longer periods of scarcity which can extend for as long as eight consecutive months (Tutin & White, L.J.T., in press). Figure 2A shows the mean number of species in the phenology sample that bore ripe fruit each month and Figure 2B illustrates inter-annual variation by comparing phenology data from 1991 and 1992. The frugivore's year at Lopé can be divided into three parts: during the 3-month dry season few ripe fruit are present; for four months after the onset of rains fruit availability is consistently high; while for the remaining five months of the year there are large inter-annual variations in fruit production.

TABLE 1. Diet of gorillas at Lopé

FOOD CLASS	Percentage (and Number) of Species Eaten (Data from 1984–1995)	Percentage (and Number) of Feeding Observations (Data from 1990–1993)
Fruit	45.5 (100)	40.9 (447)
Seed	9.6 (21)	5.8 (63)
Leaf	21.8 (48)	33.8 (369)
Pith	7.3 (16)	11.5 (126)
Flower	1.4 (3)	1.2 (13)
Bark	5.5 (12)	7.1 (77)
Animal	4.6 (10)	0.8 (9)
Other*	4.6 (10)	0.3 (3)
TOTAL NUMBER	220	1092

* Includes wood, roots, soil and fungi

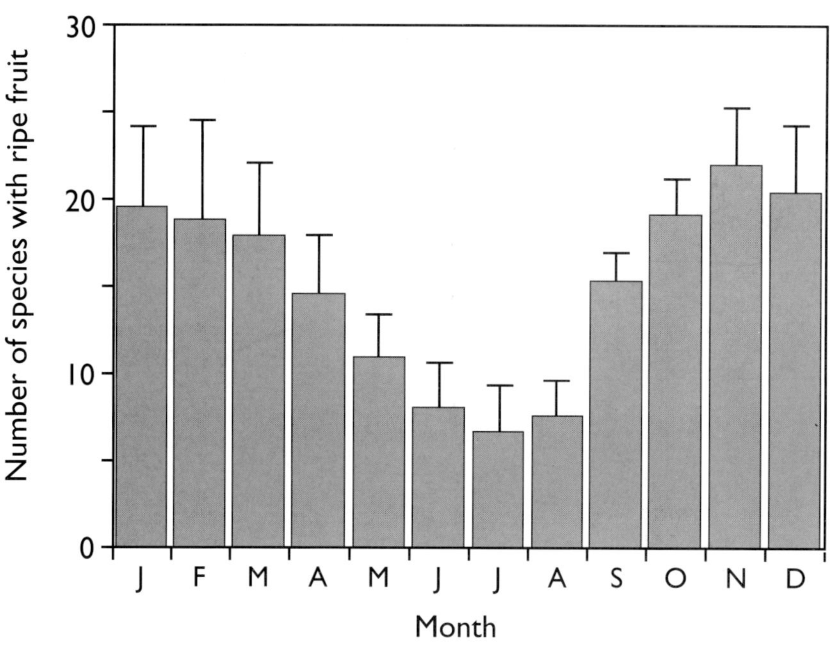

FIG. 2A. Number of fruiting species in the phenology sample, 1986–1995. Means and standard deviations.

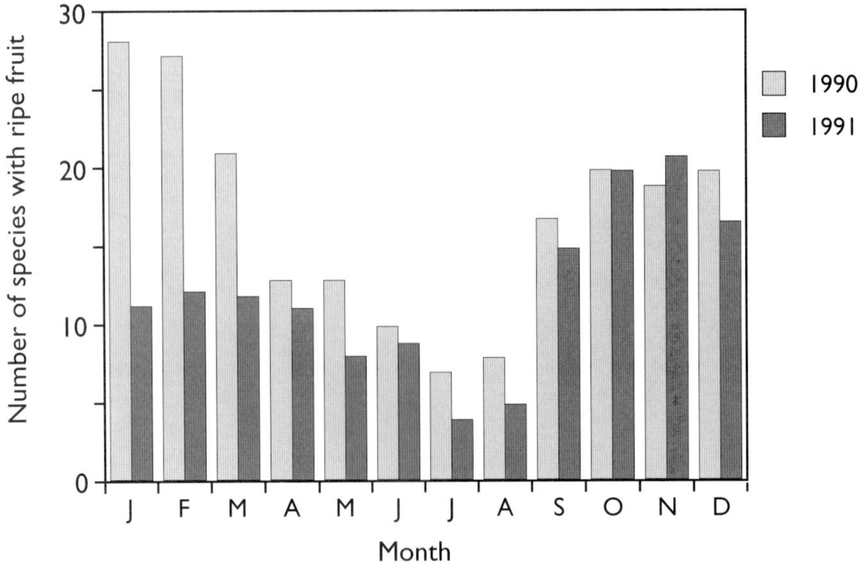

FIG. 2B. Comparison of the number of species with fruit in two consecutive years, 1990 and 1991.

When fruit is scarce, gorillas increase their consumption of pith and young leaves of common herbs but also feed on some plant parts that are completely ignored except at times of minimal fruit availability (see also Tutin *et al.*, 1997a). The four major 'fallback' foods of Lopé gorillas are listed in Table 2 along with staple foods that are eaten year round. The most important fallback food is the bark (or more correctly bast) of a large tree, *Milicia* (formerly *Chlorophora*) *excelsa* and, in the dry season, gorillas spend up to several hours each day stripping bark from branches (Rogers *et al.*, 1988; 1994). The large seeds of two species of tree are also eaten uniquely during the dry season: those of *Pentadesma butyracea* are extracted from unripe fruit and the flesh is discarded; while ripe seeds of *Detarium macrocarpum* are eaten by adult gorillas. *Detarium* seeds are inaccessible to immature gorillas who lack the necessary strength to bite through the fibrous seed coat.

Interactions between gorillas and their food plants

As the proportion of non-plant foods (ten types of insect, plus earth) in the diet is very small (Table 1), it is clear that gorillas depend totally on plants for their survival. However, gorillas' choice of foods from the diverse flora available is selective and their 209 recorded plant foods come from only 147 of the 1300+ species of plants that occur in the study area. Criteria for choice are in some cases obviously related to chemical or mechanical defences. Plants have evolved a variety of characters to protect themselves from the unwanted attentions of animals while animals riposte with morphological and physiological solutions in an on-going interactive saga.

So, which plants do gorillas choose to eat? A nutritional analysis of the food plants of the Lopé gorillas (Rogers *et al.*, 1990) found the diet to be a balance between sugary fruit, proteinaceous leaves and relatively fibrous pith. Compared to the general vegetation, gorillas avoid fatty fruit and seeds and select leaves that are high in protein and low in fibre. They eat some foods with high concentrations of phenols and condensed tannin but appear to avoid alkaloids. The relatively fibrous piths of

TABLE 2. Foods eaten by gorillas at Lopé during periods of fruit scarcity.

SPECIES	FAMILY	PART EATEN[1]	AVAILABILITY
FALLBACK FOODS[2]:			
Milicia excelsa	Moraceae	Bast & Mature leaves	Permanent
Marantachloa cordifolia	Marantaceae	Pith & Young leaves	Permanent
Pentadesma butyracea	Guttiferae	Seed	June–January
Detarium macrocarpum	Caesalpiniaceae	Seed	August–November
STAPLE FOODS[2]:			
5 species of herbs	Zingiberaceae	Pith	Permanent
4 species of herbs	Marantaceae	Young leaves & Pith	Permanent
Duboscia macrocarpa	Tiliaceae	Seeds	Permanent
Ficus, 12 species	Moraceae	Fruit	Irregular

[1] Where more than one part of a species is eaten, the major one is listed first.

[2] Fallback foods = foods only eaten when ripe fruit is scarce.   Staple foods = foods eaten year round although consumption is higher when fruit is scarce.

Zingiberaceae (principally *Aframomum* spp.) are a staple food and appear to be chosen for their high sugar content, while young leaves of Marantaceae (*Haumania, Hypselodelphys* and *Megaphrynium*) are high in protein and low in fibre. Thus, gorillas are able to balance their diet throughout the year but, when succulent fruit are available, these are preferred. Gorillas lack morphological or physiological specialisations for herbivory but their large body size, combined with an intestinal fauna that includes cellulose-digesting protozoans (Collet *et al.*, 1984), allows them to subsist on a low quality, bulky diet when necessary.

So gorillas need plants and plants have an arsenal of weapons to limit herbivory, but the story is complicated by the dependence of most plants on animals for transport of their pollen and/or seeds. Primates in general are not important pollinators although some exceptions occur (e.g. Gautier-Hion & Maisels, 1994; Kress *et al.*, 1994). The advantages of animal seed dispersal for tropical plants can be judged from the fact that the majority of plants in tropical forests have fruit with a nutritious pericarp which attracts animals (Howe & Smallwood, 1982; Janzen, 1975; 1983). At Lopé, 70 % of plant species have seeds that are dispersed by animals (White, L.J.T., 1994b).

The positive and negative impact of interactions between gorillas and their food plants are reviewed in the following sections.

Positive impact of gorilla feeding

The major positive impact of gorillas on their food plants is the dispersal of seeds away from the parent. Gorillas have large guts, a variable but long gut passage time of 17–80 hours (Milton, 1984) and travel 1–2 km each day (Tutin, 1996), all characters which make them potentially high quality seed dispersers. In addition, gorillas show two behavioural patterns that bring added benefits to at least some of the seeds they disperse. Firstly, they sleep in nests which are built each night, usually on the ground and often located in light gaps (Tutin *et al.*, 1995) and, as about half of their daily output of dung is deposited at nest-sites (Schaller, 1963), many seeds are left in a micro-habitat favourable for germination. Secondly, the 'sloppy' way that gorillas eat fruit means that they swallow large seeds which have few potential dispersers in the faunal community. Gorillas swallow the seeds of almost all of their fruit foods and only very large ones (> 7 cc in volume) are rejected. When feeding on fruit, gorillas invest little time in pre-ingestion food processing: they eat quickly and maximising intake takes priority over separating digestible from indigestible portions. Thus gorillas swallow more medium and large sized seeds than do chimpanzees of equivalent body size (Tutin & Fernandez, 1994b). Adult chimpanzees spit out seeds that exceed 4 cc volume while smaller-bodied juvenile gorillas systematically swallow seeds of four species of fruit with volumes between 4.75 and 7 cc.

Lopé gorillas disperse intact seeds of 86% of the fruit they eat. In a detailed study of gorilla seed dispersal at Lopé, Voysey (1995) looked at removal of fruit and the post-dispersal fate of seeds of four common tree species. For three of the four species, he found significantly greater germination success and seedling survival over two years, for seeds deposited in dung at nest-sites compared to those dropped below parent trees or left in dung elsewhere in the forest. For one of the species, *Cola lizae*, gorillas are the sole disperser and, at Lopé, they disperse, on average, 8–11,000 *Cola* seeds/km^2 each year. Gorillas are not good dispersers for another endemic species, *Dialium lopense*, Caesalpiniaceae, for although hundreds of thousands of seeds are swallowed by gorillas in good fruiting years, all seeds disappeared from monitored nest-sites and no germination occurred. *Dialium* seeds have a period of dormancy (during the dry season) and germinate 4–10 months after deposition, while seeds of the other three species studied, germinate within days of deposition. Thus gorilla nest-sites are advantageous microhabitats for seeds that germinate rapidly and nest-sites in light gaps provide favourable conditions for early seedling survival and growth, making gorillas seed dispersers 'par excellence' for some tree species at Lopé.

Gorillas may have other positive impacts on their food plants but these are minor compared to seed dispersal. Weaver ants (*Oecophylla longinoda*) are eaten regularly and gorillas remove entire nests from herbaceous and woody plants. Caterpillars and galls are eaten occasionally. Gorillas may, in a small way, contribute to limiting populations of insects that damage plants. Similarly, feeding by gorillas on pith of herbs has been shown to stimulate increased regeneration (Bullock, 1981), but this is unlikely to be significant in terms of the plant's life history.

Negative impacts of Gorilla feeding

Gorilla folivory is selective and rarely results in death of whole plants. When feeding on pith, individual stems of herbs are broken or uprooted, but gorillas do not uproot whole clumps. Similarly, individual leaves are eaten and even small shrubs (e.g. *Pavetta puberula*, Rubiaceae) are never completely defoliated by feeding gorillas. However, from the plant's point of view, removal of leaves or stems is likely to be negative.

Feeding on bark does impose costs on the plant and can be fatal, e.g. for *Millettia* spp. saplings which are snapped at ground level before being stripped of bark. For trees such as *Milicia excelsa*, which is a major fallback food of Lopé gorillas during the dry season, branches of a certain size are preferred and a visit of several hours by a group of gorillas results in a very severe pruning. *Milicia* trees are large and are relatively common in the study area but gorillas favour certain trees (often groves where 3–5 trees occur close together) to which they make repeated visits. Such trees can be recognised easily by the distinctive shape of their 'pruned' canopies. The loss of leaves and branches imposes costs on the trees in which gorillas feed, but we cannot assess how serious a disadvantage this is, as there may be advantages to regular pruning, possibly in terms of reducing risks of wind damage.

Feeding on fruit confers benefits through seed dispersal in most cases, but the way that gorillas harvest fruit can result in high levels of damage to tree canopies. Most vulnerable are trees that produce small fruit on terminal branches, as gorillas will break branches up to 10 cm in diameter in order gain access to the fruit. Adult gorillas in particular are understandably reluctant to venture onto small branches that might not support their weight, and they reach out and pull branches in towards the centre of the canopy. Both hands are used for picking and processing fruit, so branches are usually broken, rather than bent, to facilitate feeding. Gorilla feeding can result in up to 30% canopy death and often distinctive feeding 'nests' attest to visits by gorillas. In general trees recover quickly but for dioecious species that fruit annually, such as *Uapaca guineensis* (Euphorbiaceae), female trees are easily distinguished from males by their compact canopies resulting from the regular pruning of terminal branches.

For some small tree species, the main trunk may be broken to bring the crown to the ground for comfortable feeding. We have seen gorillas climb into fruiting *Diospyros dendo* (Ebenaceae) trees (approximately 5–8 m high and 5–10 cm dbh) and use their weight to bend the trunk towards a horizontal position; they continue walking quadrupedally along the trunk until the crown bends towards the ground and then, with unexpected agility, they swing off the trunk while retaining a firm grip with one hand. The gorilla's descent is buffered by the resistance of the tree but ends as the gorilla lands gently on the ground, surrounded by the fruit-bearing branches of the small tree. Most trees recover partially, producing new shoots from the trunk but the dramatic reduction in stature is undoubtedly a severe handicap.

Gorillas use plants to build their nests and inflict a certain amount of damage. Both herbaceous plants used for ground nests and tree branches used to build the 35% of nests in trees, are more often bent than broken to create the nest (Tutin *et al.*, 1995). However, the weight of sleeping gorillas kills most of the vegetation in the nest and herbaceous vegetation in light-gaps takes several months to recover from the flattening inflicted by the nest-site of a group of gorillas. The flattening of herbaceous vegetation favours early growth of seedlings that sprout from seeds in gorilla dung deposited at nest-sites (Voysey, 1995).

Gorillas systematically or occasionally destroy the seeds of 21 species of plant at Lopé. Systematic seed predation can be on ripe seeds of species with a non-animal mode of seed dispersal (either wind, e.g. *Pterocarpus soyauxii*, Papilionaceae; mechanical, e.g. *Haumania liebrechtsiana*, Marantaceae; or water, e.g. *Treculia africana*, Moraceae) or on immature seeds of animal-dispersed species, in which case the unripe fruit flesh is rejected. Systematic seed predation on these latter species occurs only in years when fruit is scarce on a community-wide basis. In such years (3 of 13 at Lopé) seed predation can be so extreme as to destroy the entire crops of species such as *Dialium lopense* and *Diospyros zenkeri* before maturation. Occasional seed predation occurs when feeding principally on ripe fruit of some species: typically, fruit flesh is consumed rapidly at the start of a feeding bout while seeds are swallowed, but later on, presumably as the gorilla approaches satiation, some seeds are deliberately crunched up and ingested, a process that reduces feeding rate considerably. Some of the large seeds of *Diospyros mannii* (Ebenaceae) are crunched up during feeding on ripe fruit but immature seeds of this species are protected by the dense mat of sharp irritant hairs that cover the fruit during the 7–8 months it takes for fruit to develop. When the fruit is ripe, the hairs are largely retained but become looser and can be removed by rubbing the fruit against branches. The ten large seeds (52 by 23 mm) contained in each fruit are potentially vulnerable to seed predation, but the irritant hairs provide an effective protection even against specialist predators on immature seeds such as black colobus monkeys (Tutin *et al.*, 1996).

A variation on the seed predation theme is selective copraphagy: seeds of some species are picked out of faeces and crunched up. It is possible that passage through the gut softens the testa making these reingested seeds easier to break between the molars.

Seeds swallowed by gorillas face costs as well as the benefits described in the preceding section. Gorillas swallow large numbers of seeds and a single faecal sample can contain several hundred large seeds such as *Cola lizae* or *Santiria trimera*

(Burseraceae) and over a thousand seeds of species such as *Uapaca guineensis*, *Dialium lopense* and *Diospyros* spp. This clumping of seeds and seedlings can increase vulnerability to secondary seed predation or to leaf-loss in seedlings due to insects and herbivorous mammals. Competition undoubtedly occurs between young seedlings in the crowds that germinate from gorilla dung.

Discussion

The interactions between animals and their food plants are complex but, when considered from the points of view of both sides, can shed light on likely costs and benefits that exert selective pressures. The interactions observed today are the result of thousands, or even millions, of years of co-evolution and many authors have underlined the dangers of over-simplification, or of generalisations (e.g. Herrera, 1986; Pannell & Koziol, 1987; Wheelwright & Orians, 1982). Precise reconstruction of the evolutionary history of individual interactions is rarely possible but advances can come from careful field observations and interpretation of patterns (e.g. Pannell & White, F., 1988; Tutin *et al.* 1996; White, F., 1993a).

In the case of gorillas, the diets of mountain gorillas living in montane habitats differ substantially from those of lowland gorillas in tropical rain forests in terms of the parts of plants that are eaten. The differences are, of course, due to the diversity and types of plants that are presented on the 'menu' offered by each habitat. Comparison of gorilla diets at different sites, allows some general conclusions to be reached: gorillas prefer a diet composed of succulent fruit (that provide easily digestible sugars with little fibre or digestion-inhibiting secondary compounds) and young leaves (that provide protein and are again, low in fibre and secondary compounds). Stem pith (high in sugar but also high in fibre) is eaten regularly and the fibre may be important in slowing gut passage time in order to maximise digestion of nutrients. In the absence of fruit, gorillas eat foods of lower nutritional quality and can subsist on an entirely folivorous diet.

Inter-site comparisons of gorilla diet also allow some explanation of the geographical distribution and local abundance of the genus. Plant species composition defines potential habitat and gorillas occur only in areas with a year-round abundance of herbaceous plants of nutritional value sufficient to provide their calorific needs. Gorillas, unlike chimpanzees, cannot live in the seasonal 'dry' forests of tropical Africa and maximum population densities of gorillas are found in habitats such as the montane vegetation of the Virunga Volcanoes and the swamp-forests of northern Congo, which provide abundant high quality herbaceous foods (Blake *et al.*, 1995; Fay *et al.*, 1989). Gorillas are absent from large areas of apparently suitable habitat in the Congo basin. Large rivers such as the Zaïre and Oubangui form barriers that block, or make difficult, their dispersion, but lasting results of historical changes in vegetation linked to climate change may also be involved. For example, during Pleistocene forest shrinkages in the northern Congo basin (Maley, 1989; 1993), gorillas would have been restricted to refugia and their subsequent expansion will have followed that of suitable forest habitat. Human activities, particularly burning, are known to have affected vegetation dynamics of the region (Oslisly & Deschamps, 1994; Tutin & Oslisly, 1995) and may explain some of the enigmas of present-day gorilla distribution (Groves, 1971).

The dietary differences between gorilla populations have repercussions for social organisation and behaviour and have undoubtedly contributed to the observed morphological differences. Frugivorous gorillas live in smaller groups, travel further each day and have larger home ranges than do folivorous populations (Tutin, 1996; Watts, 1996). These differences all relate to the contrasts in the distribution of fruit versus foliage in both time and space. Most fruit are produced seasonally and fruit-bearing trees tend to be widely dispersed, while foliage is available year round and is more evenly distributed (Malenky *et al.*, 1994; Rogers & Williamson, 1987; White, L.J.T. *et al.*, 1995). The ephemeral nature of fruit availability can also explain the contrast in food processing techniques used by Lopé gorillas for fruit versus foliage: when eating fruit, maximising intake is the priority and this leads to 'sloppy' feeding; while for abundant foliage, digestible portions are meticulously separated from indigestible parts by time-consuming manual processing (Tutin & Fernandez, 1994b). Thus, while diet is determined by the choice of foods available, fundamental differences between the spatial and temporal availability of fruit and foliage have far-reaching implications on social structure and behaviour of consumers (e.g. Wrangham, 1987; Wrangham *et al.*, 1993).

Of the positive and negative impacts that gorillas exert on their food plants, their impact as seed dispersers is the most significant. The negative impact of folivory is rarely fatal to the plant. Plants have evolved effective chemical, structural and behavioural adaptations that limit the damage inflicted by consumers. Plant interactions with animals began long before the appearance of gorillas, and plant defences are effective against a range of faunal 'pests'. For example, chemical and structural adaptations that reduce the palatability of leaves, plus patterns of leaf renewal that limit the availability of unprotected new leaves in either time, or quantity, serve to reduce the damage inflicted by folivores whether they be gorillas or caterpillars. The size and strength of gorillas may challenge normally effective defences and in a few cases mortality can result, e.g. from bark stripping of saplings. Damage inflicted by regular bark stripping of *Milicia excelsa*, or by gorillas breaking branches when feeding on fruit, leave visible traces. Such behaviour is unlikely to benefit plants but if costs in lost fitness are great, it is likely that better defences, perhaps via modification of the mechanical properties of bark or wood that allow bending without breaking, will evolve. Bark stripping is a common feeding pattern of elephants at Lopé, and tree species that are visited regularly tend to have either friable bark fibre that breaks off in small plaques; a capacity for rapid healing; or buttress roots: all adaptations that serve to limit the damage that an elephant can inflict, even if this was not their prime, or original, function.

Seed predation by gorillas rarely depletes crops as huge numbers of seeds are produced, especially by species that rely on wind or mechanical dispersal. However, for two tree species at Lopé, *Dialium lopense* and *Diospyros zenkeri*, gorillas can have devastating effects, destroying entire fruit crops prior to maturation. Immature and ripe seeds are fallback foods of gorillas in years when, between January and May, succulent fruit are scarce due to crop failures. *Dialium lopense* and *Diospyros zenkeri* belong to a group of species which flower in response to a low temperature cue: an environmental trigger that is normally experienced only during the dry season at Lopé and serves to synchronise flowering (Tutin & Fernandez, 1993b; Tutin & White, L.J.T., in press). In years when temperatures do not fall to 19°C during the dry season, the majority of trees of these species do not flower and a community-wide scarcity of succulent fruit is experienced between January and May. However, a few trees flower normally, perhaps because they are less sensitive than are their conspecifics to the environmental cue that triggers flowering, or because these individuals are exposed to cooler micro-climates. For such trees high costs are imposed by gorilla feeding and such factors may contribute to defining micro-habitat preferences of certain tree species.

Similarly, the strength of gorillas means that they can severely damage, or even kill, small trees by breaking major branches or trunks while feeding on fruit. For a tree living with gorillas, it is risky to produce fruit before achieving a stature that can resist the activities of an arboreal frugivore weighing up to 150 kg.

Gorillas disperse large numbers of intact seeds and undoubtedly play an important role in the dynamics of plant regeneration in tropical rain forests (Rogers *et al.*, in press; Tutin & Fernandez, 1993a; Voysey, 1995; Williamson *et al.*, 1990). The growing number of surveys of seed dispersal by animals in tropical rain forests indicates that general syndromes of fruit characters can be identified and that these correlate with broad classes of vertebrate dispersers (Chapman & Chapman, 1996; Gautier-Hion *et al.*, 1985; Howe & Smallwood, 1982; Janson, 1983; Leighton & Leighton, 1983; Pannell & Koziol, 1987). However, little evidence exists for tight co-evolution between individual plant and animal species and Herrera (1986) and others have argued convincingly that such specific relationships are improbable. Reliance on a sole vertebrate seed disperser is unlikely for evolutionary reasons (different histories, life-histories and plasticity of trees and vertebrates) and would anyway be a risky strategy (vagaries of biogeography and often conflicting interests of the two parties). In addition, it has been pointed out that seed dispersal differs in some critical ways from pollen dispersal in that, while "plants can control pollinators' movements by providing nutritional and reproductive incentives at the appropriate site (flower)...., there is no similar incentive for seed dispersers to drop seeds in appropriate places" (Wheelwright & Orians, 1982: 405).

It is true that for the majority of seeds dispersed by gorillas at Lopé, other dispersers exist. Most often these are other primates, elephants or frugivorous birds but some exceptions exist. For *Cola lizae*, gorillas are the sole disperser and for at least 12 tree species with very large seeds (>10 cc in volume), elephants are the only known disperser (White, L.J.T. *et al.*, 1993). The unusual relationship between *Cola lizae* and gorillas is obviously successful as this tree dominates the vegetation of the study area. The limited geographical range of this endemic tree and its dependence on gorillas suggest that it is a recently evolved species that thrived in the post-Pleistocene phases of forest expansion. Its dominance in the forests of the study area can be attributed to the fact that gorillas deposit so many seeds at nest-sites in light-gaps, which give a significant advantage to establishment and seedling growth during at least the first four years (Rogers *et al.*, in press).

Plants produce extraordinary numbers of seeds during their lifetimes yet only a very few will survive to reproductive age. This argues against strong selective pressures which optimise dispersal. Chance will inevitably play a major role in determining eventual success, which perhaps explains the rarity of dependence on a single species of disperser. Data on natural reproductive success among tropical trees are impossible to obtain by direct observation but new techniques in molecular genetics may soon provide a framework in which plant adaptations to, and interactions with, animals can be examined in an evolutionarily meaningful way.

While it is difficult to quantify the merits of the diverse dispersal strategies of plants by following the fate of individual seeds, the results can be assessed in a more global way. Patterns of plant biogeography with respect to major faunal boundaries can sometimes be related to types of seed dispersal (Pannell & White, F., 1988) and it is obvious for island floras that dispersal by water, birds or bats are privileged modes of transport. Frank White underlined the relevance of seed dispersal to reconstructing historical impacts on the vegetation of Africa: it is obvious that animals (particularly flying ones) provide an unparalleled potential for seed transport over long distances but terrestrial vertebrates may have been important in advancing seeds along the migratory corridors that bear witness to the history of African forests (White, F., 1983; 1993b). The other side of the coin is the possibility of identifying forest refugia by the present-day distribution of forest plants, for example, *Begonia* species or certain Caesalpiniaceae trees, which would have survived only in forest refugia and have travelled little since because of very low seed mobility (Rietkerk *et al.*, 1995; Sosef, 1991).

Frank White ended a number of his papers with advice for future directions. When reviewing the biogeography of African plants, he underlined the need for a multidisciplinary approach in order to break away from the trend to narrow specialisation, to achieve "broad syntheses, which could then serve as starting points for further intellectual advance" (1993b: 405). Taking a leaf out of Frank's book, a similar plea is needed with respect to plant-animal interactions. The review of our observations of gorillas and their food plants illustrates the complexity of the interactions between one animal species and a sub-set of plants in one small area of forest for a period of time that represents 80% of one gorilla generation and perhaps 20–50% of the generation times of most of the trees concerned. Few generalisations emerge but some pointers to future directions can be made.

Studies of animal dispersal of seeds and plant-animal interactions have focused on the morphology, size, colour and chemical composition of fruit. Other characters are likely to be important. For example, the protection of seeds from predation, infestation by insects or attack by pathogens during the period of development may produce selective pressures that conflict with ideal dispersal syndromes. Large seed size limits the number of potential dispersers and increases risks of seed predation during development but gives post-germination advantages. Elephants are excellent seed dispersers for large-seeded fruit but abscission of ripe fruit is essential as elephants cannot climb trees: a strong smell is also useful, but bright colours are irrelevant to elephants and likely to attract 'thieves' such as monkeys and birds who steal the flesh but do not move seeds away from the parent tree. We need to look not only at fruit characters in relation to potential dispersers but also at seed characteristics and the 'behaviour' of the plants. By behaviour, I mean, fruiting patterns (synchronised or not), lag between flowering and fruit ripening, size of fruit crops, timing of fruiting compared to community peaks. The behaviour of the animals can also be crucial as shown by gorillas' nesting behaviour and the long day ranges of terrestrial primates compared to arboreal species. Elephants sometimes move very long distances to feed on aggregations of seasonal fruit and almost certainly are responsible for the discontinuous, patchy distribution of *Sacoglottis gabonensis* (Humiriaceae) in the interior of Gabon (White, L.J.T., 1994c) while the presence of this lone African species of a Neotropical family on the west coast of equatorial Africa is undoubtedly due to the fact that the large seed within its fleshy pericarp, floats.

Intuitively, it would seem risky to limit the range of potential animal dispersers, a question of 'putting all the eggs in one basket' and, while Alexandre (1980) reported the disappearance of seedlings of elephant dispersed trees from a forest in Ivory Coast following the extermination of elephants by hunting, Chapman & Chapman (1995) found indications of survival in the absence of dispersers for some, but not all, of a group of large-seeded forest trees in Uganda.

In tropical Africa, at best 5–10% of the remaining natural forest vegetation lies within National Parks or Reserves. Selective logging for timber is an essential part of the economies of all countries of the region and conservation efforts are currently focused on developing sustainable forestry. For this, an understanding of forest dynamics and regeneration is essential but the area of plant-animal interactions has been neglected. For example, the current guidelines for awarding 'green' certification labels to logging companies (Read, 1994; Upton & Bass, 1995) include no recommendations for the protection, or management, of fauna within 'sustainably' managed timber concessions. I hope that enough people will heed Frank White's plea for more empirical investigation and increased inter-disciplinary contacts. Both are essential if we are to protect and manage tropical rain forests in an effective way and at the same time continue in Frank White's footsteps to advance our understanding of the historical causes and contemporary conditions that determine patterns and processes within tropical rain forest ecosystems.

Acknowledgements

Thanks go first and foremost to the Centre International de Recherches Médicales de Franceville, Gabon, for core funding for research since 1984. Many colleagues have assisted and I thank Benoît Fontaine, Rebecca Ham, Alphonse Mackanga-Missandzou, Boo Maisels, Karen McDonald, Richard Parnell, Liz Rogers, Ben Voysey and Liz Williamson and especially Michel Fernandez, Kate Abernethy and Lee White. Warm thanks go to the organisers of the Frank White Memorial Symposium particularly Camilla Huxley-Lambrick and Caroline Pannell.

References

Alexandre, D.Y. (1980). Charactère saisonnier de la fructification dans une forêt hygrophile de Côte d'Ivoire. *Rev. Écol. Appl. Protect. Nat.* **34**: 335–350.

Blake, S., Rogers, E., Fay, M.J., Ngangoué, M. and Ebéké, G. (1995). Swamp gorillas in northern Congo. *African J. Ecol.* **33**: 285–290.

Bullock, S. (1981). Dynamics of vegetative shoots of three species of *Aframomum* (Zingiberaceae) in Cameroun. *Adansonia* **20**: 383–392.

Chapman, C.A. and Chapman, L.J. (1995). Survival without dispersers: seedling recruitment under parents. *Conservation Biol.* **9**: 675–678.

Chapman, C.A. and Chapman, L.J. (1996) Frugivory and the fate of dispersed and non-dispersed seeds of six African tree species. *J. Trop. Ecol.* **12**: 491–504.

Collet, J.Y., Bourreau, E., Cooper, R.W., Tutin, C.E.G. and Fernandez, M. (1984). Experimental demonstration of cellulose digestion by *Troglodytella gorillae*, an intestinal ciliate of lowland gorillas. *Int. J. Primatol.* **5**: 328.

Emmons, L.H., Gautier-Hion, A. and Dubost, G. (1983). Community structure of the frugivorous-folivorous forest mammals of Gabon. *J. Zool. (London)* **199**: 209–222.

Fay, J.M., Agnagna, M., Moore, J. and Oko, R. (1989). Gorillas in the Likouala swamp forests of north central Congo. *Int. J. Primatol.* **10**: 477–486.

Galat, G. and Galat-Luong, A. (1985). La communaute de primates diurnes de la forêt de Tai, Côte d'Ivoire. *Terre &Vie* **40**: 3–32.

Gautier-Hion, A., Duplantier, J.M., Quris, R., Feer, F., Sourd, C., Decoux, J.P., Dubost, G., Emmons, L.H., Erard, C., Hecketsweiler, P., Moungazi, A., Roussilhon, C. and Thiolly, J.M. (1985). Fruit characters as a basis of fruit choice and seed dispersal in a tropical forest community. *Oecologia* **65**: 324–337.

Gautier-Hion, A. and Maisels, F. (1994). Mutualism between a leguminous tree and large African monkeys as pollinators. *Behav. Ecol. Sociobiol.* **34**: 203–210.

Groves, C.P. (1971). Distribution and place of origin of the gorilla. *Man* **6**: 44–51.

Hallé, N. (1987). *Cola lizae* N. Hallé (*Sterculiaceae*), nouvelle espèce du Moyen Ogooué (Gabon). *Bull. Mus. Natl. Hist. Nat., B, Adansonia*, série 4, **9**: 229–237.

Herrera, C.M. (1986). Vertebrate-dispersed plants: why they don't behave the way they should. In: A. Estrada and T. H. Fleming (editors). Frugivores and seed dispersal. Pp. 5–18. Dr. W. Junk Publishers, Dordrecht.

Howe, H.F. and Smallwood, J. (1982). The ecology of seed dispersal. *Annual Rev. Ecol. Syst.* **13**: 201–223.

Janson, C.H. (1983) Adaptations of fruit morphology to dispersal agents in a Neotropical forest. *Science* **219**: 187–189.

Janzen, D.H. (1975). Ecology of Plants in the Tropics. E. Arnold, London.

Janzen, D.H. (1983). Seed and pollen dispersal by animals: Convergence in the ecology of contamination and sloppy harvest. *Biol. J. Linn. Soc.* **20**: 103–113.

Kress, W.J., Schatz, G.E., Andrianifahanana, M. and Simons Morland, H. (1994). Pollination of *Ravenala madagascariensis* (Strelitziaceae) by lemurs in Madagascar: Evidence for an archaic coevolutionary system? *Amer. J. Bot.* **81**: 542–551.

Leighton, M. and Leighton, D.R. (1983). Vetebrate response to fruiting aseasonality within a Bornean rain forest. In: S.L. Sutton, T.C. Whitmore and A.C. Chadwick (editors). Tropical Rain Forest: Ecology and Management. Pp. 181–196. Blackwell Scientific, London.

Malenky, R.K., Kuroda, S., Vineberg, E.O. and Wrangham, R.W. (1994). The significance of terrestrial herbaceous foods for bonobos, chimpanzees and gorillas. In: R.W. Wrangham, W.C. McGrew, F.B.M. de Waal and P.G. Heltne (editors). Chimpanzee Cultures. Pp. 59–75. Harvard University Press, Cambridge, MA.

Maley, J. (1989). Late Quaternary climatic changes in African rain forest: Forest refugia and the major role of sea surface temperature variations. In: M. Leinen and M. Sarnthein (editors). Paleoclimatology and paleometeorology: Modern and past patterns of global atmospheric transport. Pp. 585–616. Kluwer Academic Publishers, Dordrecht.

Maley, J. (1993). The climatic and vegetational history of equatorial regions of Africa during the Upper Quaternary. In T. Shaw, P. Sinclair, B. Andah and A. Okpoko (editors). The Archaeology of Africa: foods, metals and towns. Pp. 43–52. Routledge, London.

Milton, K. (1984). The role of food-processing factors in primate food choice. In: P.S. Rodman and J.G.H. Cant (editors). Adaptations for Foraging in Non-human Primates. Pp. 250–279. Columbia University Press, New York.

Nishihara, T. (1992). A preliminary report on the feeding habits of western lowland gorillas (*Gorilla gorilla gorilla*) in the Ndoki Forest, northern Congo. In: N. Itoigawa, Y. Sugiyama, G.P. Sackett and R.K.R. Thompson (editors). Topics in Primatology. Volume 2. Behavior, Ecology and Conservation. Pp. 225–240. University of Tokyo Press, Tokyo.

Oslisly, R. and Deschamps, R. (1994). Découverte d'une zone d'incendie dans la forêt ombrophile du Gabon *ca* 1500 BP: Essai d'explication anthropique et implications paléoclimatiques. *Compt. Rend. Acad. Sci. Paris, série II* 318: 555–560.

Pannell, C.M. and Koziol, M.J. (1987). Ecological and phytochemical diversity of arillate seeds in *Aglaia* (Meliaceae): A study of vertebrate dispersal in tropical trees. *Phil. Trans., Ser. B* **316**: 303–333.

Pannell, C.M. and White, F. (1988). Patterns of speciation in Africa, Madagascar, and the tropical Far East: Regional faunas and cryptic evolution in vertebrate-dispersed plants. *Monogr. Syst. Bot. Missouri Bot. Gard.* **25**: 639–659.

Read, M. (1994). Truth or trickery? Timber labelling past and future. WWF, Godalming, Surrey.

Remis, M.J. (1994). Feeding ecology and positional behavior of western lowland gorillas (*Gorilla gorilla gorilla*) in the Central African Republic. Ph.D., Yale University.

Rietkerk, M., Ketner, P. and de Wilde, J.J.F.E. (1995). Caesalpinioideae and the study of forest refuges in Gabon: preliminary results. *Bull. Mus. Natl. Hist. Nat., B, Adansonia*, serie 4, **17**: 95–105.

Rogers, M.E., Maisels, F., Williamson, E.A., Fernandez, M. and Tutin, C.E.G. (1990). Gorilla diet in the Lopé Reserve, Gabon: A nutritional analysis. *Oecologia* **84**: 326–339.

Rogers, M.E., Tutin, C.E.G., Parnell, R.J., Voysey, B.C., Williamson, E.A. and Fernandez, M. (1994). Seasonal feeding on bark by gorillas: An unexpected keystone food? In: B. Thierry, J.R. Anderson, J.J. Roeder and N. Herrenschmidt (editors). Current Primatology Volume 1: Ecology and Evolution. Pp. 34–43. Université Louis Pasteur, Strasbourg.

Rogers, M.E., Voysey, B.C., McDonald, K., Parnell, R.J. and Tutin, C.E.G. (in press). Lowland gorillas and seed dispersal: The importance of nest sites.

Rogers, M.E. and Williamson, E.A. (1987). Density of herbaceous plants eaten by gorillas in Gabon: some preliminary data. *Biotropica* **19**: 278–281.

Rogers, M.E., Williamson, E.A., Tutin, C.E.G. and Fernandez, M. (1988). Effects of the dry season on gorilla diet in Gabon. *Primate Rep.* **19**: 29–34.

Schaller, G.B. (1963). The Mountain Gorilla. The University of Chicago Press, Chicago.

Sosef, M.S.M. (1991). New species of *Begonia* in Africa and their relevance to the study of glacial forest refuges. *Wageningen Agric. Univ. Pap.* **91 (4)**: 117–151.

Struhsaker, T.T. and Leland, L. (1979). Socioecology of five sympatric monkey species in the Kibale Forest, Uganda. *Adv. Study Behav.* **9**: 159–228.

Tutin, C.E.G. (1996). Ranging and social structure of lowland gorillas in the Lopé Reserve, Gabon. In: W.C. McGrew, L.F. Marchant and T. Nishida (editors). Great Ape Societies. Pp. 58–70. Cambridge University Press, Cambridge.

Tutin, C.E.G. and Fernandez, M. (1991). Responses of wild chimpanzees and gorillas to the arrival of primatologists: Behaviour observed during habituation. In: H.O. Box (editor). Primate Responses to Environmental Change. Pp. 187–197. Chapman and Hall, London.

Tutin, C.E.G. and Fernandez, M. (1993a). Composition of the diet of chimpanzees and comparisons with that of sympatric lowland gorillas in the Lopé Reserve, Gabon. *Amer. J. Primatol.* **30**: 195–211.

Tutin, C.E.G. and Fernandez, M. (1993b). Relationships between minimum temperature and fruit production in some tropical forest trees in Gabon. *J. Trop. Ecol.* **9**: 241–248.

Tutin, C.E.G. and Fernandez, M. (1994a). Faecal analysis as a method of describing diets of apes: examples from sympatric gorillas and chimpanzees at Lopé, Gabon. *Tropics* **2**: 189–198.

Tutin, C.E.G. and Fernandez, M. (1994b). Comparison of food processing by sympatric apes in the Lopé Reserve, Gabon. In: B. Thierry, J.R. Anderson, J.J. Roeder and N. Herenschmidt (editors). Current Primatology Volume 1. Ecology and Evolution. Pp. 29–36. Université Louis Pasteur, Strasbourg.

Tutin, C.E.G., Fernandez, M., Rogers, M.E., Williamson, E.A. and McGrew, W.C. (1991b). Foraging profiles of sympatric lowland gorillas and chimpanzees in the Lopé Reserve, Gabon. *Phil. Trans., Ser. B* **334**: 179–186.

Tutin, C.E.G., Ham, R.M., White, L.J.T. and Harrison, M.J.S. (1997a). The primate community of the Lopé Reserve, Gabon: Diets, responses to fruit scarcity and effects on biomass. *Amer. J. Primatol.* **42**: 1–24.

Tutin, C.E.G. and Oslisly, R. (1995). *Homo, Pan* and *Gorilla*: Co-existence over 60,000 years at Lopé in central Gabon. *J. Human Evol.* **28**: 597–602.

Tutin, C.E.G., Parnell, R.J. and White, F. (1996). Protecting seeds from primates: examples from *Diospyros* spp. in the Lopé Reserve, Gabon. *J. Trop. Ecol.* **12**: 371–384.

Tutin, C.E.G., Parnell, R.J., White, L.J.T. and Fernandez, M. (1995). Nest building by lowland gorillas in the Lopé Reserve, Gabon: environmental influences and implications for censusing. *Int. J. Primatol.* **16**: 53–76.

Tutin, C.E.G. and White, L.J.T. (in press). Primates, phenology and frugivory: present, past and future patterns in the Lopé Reserve, Gabon. In: D.M. Newbery, H.H.T. Prins and N. Brown (editors). Dynamics of Populations and Communities in the Tropics. Brit. Ecol. Soc. Symp. 37. Blackwell Scientific, Oxford.

Tutin, C.E.G., White, L.J.T. and Mackanga-Missandzou, A. (1997b) The use by rain forest mammals of natural forest fragments in an equatorial African savanna. *Conservation Biol.* **11**: 1190–1203.

Tutin, C.E.G., White, L.J.T., Williamson, E.A., Fernandez, M. and McPherson, G. (1994). List of plant species identified in the northern part of the Lopé Reserve, Gabon. *Tropics* **3**: 249–276.

Tutin, C.E.G., Williamson, E.A., Rogers, M.E. and Fernandez, M. (1991a). A case study of a plant-animal relationship: *Cola lizae* and lowland gorillas in the Lopé Reserve, Gabon. *J. Trop. Ecol.* **7**: 181–199.

Upton, C. and Bass, S. (1995). The Forest Certification Handbook. Earthscan Publications Ltd., London.

Vedder, A.L. (1984). Movement patterns of a group of free-ranging mountain gorillas (*Gorilla gorilla beringei*) and their relation to food availability. *Amer. J. Primatol.* **7**: 73–89.

Voysey, B.C. (1995). Seed dispersal by gorillas in the Lopé Reserve, Gabon. Ph.D., Edinburgh.

Watts, D.P. (1984). Composition and variability of mountain gorilla diets in the central Virungas. *Amer. J. Primatol.* **7**: 323–356.

Watts, D.P. (1990). Ecology of gorillas and its relation to female transfer in mountain gorillas. *Int. J. Primatol.* **11**: 21–45.

Watts, D.P. (1996). Comparative socio-ecology of gorillas. In: W.C. McGrew, L.F. Marchant and T. Nishida (editors). Great Ape Societies. Pp. 16–28. Cambridge University Press, Cambridge.

Wheelwright, N. and Orians, G. (1982). Seed dispersal by animals: contrasts with pollen dispersal, problems of terminology, and constraints on coevolution. *Amer. Naturalist* **119**: 402–413.

White, F. (1983). The Vegetation of Africa: a descriptive memoir to accompany the UNESCO/AETFAT/UNSO vegetation map of Africa. *Natural Resources Research* **20**. UNESCO, Paris.

White, F. (1993a). The AETFAT chorological classification of Africa: history, methods and applications. *Bull. Jard. Bot. Belg.* **62**: 225–281.

White, F. (1993b). Refuge theory, ice-age aridity and the history of tropical biotas: an essay in plant geography. *Fragm Florist. Geobot. Suppl.* **2**: 385–409.

White, L.J.T. (1994a). Biomass of rain forest mammals in the Lopé Reserve, Gabon. *J. Anim. Ecol.* **63**: 499–512.

White, L.J.T. (1994b). Patterns of fruit-fall phenology in the Lopé Reserve, Gabon. *J. Trop. Ecol.* **10**: 289–312.

White, L.J.T. (1994c). *Sacoglottis gabonensis* fruiting and the seasonal movement of elephants in the Lopé Reserve, Gabon. *J. Trop. Ecol.* **10**: 121–125.

White, L.J.T. (in press). Forest-savanna dynamics and the origins of 'Marantaceae Forest' in the Lopé Reserve, Gabon. In: B. Weber, A. Vedder, H. Simons-Morland and L.J.T. White (editors). African rain forest ecology and conservation. Yale University Press, New Haven.

White, L.J.T. and Abernethy, K.A. (1996). Guide de la végétation de la Réserve de la Lopé, Gabon. ECOFAC, Libreville, Gabon.

White, L.J.T., Oslisly, R., Abernethy, K.A. and Maley, J. (in press). *Aucoumea klaineana*: A Holocene success story now in decline? In: Proceedings of ECOFIT symposium. ORSTOM, Paris.

White, L.J.T., Rogers, M.E., Tutin, C.E.G., Williamson, E.A. and Fernandez, M. (1995). Herbaceous vegetation in different forest types in the Lopé Reserve, Gabon: Implications for keystone food availability. *African J. Ecol.* **33**: 124–141.

White, L.J.T., Tutin, C.E.G. and Fernandez, M. (1993). Group composition and diet of forest elephants, *Loxodonta africana cyclotis*, Matschie 1900, in the Lopé Reserve, Gabon. *African J. Ecol.* **31**: 181–199.

Williamson, E.A., Tutin, C.E.G., Rogers, M.E. and Fernandez, M. (1990). Composition of the diet of lowland gorillas at Lopé in Gabon. *Amer. J. Primatol.* **21**: 265–277.

Wrangham, R.W. (1987). Evolution of social structure. In: B.B. Smuts, D.L. Cheney, R.M. Seyfarth, R.W. Wrangham and T.T. Struhsaker (editors). Primate Societies. Pp. 282–296. University of Chicago Press, Chicago.

Wrangham, R.W., Gittleman, J.L. and Chapman, C.A. (1993). Constraints on group size in primates and carnivores: Population density and day range as assays of exploitation competition. *Behav. Ecol. Sociobiol.* **32**: 199–209.